# KINETIC METHODS IN ANALYTICAL CHEMISTRY

# KINETIC METHODS IN ANALYTICAL CHEMISTRY

*By*
*Prof. Ketty Edberg*

2016

SBS Publishers & Distributors Pvt. Ltd.
New Delhi

ISBN 13 : 9789380090962

First Published in 2016



Published by:

SBS PUBLISHERS & DISTRIBUTORS PVT. LTD.

2/9, Ground Floor, Ansari Road, Darya Ganj,

New Delhi - 110002,

INDIA

Tel: 0091.11.23289119 / 41563911

Email: mail@sbspublishers.com

www.sbspublishers.com

# Preface

Analytical chemistry is the study of the separation, identification, and quantification of the chemical components of natural and artificial materials. Qualitative analysis gives an indication of the identity of the chemical species in the sample, and quantitative analysis determines the amount of certain components in the substance. The separation of components is often performed prior to analysis. A kinetic method which utilizes a catalyzed reaction for analysis. Analytical chemistry is also focused on improvements in experimental design, chemometrics, and the creation of new measurement tools to provide better chemical information. Analytical chemistry has applications in forensics, bio-analysis, clinical analysis, environmental analysis, and materials analysis. Kinetic methods of analysis may be applied to the identification of both relatively large and small quantities of materials, catalytic reactions, in which the material to be identified is either consumed during the reaction or serves as its catalyst, are used in the latter case. Most kinetic methods of analysis take advantage of selective catalysts. The most common case is the use of enzymes for the determination of organic and biological molecules. However, catalytic reactions for inorganic species are also commonly performed.

Editor

# Contents

# Chapter 1

# DIASTEREOSELECTIVE THREE-COMPONENT REACTIONS OF CHIRAL NICKEL(II) GLYCINATE FOR CONVENIENT SYNTHESIS OF NOVEL α-AMINO-β-SUBSTITUTED-γ,γ-DISUBSTITUTED BUTYRIC ACIDS

Rui Zhou [1], Li Guo [1], Cheng Peng [2], Gu He [1,*] , Liang Ouyang [1] and Wei Huang [1,2,*]

[1] State Key Laboratory of Biotherapy, West China Hospital, and West China School of Pharmacy, Sichuan University, Chengdu 610041, Sichuan, China; E-Mails: sklb_zhourui@126.com (R.Z.); guoli@scu.edu.cn (L.G.); ouyangliang@scu.edu.cn (L.O.)

[2] State Key Laboratory Breeding Base of Systematic Research, Development and Utilization of Chinese Medicine, Chengdu University of Traditional Chinese Medicine, Chengdu 610041, Sichuan, China; E-Mail: pengchengchengdu@126.com

## ABSTRACT

The convenient, high yielding and diastereoselective synthesis of α-amino-β-substituted-γ,γ-disubstituted butyric acid derivatives was carried out by a three-component tandem reaction of a chiral equivalent of nucleophilic glycine. The reaction was performed smoothly under mild conditions and enabled the construction of two or three adjacent chiral centers in one step, thus affording a novel and convenient route to α-amino-β-substituted-γ,γ-disubstituted butyric acid derivatives.

# INTRODUCTION

Chiral α-amino-γ,γ-disubstituted fragments are frequently found in various bioactive compounds, such as anti-infective agents (compound 1), anti-tuberculosis agents (compound 2), modulators of RNA binding proteins (compound 3) and compositions for specific inhibition of protein splicing by small molecules, and used in the treatment of tuberculosis and other conditions (compound 4) (Figure 1) [1,2,3,4]. Catalytic diastereoselective synthesis of these chiral α-amino-β-substituted butyric acid derivatives rely on many reactions, for example, addition of α,β-unsaturated acyloxazolidinones, then the removal of the oxazolidinone portions [5], cycloaddition of chiral nitrones with (E)-1,4-dichlorobut-2-ene, followed by acid-catalyzed hydrolysis and then by amide hydrolysis [6], but the Michael addition should be considered the main method to get these products when a chiral equivalent of glycine is used. Indeed, several examples of such reactions using chiral auxiliaries have been reported [7,8,9,10,11,12,13,14,15]. However, to our knowledge, there are no reports about the synthesis of chiral α-amino-β-substituted-γ,γ-disubstituted butyric acid derivatives.

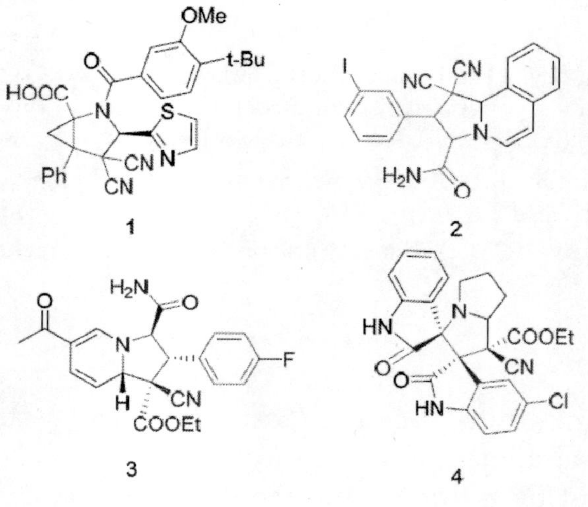

**Figure-1.** Structures of some biologically important compounds containing α-amino-β-substituted γ,γ- disubstituted butyric acid motifs.

The chiral Ni(II) complex of the Schiff base of glycine (abbreviated as (S)-BPB) is commonly used in the asymmetric syntheses of unnatural amino acids. Product mixtures with a high excess of the (S)-amino acid diastereomer are always generated by the addition using (S)-BPB as a ligand [16,17,18]. The products can be easily isolated by column chromatography, and decomposed by acid to get chiral pure amino acids. Moreover, the recovery of (S)-BPB can be high (up to 85%). To the best of our knowledge, a variety of glutamic acid and proline derivatives with a high ee values can be synthesized through Michael additions of activated olefins to Ni(II) glycinate [19,20,21]. Recently, Liu et al. reported the efficient synthesis of β-substituted α,γ-diaminobutyric acid derivatives using asymmetric Michael addition reactions of chiral nickel(II) glycinate with nitroalkenes [22,23,24,25,26,27], Schneider et al. reported the stereoselectivity synthesis of γ-carboxyglutamic acids using asymmetric Michael addition reactions of chiral copper(II) glycinate with di-tert-butyl methylenemalonate [28]. This report focuses on the synthesis of α-amino-β-substituted-γ,γ-disubstituted butyric acid derivatives through the reaction of aromatic aldehydes, a chiral Ni(II) glycinate complex, and an α-carbanion of two electron-withdrawing groups (malononitrile or ethyl cyanoacetate) as a continuation of our previous research on new methods for the preparation of potentially bioactive compounds by multi-component reactions [29,30,31,32,33]. In the process, two carbon-carbon bonds were constructed and two or three chiral centers were generated in a convenient one-pot reaction with a high stereoselectivity.

## RESULTS AND DISCUSSION

The Michael addition reaction was considered as an effective way to get the products. Firstly, the optimization of the reaction conditions was undertaken using a model reaction of chiral nickel(II) glycinate with 2-benzylidenemalononitrile (Table 1). The reaction with 1,8-diazabicyclo[5.4.0]undec-7-ene (DBU), triethylamine(TEA), 4-methylmorpholine (NMM) and piperidine gave a little lower diastereoselectivities (Entries 1–5, Table 1) than Hunig's base (DIEA) did, and all the reactions gave satisfactory yields, except the one in NMM.

**Table-1.** Optimization of the reaction conditions [a].

(S)-5 + 6a + malononitrile —Solvent, Base→ (S,2S,3R)-7a

| Entry | Base | Solvent | Yield (%) [b] | *de* [c] |
|---|---|---|---|---|
| 1 | DBU [d] | $CH_3CN$ | 94 | 88% |
| 2 | TEA | $CH_3CN$ | 97 | 94% |
| 3 | DIEA | $CH_3CN$ | 94 | 96% |
| 4 | NMM | $CH_3CN$ | 63 | 92% |
| 5 | Piperidine | $CH_3CN$ | 89 | 96% |
| 6 | DIEA | DMF | 99 | 88% |
| 7 | DIEA | EA | 98 | 80% |
| 8 | DIEA | MeOH | 99 | 88% |
| 9 | DIEA | DCM | 97 | 58% |
| 10 | DIEA | dioxane | 91 | 97% |
| 11 | DIEA | $CHCl_3$ | 59 | 90% |
| 12 | DIEA | DMSO | 91 | 76% |

[a] All the reactions were conducted at ambient temperature;

[b] Yield of the major products after silica gel column chromatography; [c] Determined by HPLC analysis;

[d] DBU was used in 0.15 equiv.; and other bases were used in 3 equiv.

The reaction proceeded smoothly in most of the solvents tested, although the one in chloroform gave a bad yield and the one in dichloromethane (DCM) gave poor diastereoselectivity. Good diastereoselectivities and yields were observed with the use of acetonitrile, N,N-dimethylformamide (DMF), ethyl acetate (EA), methanol, dioxane and dimethyl sulfoxide (DMSO).

Above all, diastereoselectivity was not obviously influenced by the kind of the bases used, and polar solvents seemed to be better than nonpolar ones. As 2-benzylidenemalononitrile can be easily generated from benzaldehyde and malononitrile under alkaline conditions, domino reaction of these three components was thought to be feasible. In fact, TLC showed that when these three components were mixed together under basic conditions, benzaldehyde first reacted quickly with malononitrile, then added to the nickel(II) glycinate and the product 7a appeared. The results showed no big difference with those in Table 1, so the substrate scope was investigated using DIEA as the base and dioxane as the solvent (entry 11, Table 1) without further optimization.

The aromatic aldehydes with substituents at different positions were introduced into this reaction (Table 2). Whether functionalized with either electron-withdrawing or electron-donating groups, these aldehydes gave the products in good to high yields. As the result obtained with malononitrile was inspiring, ethyl cyanoacetate was introduced into the reaction, and gave a satisfactory result, so the reactions with malononitrile and ethyl cyanoacetate could be looked as two series.

The results of the ethyl cyanoacetate series seemed a little better than the malononitrile series on average, despite the fact three chiral centers are newly generated. In both series, the naphthyl-functionalized aldehydes had the best diastereoselectivities (Table 2, enties 9 and 10), and ortho-functionalized aromatic aldehydes gave relatively high yields and diastereoselectivities (Table 2, entries 12, 14 and 18). The results were quite different in this two series when t-Bu- and 3-Cl-substituted substrates were involved (Table 2, entries 2 and 22, 3 and 15). However, furaldehyde and thienaldehyde were not tolerated (Table 2, enties 13 and 14). To elucidate the relative and absolute configurations of the products, X-ray single crystal structures of (S,2S,3R)-7a (CCDC 951535) and (S,2S,3R,4S)-7q (CCDC 949234) are given below (Figure 2).

To further confirm the structure, diastereoselectivity and regioselectivity, detailed NMR spectral and X-ray analyses were carried out. The structures proposed for all products were in agreement with their NMR spectra, as discussed for compounds 7a and 7q as examples.

In the $^{1}$H-NMR spectrum of 7a and 7q, the α-C proton of glycine exhibited double(d) peaks at δ 4.57 (d, J = 4 Hz, 1H) and δ 4.60 (d, J = 3.6 Hz, 1H), respectively. The α-C proton of malononitrile in 7a appeared as a doublet at δ 5.19 (d, J = 12 Hz, 1H), and the corresponding proton of ethyl cyanoacetate in 7q appeared as a doublet at δ 4.51 (d, J = 12.1 Hz, 1H). The relative configuration of these structures should be as same as compound 7a and 7q shown in

Figure 2a,c, the configurations were further confirmed by the X-ray study of single crystals (Figure 2b,d). The $^{13}C$-NMR of compound 4b supported the proposed structure as well.

A plausible mechanism for the high diastereoselectivity of the reaction could be explained as follows (Scheme 1): malononitrile or cyanide ethyl acetate first reacted with aromatic aldehyde, and the intermediate formed continued to react with the complex. When (S)-N-benzylproline was used, the benzyl group was on a certain side of this complex, so the steric hindrance was large on this side, and the intermediate would prefer attacking from the other side. Still, steric hindrance from the phenyl groups of the intermediate could contribute to the diasteroselectivity, this may explain why naphthaldehydes provided a high de value. As the diastereoselectivity was mainly controlled by the substrates, the reaction was easy to carry, making it a convenient way to get chiral α-amino-β-substituted-γ,γ-disubstituted butyric acid derivatives.

**Table-2.** Asymmetric Michael reactions of chiral nickel(II) glycinate (S)-5 with aromatic aldehydes and α-carbanions [a].

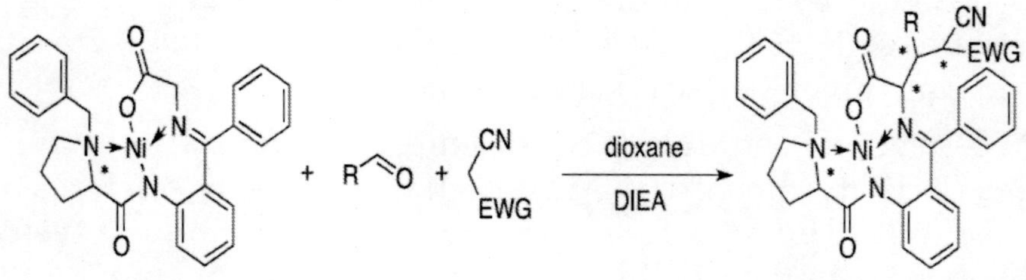

(S)-5 + R–CHO + CN/EWG —dioxane, DIEA→ (S,2S,3R,4S)-7

| Entry | Product | R | EWG | Yield (%) [b] | *de* [c] |
|---|---|---|---|---|---|
| 1 | (*S*,2*S*,3*R*)-**7a** | Ph | CN | 91 | 97% |
| 2 | (*S*,2*S*,3*R*)-**7b** | 4-(*t*-Bu)-$C_6H_4$ | CN | 86 | >99% |
| 3 | (*S*,2*S*,3*R*)-**7c** | 3-Cl-$C_6H_4$ | CN | 52 | 90% |
| 4 | (*S*,2*S*,3*R*)-**7d** | 4-F-$C_6H_4$ | CN | 83 | 93% |
| 5 | (*S*,2*S*,3*R*)-**7e** | 4-Br-$C_6H_4$ | CN | 44 | 98% |
| 6 | (*S*,2*S*,3*R*)-**7f** | 3,4-di-Cl-$C_6H_3$ | CN | 84 | 97% |
| 7 | (*S*,2*S*,3*R*)-**7g** | 3-Br-$C_6H_4$ | CN | 38 | 95% |
| 8 | (*S*,2*S*,3*R*)-**7h** | 3-OMe-$C_6H_4$ | CN | 82 | >99% |
| 9 | (*S*,2*S*,3*R*)-**7i** | 2-naphthyl | CN | 80 | 98% |
| 10 | (*S*,2*S*,3*R*)-**7j** | 1-naphthyl | CN | 26 | 98% |
| 11 | (*S*,2*S*,3*R*)-**7k** | 3-OH-$C_6H_4$ | CN | 46 | 98% |
| 12 | (*S*,2*S*,3*R*)-**7l** | 2-F-4-Br-$C_6H_3$ | CN | 90 | >99% |
| 13 | (*S*,2*S*,3*R*,4*S*)-**7m** | Ph | COOEt | 78 | 98% |
| 14 | (*S*,2*S*,3*R*,4*S*)-**7n** | 2-Br-$C_6H_4$ | COOEt | 88 | >99% |
| 15 | (*S*,2*S*,3*R*,4*S*)-**7o** | 3-Cl-$C_6H_4$ | COOEt | 89 | 97% |
| 16 | (*S*,2*S*,3*R*,4*S*)-**7p** | 4-F-$C_6H_4$ | COOEt | 75 | 96% |
| 17 | (*S*,2*S*,3*R*,4*S*)-**7q** | 3,4-di-Cl-$C_6H_3$ | COOEt | 82 | 98% |
| 18 | (*S*,2*S*,3*R*,4*S*)-**7r** | 2,4-di-Cl-$C_6H_3$ | COOEt | 96 | 98% |
| 19 | (*S*,2*S*,3*R*,4*S*)-**7s** | 4-$CH_3$-$C_6H_4$ | COOEt | 76 | >99% |
| 20 | (*S*,2*S*,3*R*,4*S*)-**7t** | 4-$OCH_3$-$C_6H_4$ | COOEt | 77 | 97% |
| 21 | (*S*,2*S*,3*R*,4*S*)-**7u** | 4-$NO_2$-$C_6H_4$ | COOEt | 69 | 97% |
| 22 | (*S*,2*S*,3*R*,4*S*)-**7v** | 4-(*t*-Bu)-$C_6H_4$ | COOEt | 51 | 97% |
| 23 | (*S*,2*S*,3*R*,4*S*)-**7w** | 1-naphthyl | COOEt | 67 | >99% |
| 24 | (*S*,2*S*,3*R*)-**7x** | 3-Br-thienyl | CN | NR [d] | NR [d] |
| 25 | (*S*,2*S*,3*R*)-**7y** | 4-Me-Furyl | CN | NR [d] | NR [d] |

[a] All the reactions were conducted at ambient temperature, 3 equiv. of all the bases were used;

[b] Yield of the major products after silica gel column chromatography;

[c] Determined by HPLC analysi

[d] Not Reaction.

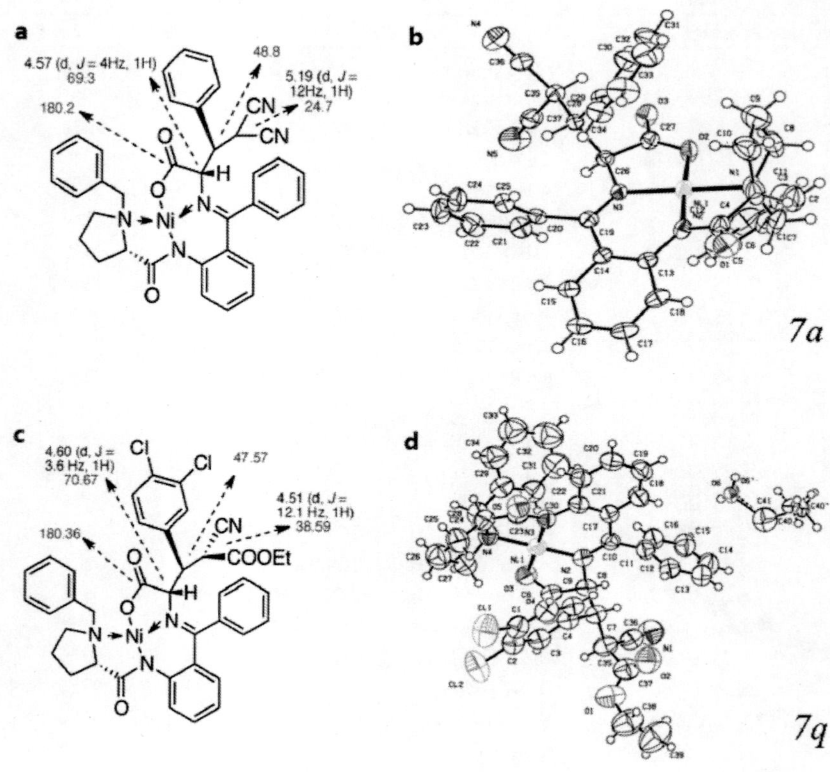


**Figure 2.** (**a**) Selected $^{1}$H- and $^{13}$C-NMR chemical shifts of (*S*,2*S*,3*R*)-**7a**; (**b**) Single crystal X-ray diffraction study of (*S*,2*S*,3*R*)-**7a**; (**c**) Selected $^{1}$H- and $^{13}$C-NMR chemical shifts of (*S*,2*S*,3*R*,4*S*)-**7q** and (**d**) Single crystal X-ray diffraction study of (*S*,2*S*,3*R*,4*S*)-**7q**.

With high diastereoselectivities and mild reaction conditions, the synthesis of (2S,3R)-8a (Scheme 2) was completed by optimizing the metal complex decomposition and Fmoc-protection conditions. Typically, the compound (S,2S,3R)-7a was decomposed by heating a suspension in methanol/6N HCl. However, we found that one of the nitrile groups was partly hydrolyzed in this process, so suitable conditions were sought to ensure that the nitrile groups remain inert. When we stirred (S,2S,3R)-7a in THF with a 3N concentration HCl at ambient temperature, the complex was decomposed and the nitrile group preserved (Scheme 2). The chiral ligand (S)-BPB can be easily recovered quantitatively. (2S,3R)-8a was synthesized after (S,2S,3R)-

7a was decomposed, the (S)-BPB was extracted with ethyl acetate (EA) and the α-amino-β-substituted γ,γ-disubstituted butyric acid product was protected by a Fmoc group. Ultimately, the yield of (S,2R)-5a from (S,2S,3R)-7a was 62% over two steps.

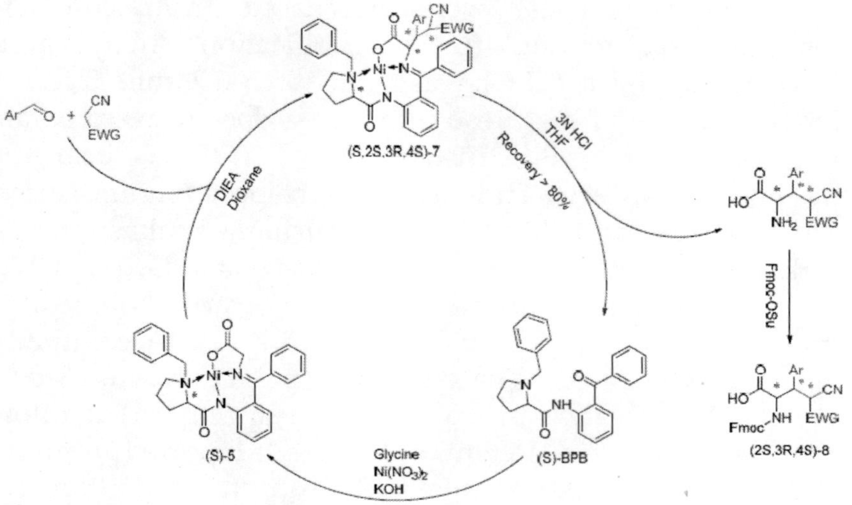

Scheme 1. Asymmetric domino reactions of chiral nickel(II) glycinate (*S*)-5 with aromatic aldehydes and α-carbanion.

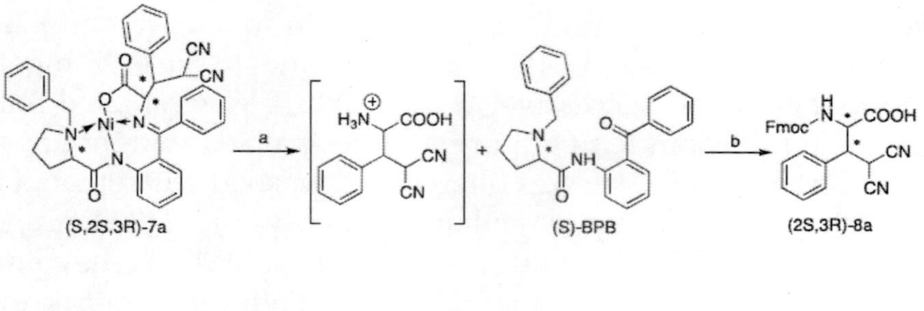

Scheme-2 Decomposition of Ni(II) complex 7a to release product 8a and recovery of the (*S*)-BPB.

# EXPERIMENTAL

## 3.1. General

The reagents (chemicals) were purchased from commercial sources, and used without further purification. Analytical thin layer chromatography (TLC) was $GF_{254}$ (0.15–0.2 mm thickness). The mass spectra and high resolution mass spectra were obtained using Bruker microTOF-Q instrument or TOF-MS instrument. The $^1H$- and $^{13}C$-NMR spectra have been respectively measured in $CDCl_3$ or DMSO-$d_6$ at 400 and 100 MHz using a Bruker Avance III 400 MHz instrument with TMS as an internal standard. Analytical high performance liquid chromatography was carried out using the Waters Alliance 2695 HPLC, using the Chiralpak IA column. The loading loop was 10 μL. The eluting employed was an isocratic mixture of n-hexane and i-propanol (50:50 respectively) at a flow of 1 mL/min unless stated. Retention times are reported in minutes. The enantiomeric excess was calculated from the integration of the absorption peaks at 220 nm.

### Typical Procedure for the Synthesis of (S)-Nickel(II) Complex (5) [18]

(S)-BPB (1 g, 2.60 mmol), $Ni(NO_3)_2 \cdot 6H_2O$ (1.52 g, 5.21 mmol) and glycine (976 mg, 13.0 mmol) were stirred in MeOH (50 mL). Then NaH (55%–65% in oil, 1.04 g, 26 mmol) and KOH (437 mg, 7.81 mmol) were added successively. The mixture was refluxed for 2 h then cooled to room temperature and neutralized with acetic acid. After 12 h the precipitate was filtered and washed with ethanol (100 mL), followed by stirring in methane/water (v/v) 1:2 (200 mL), then filtered to form a red crystalline solid (1.27 g, yield 98%). The complex was sufficiently pure for further use without additional purification.

### General Procedure for the Synthesis of Ni(II) (7)

The nickel(II) complex of glycine (S)-5 (1.0 equiv.) was dissolved in dioxane, and DIEA (3.0 equiv.), aromatic aldehyde (1.2 equiv.) and malononitrile/ethyl cyanoacetate (1.2 equiv.) was added at room temperature. The mixture was then stirred at room temperature for 12 h, then poured into 10% citric acid solution, extracted with $CH_2Cl_2$ (three times), dried with anhydrous $Na_2SO_4$, concentrated, and purified on silica (petroleum ether/ethyl acetate = 1/1) to give 7 as a red solid.

### *Ni(II)-(S)-BPB/(2S,3R)-2-Amino-4,4'-dicyano-3-phenylbutyric Acid Schiff Base Complex (7a)*

Yield = 91%, m.p. 202–204 °C. $[\alpha]_D^{18}$ = +1602 (ca. 0.2 g/100 mL, $CHCl_3$). $^1$H-NMR ($CDCl_3$) δ 8.25 (d, J= 8 Hz, 1H), 7.98 (d, J = 8 Hz, 2H), 7.68–7.63 (m, 6H), 7.39 (d, J = 8 Hz, 1H), 7.30–7.26 (m, 4H), 7.18–7.14 (m, 3H), 6.70 (d, J = 4 Hz, 1H), 5.19 (d, J = 12 Hz, 1H), 4.57 (d, J = 4 Hz, 1H), 4.15 (d, J = 12 Hz, 1H), 3.42 (d, J = 12 Hz, 1H), 3.29–3.21 (m, 2H), 2.97–2.91 (m, 1H), 2.32–2.20 (m, 1H), 2.06–2.02 (m, 1H), 1.95–1.91 (m, 1H), 1.85–1.83 (m, 1H), 1.56–1.50 (m, 1H). $^{13}$C-NMR ($CDCl_3$) δ 180.2, 176.2, 173.5, 143.4, 133.9, 133.5, 133.2, 132.6, 131.4, 130.9, 130.6, 130.1, 129.9, 129.5, 128.9, 128.9, 127.5, 127.0, 125.5, 123.3, 120.8, 111.5, 111.3, 70.5, 69.7, 63.9, 57.4, 48.8, 29.7, 24.7, 23.1. ESI-MS (m/z): calcd. 652.2, found 652.2 ($[M+H]^+$); HRMS (m/z): calcd. $C_{37}H_{32}N_5NiO_3$ for 652.1859, found 652.1857 ($[M+H]^+$). HPLC (Chiralpak IA, n-hexane/i-propanol = 50/50, flow rate 1.0 mL/min, λ = 220 nm), $t_{major}$ = 32.3 min, $t_{minor}$ = 13.4 min, de = 97%.

### *Ni(II)-(S)-BPB/(2S,3R)-2-Amino-4,4'-dicyano-3-(4-tert-butyl)-phenylbutyric Acid Schiff Base Complex (7b)*

Yield = 86%, m.p. 197–199 °C. $[\alpha]_D^{18}$ = +1602 (ca. 0.2 g/100 mL, $CHCl_3$). $^1$H-NMR ($CDCl_3$) δ 8.29 (d, J= 8.7 Hz, 1H), 7.95 (d, J = 7.5 Hz, 2H), 7.68 (dt, J = 13.0, 6.8 Hz, 3H), 7.60 (d, J = 8.1 Hz, 2H), 7.41 (d, J= 7.2 Hz, 1H), 7.29 (dd, J = 14.5, 7.1 Hz, 3H), 7.19 (d, J = 7.9 Hz, 3H), 7.16–7.09 (m, 2H), 6.70 (d, J = 2.9 Hz, 2H), 5.16 (d, J = 12.0 Hz, 1H), 4.58 (d, J = 3.4 Hz, 1H), 4.16 (d, J = 12.7 Hz, 1H), 3.79 (s, 1H), 3.49 (d, J = 12.7 Hz, 1H), 3.28–3.13 (m, 2H), 3.05–2.90 (m, 1H), 2.29–2.13 (m, 1H), 2.12–2.00 (m, 1H), 1.92 (dt, J = 27.0, 8.6 Hz, 1H), 1.78 (dt, J = 18.9, 9.3 Hz, 1H), 1.54 (s, 1H), 1.46–1.31 (m, 9H). $^{13}$C-NMR ($CDCl_3$) δ 180.2, 176.3, 173.3, 160.9, 143.2, 133.9, 133.4, 133.3, 133.2, 131.4, 130.8, 130.5, 129.4, 128.9, 128.8, 127.4, 127.0, 125.6, 124.1, 123.2, 120.8, 115.2, 111.6, 111.3, 70.6, 69.8, 64.0, 57.6, 55.4, 48.1, 30.5, 24.7, 22.9. ESI-MS (m/z): calcd.

708.2, found 708.3 ([M+H]$^+$); HRMS (m/z): calcd. $C_{41}H_{40}N_5NiO_3$ for 708.2485, found 708.2482 ([M+H]$^+$). HPLC (Chiralpak IA, n-hexane/i-propanol = 50/50, flow rate 1.0 mL/min, λ = 220 nm), $t_{major}$ = 30.9 min, $t_{minor}$ = 17.3 min, de > 99%.

### *Ni(II)-(S)-BPB/(2S,3R)-2-Amino-4,4'-dicyano-3-(3-chlorophenyl) Butyric Acid Schiff Base Complex (7c)*

Yield = 52%, m.p. 211–213 °C. $[\alpha]_D^{18}$ = +1622 (ca. 0.2 g/100 mL, $CHCl_3$). $^1$H-NMR ($CDCl_3$) δ 8.27 (d, J= 8.7 Hz, 1H), 8.00 (d, J = 7.5 Hz, 2H), 7.75–7.60 (m, 4H), 7.55 (d, J = 15.7 Hz, 1H), 7.42–7.34 (m, 2H), 7.32 (dd, J = 14.6, 6.9 Hz, 3H), 7.22–7.09 (m, 4H), 6.70 (d, J = 3.7 Hz, 2H), 5.14 (d, J = 11.9 Hz, 1H), 4.56 (d, J = 3.8 Hz, 1H), 4.15 (d, J = 12.7 Hz, 1H), 3.43 (d, J = 12.6 Hz, 1H), 3.25 (ddd, J = 16.2, 10.7, 5.6 Hz, 2H), 3.02 (dd, J = 9.8, 5.5 Hz, 1H), 2.25 (dd, J = 18.6, 10.0 Hz, 1H), 2.10–1.98 (m, 3H), 1.70–1.63 (m, 1H). $^{13}$C-NMR ($CDCl_3$) δ 180.2, 175.9, 173.7, 143.4, 136.3, 134.6, 133.9, 133.4, 133.4, 133.2, 131.3, 131.0, 131.0, 130.6, 130.4, 129.5, 129.0, 128.8, 127.4, 127.0, 125.4, 123.3, 120.8, 111.2, 110.9, 70.5, 69.4, 64.0, 57.6, 48.4, 30.7, 29.6, 24.6, 23.0. ESI-MS (m/z): calcd. 686.1, found 686.2 ([M+H]$^+$). HRMS (m/z): calcd. $C_{37}H_{31}ClN_5NiO_3$ for 686.1469, found 686.1475 ([M+H]$^+$). HPLC (Chiralpak IA, n-hexane/i-propanol = 50/50, flow rate 1.0 mL/min, λ = 220 nm), $t_{major}$ = 31.0 min, $t_{minor}$ = 13.0 min, de = 90%.

### *Ni(II)-(S)-BPB/(2S,3R)-2-Amino-4,4'-dicyano-3-(4-fluorophenyl)butyric Acid Schiff Base Complex (7d)*

Yield = 83%, m.p. 207–209 °C. $[\alpha]_D^{18}$ = +1734 (ca. 0.2 g/100 mL, $CHCl_3$). $^1$H-NMR ($CDCl_3$) δ 8.25 (d, J= 8.6 Hz, 1H), 8.00 (d, J = 7.4 Hz, 2H), 7.73–7.62 (m, 3H), 7.34 (ddd, J = 14.8, 13.7, 7.2 Hz, 7H), 7.21–7.09 (m, 3H), 6.73–6.66 (m, 2H), 5.16 (d, J = 12.0 Hz, 1H), 4.55 (d, J = 3.7 Hz, 1H), 4.15 (d, J = 13.1 Hz, 1H), 3.48–3.37 (m, 1H), 3.32–3.22 (m, 2H), 3.02–2.90 (m, 1H), 2.33–2.23 (m, 1H), 2.13–2.05 (m, 1H), 1.97 (dd, J = 13.3, 7.4 Hz, 2H), 1.67–1.61 (m, 1H). $^{13}$C-NMR ($CDCl_3$) δ 180.2, 176.1, 173.6, 165.1, 162.6, 143.3, 133.9, 133.4, 133.3, 133.2, 131.3, 131.2, 130.9, 130.6, 129.5, 129.0, 128.8, 128.4, 128.4, 127.3, 127.0, 125.4, 123.3, 120.8, 117.1, 116.9, 111.3, 111.0, 70.5, 69.6, 64.0, 57.5, 48.1, 30.6, 24.7, 22.9. ESI-MS (m/z): calcd. 670.2, found 670.2 ([M+H]$^+$). HRMS (m/z): calcd. $C_{37}H_{31}FN_5NiO_3$ for 670.1764, found 670.1801 ([M+H]$^+$). HPLC

(Chiralpak IA, n-hexane/i-propanol = 50/50, flow rate 1.0 mL/min, $\lambda$ = 220 nm), $t_{major}$ = 26.0 min, $t_{minor}$ = 10.1 min, de = 93%.

## Ni(II)-(S)-BPB/(2S,3R)-2-amino-4,4'-dicyano-3-(4-bromophenyl) Butyric Acid Schiff Base Complex (7e)

Yield = 44%, m.p. 197–199 °C. $[\alpha]_D^{18}$ = +1972 (ca. 0.2 g/100 mL, $CHCl_3$). $^1$H-NMR ($CDCl_3$) δ 8.23 (d, J= 8.7 Hz, 1H), 8.01 (d, J = 7.4 Hz, 2H), 7.78–7.58 (m, 3H), 7.38 (d, J = 6.8 Hz, 1H), 7.29 (dd, J = 14.9, 7.4 Hz, 3H), 7.25–7.08 (m, 7H), 6.70 (d, J = 13.7 Hz, 2H), 5.14 (d, J = 12.0 Hz, 1H), 4.53 (d, J = 3.2 Hz, 1H), 4.15 (d, J = 12.7 Hz, 1H), 3.40 (d, J = 12.6 Hz, 1H), 3.31–3.18 (m, 2H), 2.99 (dd, J = 10.5, 6.0 Hz, 1H), 2.23 (dt, J = 24.1, 12.1 Hz, 1H), 2.13–2.02 (m, 1H), 2.02–1.87 (m, 2H), 1.55 (dd, J = 12.1, 6.5 Hz, 1H). $^{13}$C-NMR ($CDCl_3$) δ 180.2, 176.3, 173.3, 160.9, 143.2, 133.9, 133.4, 133.3, 133.2, 131.4, 130.8, 130.5, 129.4, 128.9, 128.8, 127.4, 127.0, 125.6, 124.1, 123.2, 120.8, 115.2, 111.6, 111.3, 70.6, 69.8, 64.0, 57.6, 55.4, 48.1, 30.5, 24.7, 22.9. ESI-MS (m/z): calcd. 730.1, found 730.2 ($[M+H]^+$). HRMS (m/z): calcd. $C_{37}H_{31}BrN_5NiO_3$ for 730.0966, found 730.0978 ($[M+H]^+$). HPLC (Chiralpak IA, n-hexane/i-propanol = 50/50, flow rate 1.0 mL/min, $\lambda$ = 220 nm), $t_{major}$ = 32.0 min, $t_{minor}$ = 15.3 min, de = 98%.

## Ni(II)-(S)-BPB/(2S,3R)-2-Amino-4,4'-dicyano-3-(3,4-dichlorophenyl) Butyric Acid Schiff Base Complex (7f)

Yield = 84%, m.p. 213–215 °C. $[\alpha]_D^{18}$ = +1660 (ca. 0.2 g/100 mL, $CHCl_3$). $^1$H-NMR ($CDCl_3$) δ 8.27 (d, J= 8.7 Hz, 1H), 8.01 (d, J = 7.4 Hz, 2H), 7.69 (dd, J = 17.4, 7.8 Hz, 4H), 7.44 (s, 1H), 7.38 (d, J = 6.3 Hz, 1H), 7.31 (t, J = 7.4 Hz, 2H), 7.18 (dd, J = 15.9, 8.3 Hz, 2H), 7.09 (t, J = 7.8 Hz, 2H), 6.71 (d, J = 14.3 Hz, 2H), 5.15 (d, J = 11.9 Hz, 1H), 4.53 (s, 1H), 4.12 (dd, J = 17.0, 9.9 Hz, 1H), 3.42 (d, J = 12.6 Hz, 1H), 3.30 (t, J = 8.3 Hz, 1H), 3.22 (d, J = 11.9 Hz, 1H), 2.99 (d, J = 5.8 Hz, 1H), 2.33 (dd, J = 20.4, 8.8 Hz, 1H), 2.02 (qd, J = 13.8, 6.8 Hz, 3H), 1.77–1.66 (m, 1H). $^{13}$C-NMR ($CDCl_3$) δ 180.2, 175.8, 173.8, 143.4, 135.0, 134.6, 134.0, 133.5, 133.3, 132.7, 131.7, 131.3, 131.1, 130.7, 129.6, 129.0, 128.9, 127.2, 126.9, 125.2, 123.3, 120.8, 111.1, 110.6, 70.5, 69.3, 64.1, 57.8, 48.0, 30.7, 24.4, 22.8. ESI-MS (m/z): calcd. 720.1, found 720.2 ($[M+H]^+$); HRMS (m/z): calcd. $C_{37}H_{30}Cl_2N_5NiO_3$ for 720.1079, found 720.1090 ($[M+H]^+$). HPLC (Chiralpak IA, n-hexane/i-propanol = 50/50, flow rate 1.0 mL/min, λ = 220 nm), $t_{major}$ = 26.0 min, $t_{minor}$ = 12.5 min, de = 97%.

## *Ni(II)-(S)-BPB/(2S,3R)-2-Amino-4,4'-dicyano-3-(3-bromophenyl) Butyric Acid Schiff Base Complex (7g)*

Yield = 38%, m.p. 208–210 °C. $[\alpha]_D^{18}$ = +1882 (ca. 0.2 g/100 mL, $CHCl_3$). $^1$H-NMR ($CDCl_3$) δ 8.28 (d, J= 8.7 Hz, 1H), 8.00 (d, J = 7.4 Hz, 2H), 7.80 (d, J = 8.0 Hz, 1H), 7.75–7.60 (m, 3H), 7.49 (t, J = 7.8 Hz, 2H), 7.39 (d, J = 6.7 Hz, 1H), 7.31 (t, J = 7.6 Hz, 2H), 7.24–7.08 (m, 4H), 6.70 (d, J = 4.0 Hz, 2H), 5.14 (d, J = 11.9 Hz, 1H), 4.55 (d, J = 3.7 Hz, 1H), 4.20–4.09 (m, 1H), 3.42 (d, J = 12.6 Hz, 1H), 3.25 (ddd, J = 15.7, 10.7, 5.6 Hz, 2H), 3.03 (dd, J = 10.0, 5.6 Hz, 1H), 2.25 (dt, J = 16.4, 8.4 Hz, 1H), 2.15–1.91 (m, 4H).$^{13}$C-NMR ($CDCl_3$) δ 180.2, 176.3, 173.3, 160.9, 143.2, 133.9, 133.4, 133.3, 133.2, 131.4, 130.8, 130.5, 129.4, 128.9, 128.8, 127.4, 127.0, 125.6, 124.1, 123.2, 120.8, 115.2, 111.6, 111.3, 70.6, 69.8, 64.0, 57.6, 55.4, 48.1, 30.5, 24.7, 22.9. ESI-MS (m/z): calcd. 730.1, found 730.2 ($[M+H]^+$); HRMS (m/z): calcd. $C_{37}H_{31}BrN_5NiO_3$ for 730.0966, found 730.0966 ($[M+H]^+$). HPLC (Chiralpak IA, n-hexane/ i-propanol = 50/50, flow rate 1.0 mL/min, λ = 220 nm), $t_{major}$ = 32.0, $t_{minor}$ = 14.0 min, de = 95%.

## *Ni(II)-(S)-BPB/(2S,3R)-2-Amino-4,4'-dicyano-3-(3-methoxyphenyl) Butyric Acid Schiff Base Complex (7h)*

Yield = 82%, m.p. 222–224 °C. $[\alpha]_D^{18}$ = +1678 (ca. 0.2 g/100 mL, $CHCl_3$). $^1$H-NMR ($CDCl_3$) δ 8.18 (d, J= 8.7 Hz, 1H), 7.93 (d, J = 7.5 Hz, 2H), 7.59 (dd, J = 13.8, 7.2 Hz, 3H), 7.44 (t, J = 7.9 Hz, 1H), 7.32 (d, J= 6.9 Hz, 1H), 7.23 (t, J = 7.5 Hz, 2H), 7.07 (dd, J = 13.9, 6.8 Hz, 4H), 6.75 (d, J = 11.2 Hz, 2H), 6.62 (d,J = 4.0 Hz, 2H), 5.09 (d, J = 12.0 Hz, 1H), 4.47 (d, J = 3.4 Hz, 1H), 4.04 (dd, J = 15.5, 9.7 Hz, 1H), 3.71 (s, 3H), 3.34 (d, J = 12.6 Hz, 1H), 3.16 (dd, J = 16.1, 6.4 Hz, 2H), 2.89 (q, J = 10.2 Hz, 1H), 2.26–2.06 (m, 1H), 2.01 (dd, J = 13.8, 7.1 Hz, 1H), 1.87 (dd, J = 18.6, 6.4 Hz, 2H), 1.49 (dd, J = 14.2, 9.1 Hz, 1H). $^{13}$C-NMR ($CDCl_3$) δ 180.1, 176.2, 173.3, 160.6, 143.3, 133.9, 133.8, 133.4, 133.3, 133.2, 131.4, 130.9, 130.9, 130.5, 129.4, 128.9, 128.8, 127.5, 127.0, 125.5, 123.2, 120.7, 115.7, 111.5, 111.2, 70.5, 69.6, 63.9, 57.4, 55.3, 48.6, 30.6, 24.6, 23.0. ESI-MS (m/z): calcd. 682.2, found 682.3 ($[M+H]^+$); HRMS (m/z): calcd. $C_{38}H_{34}N_5NiO_4$ for 682.1964, found 682.1959 ($[M+H]^+$).

HPLC (Chiralpak IA, n-hexane/i-propanol = 50/50, flow rate 1.0 mL/min, λ = 220 nm), $t_{major}$ = 36.2, $t_{minor}$ = 16.5 min, de > 99%.

## *Ni(II)-(S)-BPB/(2S,3R)-2-Amino-4,4'-dicyano-3-(2-naphthyl) Butyric Acid Schiff Base Complex (7i)*

Obtained as a red solid, yield = 80%, m.p. 187–189 °C. $[\alpha]_D^{18}$ = +1582 (ca. 0.2 g/100 mL, $CHCl_3$). $^1$H-NMR ($CDCl_3$) δ 8.19 (d, J = 8.7 Hz, 1H), 8.08 (d, J = 8.4 Hz, 1H), 7.97 (d, J = 7.5 Hz, 3H), 7.93 (d, J = 7.7 Hz, 1H), 7.81 (s, 1H), 7.70 (d, J = 7.8 Hz, 3H), 7.60 (p, J = 6.7 Hz, 2H), 7.41 (d, J = 7.0 Hz, 1H), 7.32 (d, J = 7.0 Hz, 1H), 7.26 (t, J = 7.6 Hz, 3H), 7.20–7.09 (m, 2H), 6.70 (d, J = 14.1 Hz, 2H), 5.34 (d, J = 12.0 Hz, 1H), 4.62 (d, J = 3.1 Hz, 1H), 4.12 (d, J = 7.1 Hz, 1H), 4.02 (d, J = 12.5 Hz, 1H), 3.46 (dd, J = 12.0, 3.1 Hz, 1H), 3.23 (d, J = 12.5 Hz, 1H), 3.05 (t, J = 8.5 Hz, 1H), 2.64–2.48 (m, 1H), 2.04 (s, 1H), 1.74 (d, J = 10.0 Hz, 2H), 1.24 (dd, J = 16.0, 8.8 Hz, 1H). $^{13}$C-NMR ($CDCl_3$) δ 180.1, 176.2, 173.4, 143.3, 134.1, 133.9, 133.5, 133.3, 133.2, 131.3, 130.9, 130.5, 129.8, 129.5, 128.9, 128.8, 128.5, 127.8, 127.5, 127.5, 127.3, 127.1, 125.5, 123.3, 120.7, 111.6, 111.3, 70.3, 70.0, 64.0, 57.7, 48.9, 30.0, 24.6, 22.3. ESI-MS (m/z): calcd. 702.2, found 702.3 ($[M+H]^+$); HRMS (m/z): calcd. $C_{41}H_{34}N_5NiO_3$ for 702.2015, found 702.2022 ($[M+H]^+$). HPLC (Chiralpak IA, n-hexane/i-propanol = 50/50, flow rate 1.0 mL/min, λ = 220 nm), $t_{major}$ = 31.5 min, $t_{minor}$ = 18.8 min, de = 98%.

## *Ni(II)-(S)-BPB/(2S,3R)-2-Amino-4,4'-dicyano-3-(1-naphthyl) Butyric Acid Schiff Base Complex (7j)*

Yield = 26%, m.p. 178–180 °C. $[\alpha]_D^{18}$ = +1614 (ca. 0.2 g/100 mL, $CHCl_3$). $^1$H-NMR ($CDCl_3$) δ 8.25 (d, J= 8.8 Hz, 1H), 8.14 (d, J = 8.2 Hz, 1H), 8.05 (d, J = 8.2 Hz, 1H), 7.89 (d, J = 7.5 Hz, 2H), 7.80–7.65 (m, 4H), 7.57 (d, J = 7.3 Hz, 1H), 7.55–7.43 (m, 3H), 7.41 (d, J = 7.0 Hz, 1H), 7.29–7.21 (m, 3H), 7.21–7.07 (m, 3H), 6.80–6.67 (m, 2H), 5.36 (d, J = 11.8 Hz, 1H), 4.74 (d, J = 2.6 Hz, 1H), 4.20 (dd, J = 11.8, 2.2 Hz, 1H), 4.12 (q, J = 7.1 Hz, 1H), 3.97 (d, J = 12.6 Hz, 1H), 3.25 (d, J = 12.6 Hz, 1H), 2.95 (t, J = 8.7 Hz, 1H), 2.50 (dt, J = 11.4, 5.8 Hz, 1H), 1.86 (dd, J = 12.9, 9.2 Hz, 1H), 1.75 (dt, J = 22.7, 9.6 Hz, 2H), 0.97 (dt, J= 14.6, 7.4 Hz, 1H). $^{13}$C-NMR ($CDCl_3$) δ 179.5, 176.1, 173.6, 143.4, 134.4, 134.0, 133.5, 133.4, 133.1, 133.0, 131.2, 130.6, 130.4, 130.1, 129.3, 129.2, 128.8, 128.7, 127.2, 127.1, 126.9, 126.7, 126.1, 126.0, 125.1, 123.0, 122.5, 120.5, 111.6,

111.1, 71.4, 70.3, 63.7, 57.2, 43.5, 30.2, 25.7, 22.9. ESI-MS (m/z): calcd. 702.2, found 702.3 ($[M+H]^+$); HRMS (m/z): calcd. $C_{41}H_{34}N_5NiO_3$ for 702.2015, found 702.2019 ($[M+H]^+$). HPLC (Chiralpak IA, n-hexane/i-propanol = 50/50, flow rate 1.0 mL/min, λ = 220 nm), $t_{major}$ = 55.7 min, $t_{minor}$ = 20.4 min, de = 98%.

### *Ni(II)-(S)-BPB/(2S,3R)-2-Amino-4,4'-dicyano-3-(3-hydroxyphenyl) Butyric Acid Schiff Base Complex (7k)*

Yield = 46%, m.p. 222–224 °C. $[\alpha]_D^{18}$ = +1775 (ca. 0.2 g/100 mL, $CHCl_3$). $^1$H-NMR (DMSO) δ 9.89 (s, 1H), 8.42 (d, J = 7.5 Hz, 2H), 8.11 (d, J = 8.7 Hz, 1H), 7.78 (d, J = 5.3 Hz, 1H), 7.69 (s, 3H), 7.54 (d, J = 4.9 Hz, 1H), 7.46 (t, J = 7.8 Hz, 1H), 7.37 (t, J = 7.5 Hz, 2H), 7.13 (dd, J = 16.4, 6.8 Hz, 3H), 6.93–6.80 (m, 2H), 6.73 (t, J = 7.6 Hz, 1H), 6.63 (d, J = 8.3 Hz, 1H), 5.40 (d, J = 12.4 Hz, 1H), 4.40 (d, J = 3.6 Hz, 1H), 4.09 (q, J = 7.1 Hz, 1H), 3.91 (d, J = 12.2 Hz, 1H), 3.15 (dd, J = 12.3, 3.5 Hz, 1H), 2.96–2.84 (m, 1H), 2.56 (s, 2H), 2.23 (dd, J = 15.2, 8.6 Hz, 1H), 2.17–2.07 (m, 1H), 1.99 (dd, J = 25.2, 14.2 Hz, 2H), 1.70 (d, J = 6.5 Hz, 1H). $^{13}$C-NMR (DMSO) δ 180.0, 174.8, 171.9, 158.2, 143.2, 134.5, 134.1, 133.5, 133.1, 131.8, 131.5, 130.2, 130.1, 129.7, 128.9, 128.4, 128.1, 127.8, 127.4, 125.1, 122.8, 119.8, 116.4, 113.1, 112.1, 69.7, 69.6, 63.2, 57.7, 47.5, 30.4, 25.2, 22.6. ESI-MS (m/z): calcd. 668.2, found 668.2 ($[M+H]^+$); HRMS (m/z): calcd. $C_{37}H_{32}N_5NiO_4$ for 668.1808, found 668.1819 ($[M+H]^+$). HPLC (Chiralpak IA, n-hexane/i-propanol = 50/50, flow rate 1.0 mL/min, λ = 220 nm), $t_{major}$ = 14.7 min, $t_{minor}$ = 7.1 min, de = 98%.

### *Ni(II)-(S)-BPB/(2S,3R)-2-Amino-4,4'-dicyano-3-(2-fluoro-4-bromophenyl) Butyric Acid Schiff Base Complex (7l)*

Yield = 90%, m.p. 222–224 °C. $[\alpha]_D^{18}$ = +1624 (ca. 0.2 g/100 mL, $CHCl_3$). $^1$H-NMR ($CDCl_3$) δ 8.32 (d, J= 8.7 Hz, 1H), 8.01 (d, J = 7.5 Hz, 2H), 7.66 (d, J = 6.2 Hz, 3H), 7.60 (d, J = 8.6 Hz, 2H), 7.35 (d, J = 5.2 Hz, 1H), 7.29 (dd, J = 13.5, 6.0 Hz, 3H), 7.24–7.11 (m, 3H), 6.68 (s, 2H), 5.23 (d, J = 12.0 Hz, 1H), 4.53 (d, J = 3.3 Hz, 1H), 4.20–4.05 (m, 1H), 3.80 (d, J = 9.4 Hz, 1H), 3.45 (d, J = 12.6 Hz, 1H), 3.30 (t, J = 8.5 Hz, 1H), 2.96 (q, J = 10.3 Hz, 1H), 2.44–2.25 (m, 1H), 2.25–2.08 (m, 1H), 2.02–1.89 (m, 2H), 1.79–1.64 (m, 1H). $^{13}$C-NMR ($CDCl_3$) δ 180.2, 176.3, 173.3, 160.9, 143.2, 133.9, 133.4, 133.3, 133.2, 131.4, 130.8, 130.5, 129.4, 128.9, 128.8, 127.4, 127.0, 125.6, 124.1, 123.2, 120.8, 115.2, 111.6, 111.3, 70.6,

69.8, 64.0, 57.6, 55.4, 48.1, 30.5, 24.7, 22.9. ESI-MS (m/z): calcd. 748.1, found 748.1 ($[M+H]^+$); HRMS (m/z): calcd. $C_{37}H_{30}BrFN_5NiO_3$ for 748.0869, found 748.0881 ($[M+H]^+$). HPLC (Chiralpak IA, n-hexane/i-propanol = 50/50, flow rate 1.0 mL/min, λ = 220 nm), $t_{major}$ = 36.2 min, $t_{minor}$ = 15.3 min, de > 99%.

## *Ni(II)-(S)-BPB/(2S,3R,4S)-2-Amino-4-cyano-5-ethoxy-5-oxo-3-phenylpentanoic Acid Schiff Base Complex (7m)*

Yield = 78%, m.p. 192.2–193.5 °C. $[\alpha]_D^{20}$ = +2323 (ca. 0.03 g/100 mL, $CH_2Cl_2$). $^1$H-NMR ($CDCl_3$) δ 8.24 (d, J = 8.6 Hz, 1H), 7.99 (d, J = 7.1 Hz, 2H), 7.74–7.58 (m, 3H), 7.53 (s, 3H), 7.40 (d, J = 7.3 Hz, 1H), 7.35–7.26 (m, 4H), 7.22–7.08 (m, 3H), 6.70 (q, J = 7.7 Hz, 2H), 4.63 (s, 1H), 4.57 (d, J = 12.0 Hz, 1H), 4.18 (d, J = 12.6 Hz, 1H), 3.85 (q, J = 6.9 Hz, 2H), 3.39 (t, J = 12.9 Hz, 2H), 3.22 (t, J = 8.4 Hz, 1H), 2.93 (dt, J = 9.3, 4.6 Hz, 1H), 2.17 (dt, J = 16.0, 8.1 Hz, 1H), 2.02 (dd, J = 12.6, 6.5 Hz, 1H), 1.94 (dd, J = 18.3, 8.6 Hz, 1H), 1.82 (dt, J = 19.5, 6.8 Hz, 1H), 1.47 (ddd, J = 19.2, 12.4, 6.7 Hz, 1H), 0.90 (t, J = 6.9 Hz, 3H). $^{13}$C-NMR ($CDCl_3$) δ 180.28, 176.58, 173.02, 164.34, 143.18, 134.17, 133.89, 133.71, 133.30, 132.85, 131.46, 130.59, 130.36, 129.32, 129.19, 129.12, 128.88, 128.80, 127.70, 127.12, 125.83, 123.20, 120.67, 114.69, 71.09, 70.50, 63.81, 62.42, 57.36, 48.45, 38.92, 30.62, 23.06, 13.48. HRMS (m/z): calcd. $C_{39}H_{36}N_4NaNiO_5^+$ for 721.1931, found 721.1931 ($[M+Na]^+$). HPLC (Chiralpak IA, n-hexane/i-propanol = 50/50, flow rate 1.0 mL/min, λ = 220 nm), $t_{major}$ = 31.7 min, $t_{minor}$ = 8.0 min, de = 98%.

## *Ni(II)-(S)-BPB/(2S,3R,4S)-2-Amino-3-(2-bromophenyl)-4-cyano-5-ethoxy-5-oxopentanoic Acid Schiff Base Complex (7n)*

Yield = 88%, m.p. 191.3–192.1 °C. $[\alpha]_D^{20}$ = +2120 (ca. 0.03 g/100 mL, $CH_2Cl_2$). $^1$H-NMR ($CDCl_3$) δ 8.44 (d, J = 8.7 Hz, 1H), 7.96 (d, J = 7.3 Hz, 2H), 7.88 (d, J = 7.8 Hz, 1H), 7.69–7.61 (m, 3H), 7.58–7.53 (m, 1H), 7.53–7.44 (m, 2H), 7.43–7.36 (m, 1H), 7.29 (dd, J = 10.5, 4.9 Hz, 3H), 7.16 (dt, J = 13.7, 5.3 Hz, 2H), 6.73–6.63 (m, 2H), 4.62 (d, J = 3.0 Hz, 1H), 4.49 (d, J = 12.2 Hz, 1H), 4.17 (d, J = 12.6 Hz, 1H), 4.06 (dd, J = 12.2, 3.0 Hz, 1H), 3.89 (qd, J = 7.1, 2.3 Hz, 2H), 3.43 (d, J = 12.6 Hz, 1H), 3.25 (t, J = 8.6 Hz, 1H), 2.85-2.75 (m, 1H), 2.19 (ddd, J = 19.3, 13.1, 7.3 Hz, 2H), 1.92 (dt, J = 11.2, 8.1 Hz, 1H), 1.70 (dt, J = 13.6, 7.5 Hz,

1H), 1.46 (dt, J = 18.8, 6.4 Hz, 1H), 0.95 (t, J = 7.1 Hz, 3H). $^{13}$C-NMR ($CDCl_3$) δ 180.14, 176.38, 174.21, 163.71, 143.31, 134.68, 134.06, 133.87, 133.70, 133.24, 133.05, 131.42, 130.80, 130.31, 130.03, 129.76, 128.87, 128.79, 128.55, 128.06, 127.12, 127.05, 125.84, 122.92, 120.52, 113.84, 71.89, 70.74, 63.75, 62.67, 57.14, 46.61, 39.74, 30.89, 23.02, 13.50, 0.01. HRMS (m/z): calcd. $C_{39}H_{35}BrN_4NaNiO_5^+$for 799.1037, found 799.1034 ([M+Na]$^+$). HPLC (Chiralpak IA, n-hexane/i-propanol = 50/50, flow rate 1.0 mL/min, λ = 220 nm), $t_{major}$ =42.8 min, $t_{minor}$ =10.6 min, de > 99%.

### *Ni(II)-(S)-BPB/(2S,3R,4S)-2-Amino-(3-chlorophenyl)-4-cyano-5-ethoxy-3-5-oxopentanoic Acid Schiff Base Complex (7o)*

Yield = 89%, m.p. 191.2–193.4 °C. $[\alpha]_D^{20}$ = +2376 (ca. 0.03 g/100 mL, $CH_2Cl_2$). $^1$H-NMR ($CDCl_3$) δ 8.27 (d, J = 8.7 Hz, 1H), 8.00 (d, J = 7.4 Hz, 2H), 7.70 (dd, J = 11.3, 4.9 Hz, 1H), 7.67–7.59 (m, 2H), 7.53 (d, J = 8.2 Hz, 1H), 7.46 (t, J = 7.8 Hz, 1H), 7.42–7.34 (m, 2H), 7.31 (t, J = 7.6 Hz, 2H), 7.21–7.13 (m, 3H), 7.11 (d, J = 7.2 Hz, 1H), 6.76–6.64 (m, 2H), 4.62 (d, J = 3.7 Hz, 1H), 4.53 (d, J = 12.2 Hz, 1H), 4.19 (d, J = 12.6 Hz, 1H), 3.89 (q, J = 7.1 Hz, 2H), 3.42 (d, J = 12.6 Hz, 1H), 3.33 (dd, J = 12.2, 3.7 Hz, 1H), 3.26 (dd, J = 9.3, 7.7 Hz, 1H), 3.02 (dd, J = 10.6, 5.8 Hz, 1H), 2.23 (td, J = 17.0, 7.6 Hz, 1H), 2.09 (dd, J= 13.4, 7.4 Hz, 1H), 1.99 (dd, J = 10.8, 6.3 Hz, 1H), 1.93 (dd, J = 14.0, 7.2 Hz, 1H), 1.62 (d, J = 12.7 Hz, 1H), 0.96 (t, J = 7.1 Hz, 3H). $^{13}$C-NMR () δ 180.29, 176.27, 173.26, 164.17, 143.30, 136.44, 135.58, 133.91, 133.62, 133.34, 133.04, 131.45, 130.68, 130.43, 129.42, 129.26, 128.92, 128.83, 127.58, 127.06, 125.66, 123.25, 120.68, 114.35, 70.79, 70.54, 63.88, 62.61, 57.57, 48.07, 38.81, 30.76, 23.03, 13.51. HRMS (m/z): calcd. $C_{39}H_{35}ClN_4NaNiO_5^+$for 755.1542, found 755.1541 ([M+Na]$^+$). HPLC (Chiralpak IA,n-hexane/ i-propanol = 50/50, flow rate 1.0 mL/min, λ = 220 nm), $t_{major}$ = 26.6 min, $t_{minor}$ = 8.1 min, de = 97%.

### *Ni(II)-(S)-BPB/(2S,3R,4S)-2-Amino-4-cyano-5-ethoxy-3-(4-fluorophenyl)-5-oxopentanoic Acid Schiff Base Complex (7p)*

Yield = 75%, m.p. 192.3–193.5 °C. $[\alpha]_D^{20}$ = +2250 (ca. 0.03 g/100 mL, $CH_2Cl_2$). $^1$H-NMR ($CDCl_3$) δ 8.25 (d, J = 8.7 Hz, 1H), 8.00 (d, J = 7.5

Hz, 2H), 7.69 (t, J = 7.0 Hz, 1H), 7.67–7.59 (m, 2H), 7.39 (d, J = 7.3 Hz, 1H), 7.30 (dd, J = 14.3, 6.6 Hz, 4H), 7.22(d, J = 8.5, 2H), 7.16 (t, J = 7.3 Hz, 3H), 6.70 (q, J = 8.2 Hz, 2H), 4.62 (d, J = 3.4 Hz, 1H), 4.52 (d, J = 12.2 Hz, 1H), 4.19 (d, J = 12.6 Hz, 1H), 3.88 (q, J = 7.1 Hz, 2H), 3.41 (d, J = 12.6 Hz, 1H), 3.36 (dd, J = 12.2, 3.5 Hz, 1H), 3.29–3.22 (m, 1H), 2.99 (dd, J = 10.1, 5.7 Hz, 1H), 2.25 (td, J = 16.7, 7.6 Hz, 1H), 2.09 (dt, J = 16.1, 8.6 Hz, 1H), 1.96 (dt, J = 14.1, 9.6 Hz, 2H), 1.66–1.55 (m, 1H), 0.95 (t, J = 7.1 Hz, 3H). $^{13}$C-NMR ($CDCl_3$) δ 180.32, 176.43, 173.22, 164.62, 164.23, 162.15, 143.22, 133.90, 133.68, 133.33, 132.97, 131.43, 130.65, 130.44, 130.04, 130.00, 129.23, 128.93, 128.83, 127.58, 127.09, 125.70, 123.22, 120.70, 116.40, 116.19, 114.44, 99.99, 70.99, 70.48, 63.91, 62.55, 57.48, 47.82, 38.99, 30.68, 22.94, 13.55. HRMS (m/z): calcd. $C_{39}H_{35}FN_4NaNiO_5^+$for 739.1837, found 739.1837 ($[M+Na]^+$). HPLC (Chiralpak IA, n-hexane/i-propanol = 50/50, flow rate 1.0 mL/min, λ = 220 nm), $t_{major}$ = 53.2 min, $t_{minor}$ = 6.9 min, de = 96%.

### *Ni(II)-(S)-BPB/(2S,3R,4S)-2-Amino-4-cyano-3-(3,4-dichlorophenyl)-5-ethoxy-5-oxopentanoic Acid Schiff Base Complex (7q)*

Yield = 82%, m.p. 192.7–194.7 °C. $[\alpha]_D^{20}$ = +2353 (ca. 0.03 g/100 mL, $CH_2Cl_2$). $^1$H-NMR ($CDCl_3$) δ 8.27 (d, J = 8.7 Hz, 1H), 8.02 (d, J = 7.6 Hz, 2H), 7.70 (t, J = 7.1 Hz, 1H), 7.63 (dd, J = 13.5, 8.8 Hz, 3H), 7.44 (s, 1H), 7.38 (d, J = 7.4 Hz, 1H), 7.31 (t, J = 7.6 Hz, 2H), 7.20–7.11 (m, 3H), 7.08 (d, J = 7.6 Hz, 1H), 6.78–6.64 (m, 2H), 4.60 (d, J = 3.6 Hz, 1H), 4.51 (d, J = 12.1 Hz, 1H), 4.19 (d, J = 12.5 Hz, 1H), 3.94 (q,J = 7.1 Hz, 2H), 3.41 (d, J = 12.6 Hz, 1H), 3.35–3.25 (m, 2H), 3.00 (dt, J = 10.2, 5.2 Hz, 1H), 2.30 (dt, J = 16.8, 7.7 Hz, 1H), 2.09 (dd, J = 13.3, 6.0 Hz, 1H), 2.02 (dd, J = 11.8, 7.4 Hz, 1H), 1.95 (dd, J = 13.4, 6.3 Hz, 1H), 1.67 (dt, J = 13.0, 6.4 Hz, 1H), 1.02 (t, J = 7.1 Hz, 3H). $^{13}$C-NMR ($CDCl_3$) δ 180.36, 176.15, 173.45, 164.02, 143.31, 134.66, 133.94, 133.80, 133.56, 133.37, 133.17, 131.42, 131.08, 130.75, 130.50, 129.30, 128.96, 128.86, 127.48, 127.04, 125.53, 123.27, 120.75, 114.12, 70.67, 70.54, 64.03, 62.82, 57.78, 47.57, 38.59, 30.75, 22.83, 13.59. HRMS (m/z): calcd. $C_{39}H_{34}Cl_2N_4NaNiO_5^+$for 789.1152, found 789.1151 ($[M+Na]^+$). HPLC (Chiralpak IA, n-hexane/i-propanol = 50/50, flow rate 1.0 mL/min, λ = 220 nm), $t_{major}$ = 44.5 min, $t_{minor}$ = 7.6 min, de = 98%.

### ***Ni(II)-(S)-BPB/(2S,3R,4S)-2-Amino-4-cyano-3-(2,4-dichlorophenyl)-5-ethoxy-5-oxopentanoic Acid Schiff Base Complex (7r)***

Yield = 96%, m.p. 192.6–194.5 °C. $[\alpha]_D^{20}$ = +1960 (ca. 0.03 g/100 mL, $CH_2Cl_2$). $^1$H-NMR ($CDCl_3$) δ 8.42 (d, J = 8.8 Hz, 1H), 7.99 (d, J = 7.6 Hz, 2H), 7.71 (s, 1H), 7.69–7.59 (m, 3H), 7.44 (s, 2H), 7.41 (d, J = 3.5 Hz, 1H), 7.30 (dd, J = 13.9, 6.7 Hz, 3H), 7.19–7.11 (m, 2H), 6.72–6.63 (m, 2H), 4.60 (d, J = 3.1 Hz, 1H), 4.48 (d, J = 12.2 Hz, 1H), 4.18 (d, J = 12.6 Hz, 1H), 4.05 (dd, J = 12.2, 3.1 Hz, 1H), 3.99–3.88 (m, 2H), 3.43 (d, J = 12.6 Hz, 1H), 3.34–3.22 (m, 1H), 2.90 (dt, J = 11.0, 5.6 Hz, 1H), 2.33 (dt, J = 16.8, 7.7 Hz, 1H), 2.16 (td, J = 13.6, 7.7 Hz, 1H), 2.00 (dt, J = 10.9, 7.6 Hz, 1H), 1.88 (dt, J = 14.2, 7.4 Hz, 1H), 1.69–1.53 (m, 1H), 1.01 (t, J = 7.1 Hz, 3H). $^{13}$C-NMR ($CDCl_3$) δ 180.23, 176.30, 174.38, 163.63, 143.32, 137.69, 135.70, 134.05, 133.59, 133.31, 133.17, 131.56, 131.39, 130.83, 130.43, 130.19, 130.14, 129.65, 128.91, 128.83, 128.28, 127.08, 125.58, 122.92, 120.58, 113.73, 71.39, 70.71, 63.95, 62.85, 57.45, 43.63, 39.21, 30.98, 22.89, 13.56. HRMS (m/z): calcd. $C_{39}H_{34}Cl_2N_4NaNiO_5^+$for 789.1152, found 789.1151 ($[M+Na]^+$). HPLC (Chiralpak IA, n-hexane/i-propanol = 50/50, flow rate 1.0 mL/min, λ = 220 nm), $t_{major}$ = 77.3, $t_{minor}$ = 8.8 min, de = 98%.

### ***Ni(II)-(S)-BPB/(2S,3R,4S)-2-Amino-4-cyano-5-ethoxy-5-oxo-3-(p-tolyl)pentanoic Acid Schiff Base Complex (7s)***

Yield = 76%, m.p. 182.3–183.7 °C. $[\alpha]_D^{20}$ = +2570 (ca. 0.03 g/100 mL, $CH_2Cl_2$). $^1$H-NMR ($CDCl_3$) δ 8.24 (d, J = 8.7 Hz, 1H), 8.00 (d, J = 7.3 Hz, 2H), 7.68 (dt, J = 9.7, 4.3 Hz, 1H), 7.61 (dd, J = 9.1, 5.4 Hz, 2H), 7.39 (d, J = 7.4 Hz, 1H), 7.31 (dd, J = 16.1, 8.0 Hz, 4H), 7.21–7.10 (m, 5H), 6.75–6.65 (m, 2H), 4.60 (d, J = 3.7 Hz, 1H), 4.54 (d, J = 12.2 Hz, 1H), 4.19 (d, J = 12.6 Hz, 1H), 3.86 (q, J = 7.1 Hz, 2H), 3.41 (d,J = 12.6 Hz, 1H), 3.33 (dd, J = 12.2, 3.7 Hz, 1H), 3.26–3.19 (m, 1H), 3.01–2.93 (m, 1H), 2.43 (s, 3H), 2.20 (dt, J = 16.7, 7.9 Hz, 1H), 2.04 (dt, J = 13.1, 6.8 Hz, 1H), 1.95 (dd, J = 11.1, 7.2 Hz, 1H), 1.83 (tt, J = 15.3, 7.6 Hz, 1H), 1.49 (tt, J = 12.8, 6.4 Hz, 1H), 0.94 (t, J = 7.1 Hz, 3H). $^{13}$C-NMR ($CDCl_3$) δ 180.19, 176.62, 172.87, 164.36, 143.12, 138.95, 133.86, 133.71, 133.31, 132.80, 131.45, 130.97, 130.55, 130.30, 129.97, 129.17, 128.87, 128.79, 127.70, 127.11, 125.83, 123.17, 120.65, 114.76, 71.12, 70.55, 63.87, 62.37, 57.40, 48.10, 38.85, 30.47, 22.78, 21.27, 13.51. HRMS (m/z): calcd.

$C_{40}H_{38}N_4NaNiO_5^+$ for 735.2088, found 735.2089 ([M+Na]$^+$). HPLC (Chiralpak IA, n-hexane/i-propanol = 50/50, flow rate 1.0 mL/min, $\lambda$ = 220 nm), $t_{major}$ = 46.3 min, $t_{minor}$ = 7.5 min, de > 99%.

### Ni(II)-(S)-BPB/(2S,3R,4S)-2-Amino-4-cyano-5-ethoxy-3-(4-methoxylphenyl)-5-oxopentanoic Acid Schiff Base Complex (7t)

Yield = 77%, m.p. 188.5–189.4 °C. $[\alpha]_D^{20}$ = +2376 (ca. 0.03 g/100 mL, $CH_2Cl_2$). $^1$H-NMR ($CDCl_3$) δ 8.23 (d, J = 8.7 Hz, 1H), 8.01 (d, J = 7.5 Hz, 2H), 7.68 (t, J = 6.9 Hz, 1H), 7.65–7.58 (m, 2H), 7.38 (d, J = 7.4 Hz, 1H), 7.30 (t, J = 7.6 Hz, 2H), 7.21 (d, J = 8.1 Hz, 2H), 7.15 (dd, J = 13.2, 6.6 Hz, 3H), 7.04 (d, J = 8.5 Hz, 2H), 6.74–6.65 (m, 2H), 4.60 (d, J = 3.5 Hz, 1H), 4.52 (d, J = 12.2 Hz, 1H), 4.19 (d, J = 12.6 Hz, 1H), 3.91–3.86 (m, 2H), 3.85 (s, 3H), 3.40 (d, J = 12.6 Hz, 1H), 3.32 (dd, J = 12.2, 3.5 Hz, 1H), 3.24 (t, J= 8.5 Hz, 1H), 3.04–2.96 (m, 1H), 2.21 (dt, J = 16.3, 7.6 Hz, 1H), 2.08 (dd, J = 13.3, 6.2 Hz, 1H), 1.98 (dd, J = 11.0, 6.9 Hz, 1H), 1.90 (dd, J = 13.5, 6.8 Hz, 1H), 1.56–1.47 (m, 1H), 0.95 (t, J = 7.1 Hz, 3H).$^{13}$C-NMR ($CDCl_3$) δ 180.28, 176.63, 172.88, 164.42, 160.34, 143.15, 133.85, 133.73, 133.37, 132.79, 131.46, 130.56, 130.32, 129.17, 128.88, 128.79, 127.67, 127.13, 125.85, 125.78, 123.18, 120.65, 114.73, 114.59, 71.20, 70.59, 63.92, 62.37, 57.55, 55.34, 47.89, 38.97, 30.62, 22.94, 13.56. HRMS (m/z): calcd. $C_{40}H_{38}N_4NaNiO_6^+$ for 751.2037, found 751.2037 ([M+Na]$^+$). HPLC (Chiralpak IA, n-hexane/i-propanol = 50/50, flow rate 1.0 mL/min, $\lambda$ = 220 nm), $t_{major}$ = 71.2 min, $t_{minor}$ = 8.5 min, de = 97%.

### Ni(II)-(S)-BPB/(2S,3R,4S)-2-Amino-4-cyano-5-ethoxy-3-(4-nitrophenyl)-5-oxopentanoic Acid Schiff Base Complex (7u)

Yield = 69%, m.p. 206.5–208.6 °C. $[\alpha]_D^{20}$ = +2163 (ca. 0.03 g/100 mL, $CH_2Cl_2$). $^1$H-NMR ($CDCl_3$) δ 8.40 (d, J = 8.5 Hz, 2H), 8.28 (d, J = 8.7 Hz, 1H), 7.98 (d, J = 7.5 Hz, 2H), 7.72 (t, J = 7.1 Hz, 1H), 7.69–7.60 (m, 2H), 7.48 (d, J = 8.4 Hz, 2H), 7.41 (d, J = 7.4 Hz, 1H), 7.30 (t, J = 7.6 Hz, 2H), 7.21–7.11 (m, 3H), 6.75–6.67 (m, 2H), 4.67 (d, J = 3.5 Hz, 1H), 4.62 (d, J = 12.1 Hz, 1H), 4.17 (d, J = 12.6 Hz, 1H), 3.96–3.86 (m, 2H), 3.48 (dd, J = 12.1, 3.5 Hz, 1H), 3.40 (d, J = 12.6 Hz, 1H), 3.23 (dd, J = 9.7, 7.1 Hz, 1H), 2.95–2.87 (m, 1H), 2.17 (dt, J = 17.7, 8.9 Hz, 1H), 1.97–1.86

(m, 2H), 1.68 (dd, J = 17.8, 10.7 Hz, 1H), 1.55–1.46 (m, 1H), 1.00 (t, J = 7.1 Hz, 3H). $^{13}$C-NMR ($CDCl_3$) δ 180.25, 176.03, 173.78, 163.86, 148.69, 143.31, 142.00, 134.01, 133.59, 133.32, 133.25, 131.36, 130.83, 130.64, 129.33, 129.00, 128.89, 127.47, 127.03, 125.46, 124.23, 123.26, 120.85, 113.96, 70.71, 70.32, 63.95, 62.93, 57.36, 47.97, 38.63, 30.64, 22.83, 13.61. HRMS (m/z): calcd. $C_{39}H_{35}N_5NaNiO_7^+$for 766.1782, found 766.1782 ($[M+Na]^+$). HPLC (Chiralpak IA, n-hexane/i-propanol = 50/50, flow rate 1.0 mL/min, λ = 220 nm), $t_{major}$ = 69.0 min, $t_{minor}$ = 8.6 min, de = 97%.

### *Ni(II)-(S)-BPB/(2S,3R,4S)-2-Amino-3-(4-(tert-butyl) phenyl)-4-cyano-5-ethoxy-5-oxopentanoic Acid Schiff Base Complex (7v)*

Yield = 51%, m.p. 200.8–201.7 °C. $[\alpha]_D^{20}$ = +2260 (ca. 0.03 g/100 mL, $CH_2Cl_2$). $^1$H-NMR ($CDCl_3$) δ 8.29 (d, J = 8.7 Hz, 1H), 7.95 (d, J = 7.4 Hz, 2H), 7.72–7.66 (m, 1H), 7.66–7.59 (m, 2H), 7.50 (d, J = 8.3 Hz, 2H), 7.40 (d, J = 7.4 Hz, 1H), 7.30 (t, J = 7.6 Hz, 2H), 7.17 (dt, J = 15.7, 7.6 Hz, 5H), 6.75–6.66 (m, 2H), 4.63 (d, J = 3.7 Hz, 1H), 4.53 (d, J = 12.2 Hz, 1H), 4.20 (d, J = 12.7 Hz, 1H), 3.86–3.76 (m, 2H), 3.49 (d, J = 12.7 Hz, 1H), 3.32 (dd, J = 12.2, 3.7 Hz, 1H), 3.17 (dd, J = 9.8, 7.4 Hz, 1H), 3.00 (dd, J = 10.3, 6.8 Hz, 1H), 2.18 (dt, J = 18.1, 8.3 Hz, 1H), 2.06 (dd, J = 11.2, 6.6 Hz, 1H), 1.89 (dd, J = 19.4, 8.6 Hz, 1H), 1.80 (dd, J = 10.3, 6.2 Hz, 1H), 1.48 (d, J = 8.3 Hz, 1H), 1.34 (s, 9H), 0.79 (t, J = 7.1 Hz, 3H). $^{13}$C-NMR ($CDCl_3$) δ 180.08, 176.69, 173.06, 164.52, 151.93, 143.13, 133.93, 133.77, 133.17, 132.84, 131.41, 130.88, 130.56, 130.34, 129.16, 128.85, 128.80, 127.82, 127.07, 126.29, 126.25, 125.92, 123.10, 120.69, 114.80, 70.98, 70.48, 63.49, 62.22, 56.56, 48.08, 39.24, 34.76, 31.35, 30.72, 22.77, 13.35. HRMS (m/z): calcd. $C_{43}H_{44}N_4NaNiO_5^+$for 777.2557, found 777.2557 ($[M+Na]^+$). HPLC (Chiralpak IA, n-hexane/i-propanol = 50/50, flow rate 1.0 mL/min, λ = 220 nm), $t_{major}$ = 58.6 min, $t_{minor}$ = 8.7 min, de = 97%.

### *Ni(II)-(S)-BPB/(2S,3R,4S)-2-Amino-4-cyano-5-ethoxy-3-(naphthalen-1-y)-5-oxopentanoic Acid Schiff Base Complex (7w)*

Yield = 67%, m.p. 188.4–190.3 °C. $[\alpha]_D^{20}$ = +1793 (ca. 0.03 g/100 mL, $CH_2Cl_2$). $^1$H-NMR ($CDCl_3$) δ 8.25 (d, J = 8.7 Hz, 1H), 8.00 (dd, J = 16.3,

8.0 Hz, 2H), 7.90 (d, J = 7.5 Hz, 2H), 7.71 (dd, J = 16.1, 8.1 Hz, 3H), 7.66–7.56 (m, 3H), 7.54–7.44 (m, 2H), 7.41 (d, J = 7.7 Hz, 1H), 7.23 (d, J = 7.4 Hz, 2H), 7.18–7.07 (m, 3H), 6.77 (d, J = 8.2 Hz, 1H), 6.71 (t, J = 7.5 Hz, 1H), 4.80 (s, 1H), 4.76 (d, J = 12.1 Hz, 1H), 4.30 (d, J = 10.7 Hz, 1H), 4.02 (d, J = 12.6 Hz, 1H), 3.80–3.65 (m, 2H), 3.25 (d, J = 12.5 Hz, 1H), 2.94 (t,J = 8.7 Hz, 1H), 2.51 (dt, J = 11.5, 5.9 Hz, 1H), 1.89–1.72 (m, 2H), 1.24 (dt, J = 13.6, 6.8 Hz, 2H), 0.94 (dt, J = 21.1, 7.7 Hz, 1H), 0.67 (t, J = 7.1 Hz, 3H). $^{13}$C-NMR ($CDCl_3$) δ 179.70, 176.53, 173.06, 164.07, 143.34, 134.19, 133.94, 133.72, 133.47, 133.23, 133.08, 131.36, 131.22, 130.96, 130.45, 129.86, 129.42, 128.98, 128.76, 128.69, 127.37, 126.94, 126.86, 126.59, 126.02, 125.80, 125.42, 122.98, 122.94, 120.40, 114.48, 72.69, 70.32, 63.60, 62.39, 57.23, 43.09, 39.99, 30.34, 22.94, 13.28. HRMS (m/z): calcd. $C_{43}H_{38}N_4NaNiO_5^+$for 771.2088, found 771.2088 ([M+Na]$^+$). HPLC (Chiralpak IA, n-hexane/i-propanol = 50/50, flow rate 1.0 mL/min, λ = 220 nm), $t_{major}$ = 31.7 min, $t_{minor}$ = 12.3 min, de > 99%.

## Procedure for the Synthesis of (2S,3R)-8a

In a typical procedure, 3 mol/L HCl (3.33 mL, 5.0 mmol) was added to a solution of the (S,2S,3R)-7a (1.0 mmol) dissolved in THF (13 mL). The reaction was stirred for 12 h or until the red color of the solution disappeared and was then concentrated under vacuum to half of the original volume. In the case of (S,2S,3R)-7a, the (S)-BPB was recovered from the aqueous portion by extracting with ethyl acetate (EA) and was washed with water. The organic layer was removed, and the aqueous portion was diluted with water (2 mL). The aqueous portion was transferred to a clean flask, and solid $NaHCO_3$ (336 mg, 4.0 mmol) was carefully added with stirring to neutralize the solution, followed by $Na_2$EDTA (372 mg, 1.0 mmol), and was stirred for 5 min. Additional solid $NaHCO_3$ (336 mg, 4.0 mmol) was added, followed by a solution of Fmoc-OSu (337 mg,1.0 mmol) in acetonitrile (5 mL). The reaction was stirred for 24 h under nitrogen, concentrated in vacuum to half of the original volume, adjusted to pH = 3 with 10% citric acid, and extracted with EtOAc twice. Combined organic layers were washed with brine, dried with anhydrous $MgSO_4$, concentrated, and purified on silica gel using a flash chromatography (petroleum ether/ethyl acetate = 1/2) to give (2S,3R)-8a as a white solid.

### 2-(((9H-Fluoren-9-yl)methoxy)carbonyl)-4,4-dicyano-3-phenylbutanoic Acid (8a)

$^{1}$H-NMR ($CDCl_3$) δ 8.02 (s, 1H), 7.68–7.54 (m, 3H), 7.54–7.41 (m, 2H), 7.21–6.97 (m, 6H), 6.58 (s, 1H), 4.42 (s, 1H), 4.35–4.17 (m, 2H), 4.11 (s, 1H), 4.06 (s, 1H), 3.82 (s, 1H). $^{13}$C-NMR (DMSO) δ 172.8, 162.3, 150.4, 143.2, 140.6, 140.6, 128.9, 128.8, 127.8, 127.7, 127.2, 127.2, 127.1, 126.4, 121.3, 120.1, 120.0, 109.7, 67.8, 59.7, 46.1, 25.1, 20.7. ESI-MS (m/z): calcd. 450.2, found 450.4 ($[M-H]^-$).

## CONCLUSIONS

We have reported the first asymmetric three-component reaction of chiral nickel(II) glycinate, aromatic aldehydes, and an α-carbanion of two electron-withdrawing groups (malononitrile or ethyl cyanoacetate) to give a series of novel α-amino-β-substituted γ,γ-disubstituted butyric acid derivatives. We have screened a series of reaction conditions and developed a practical system to promote the asymmetric three component reaction of chiral nicke(II) glycinate. This reaction, which constructed two carbon-carbon bonds and formed two or three chiral centers, provides a convenient synthesis of functionalized chiral Fmoc-α-amino-β-substituted γ,γ-disubstituted butyric acid derivatives. The transformation performed well with electron-deficient, electron-rich, condensed ring and sterically hindered aromatic aldehydes and addorded functionalized products. To our excitement, some of them had amazingly high diastereoselectivities, but the heteroaryl substitutes were not well tolerated. The absolute configurations of the typical products were determined. Further studies will focus on mechanistic aspects, expansion of substrate ranges, and further applications of other chiral nickel(II) complexes in important carbon-carbon bond-forming reactions.

## ACKNOWLEDGMENTS

This research has received financial support from the National

Natural Science Foundation of China (No. 81001357, 81273471 and 81303208) and the Open Research Fund of State Key Laboratory Breeding Base of Systematic Research, Development and Utilization of Chinese Medicine.

## Conflicts of Interest

The authors declare no conflict of interest.

## REFERENCES


1. Or, Y.S.; Ying, L.; Wang, C.; Long, J.; Qui, Y.L. Preparation of Substituted Pyrrolidine Derivatives for Use as Anti-Infective Agents. WO 2009003009, 31 December 2008.
2. Edwards, D.L.; Berens, M.E.; Beaudry, C. Identification of Chemical Useful for Treating e.g., Cancer, Comprising Depositing Cancer Cells on Substrate Using Guided Cell Sedimentation, Treating Cancer Cells with Library of Chemicals, and Measuring Cell Migration Rate. WO 2003102153, 11 December 2003.
3. Jones, D.T.; Harris, A.L. Identification of novel small-molecule inhibitors of hypoxia-inducible factor-1 transactivation and DNA binding. Mol. Cancer Ther. 2006, 5, 2193–2202, doi:10.1158/1535-7163.MCT-05-0443.
4. Dandia, A.; Jain, A.K.; Laxkar, A.K.; Bhati, D.S. Synthesis and stereochemical investigation of highly functionalized novel dispirobisoxindole derivatives via [3+2] cycloaddition reaction in ionic liquid.Tetrahedron 2013, 69, 2062–2069, doi:10.1016/j.tet.2012.12.021.
5. Erdbrink, H.; Peuser, I.; Gerling, U.I.M.; Lentz, D.; Koksch, B.; Czekelius, C. Conjugate hydrotrifluoromethylation of α,β-unsaturated acyl-oxazolidinones: Synthesis of chiral fluorinated amino acids.Org. Biomol. Chem. 2012, 10, 8583–8586, doi:10.1039/c2ob26810h.
6. Aouadi, K.; Jeanneau, E.; Msaddek, M.; Praly, J.P. 1,3-Dipolar cycloaddition of a chiral nitrone to (E)-1,4-dichloro-2-butene: A new efficient synthesis of (2S,3S,4R)-4-hydroxyisoleucine. Tetrahedron Lett. 2012,53, 1817–1821.
7. Tong, B.M.K.; Chiba, S. Diamine-catalyzed conjugate addition to acrylate derivatives. Org. Lett. 2011, 13, 2948–2951.
8. Dugave, C.; Cluzeau, J.; Menez, A.; Gaudry, M.; Marquet, A. Chemo-enzymic synthesis of protected cyano derivatives of glutamate. Tetrahedron Lett. 1998, 39, 5775–5778, doi:10.1016/S0040-4039(98)01207-6.
9. Heimgartner, H. 3-Amino-2H-azirines. Synthons for α,α-disubstituted α-amino acids in heterocycle and peptide synthesis. Angew. Chem. Int. Ed. 1991, 30, 238–264, doi:10.1002/anie.199102381.
10. Duthaler, R.O. Recent developments in the stereoselective synthesis

of α-amino acids. Tetrahedron 1994,50, 1539–1650, doi:10.1016/S0040-4020(01)80840-1.

11. Wirth, T. New strategies to alpha-alkylated alpha-amino acids. Angew. Chem. Int. Ed. 1997, 36, 225–227, doi:10.1002/anie.199702251.

12. Gibson, S.E.; Guillo, N.; Tozer, M.J. Towards control of chi-space: Conformationally constrained analogues of Phe, Tyr, Trp and His. Tetrahedron 1999, 55, 585–615, doi:10.1016/S0040-4020(98)00942-9.

13. Cativiela, C.; Diaz-de-Villegas, M.D. Stereoselective synthesis of quaternary alpha-amino acids. Part 1: Acyclic compounds. Tetrahedron Asymmetry 1998, 9, 3517–3599, doi:10.1016/S0957-4166(98)00391-7.

14. Cativiela, C.; Diaz-de-Villegas, M.D. Stereoselective synthesis of quaternary alpha-amino acids. Part 2. Cyclic compounds. Tetrahedron Asymmetry 2000, 11, 645–732, doi:10.1016/S0957-4166(99)00565-0.

15. Flores-Conde, M.I.; Reyes, L.; Herrera, R.; Rios, H.; Vazquez, M.A.; Miranda, R.; Tamariz, J.; Delgado, F. Highly regio- and stereoselective diels-alder cycloadditions via two-step and multicomponent reactions promoted by infrared irradiation under solvent-free conditions. Int. J. Mol. Sci. 2012, 13, 2590–2617, doi:10.3390/ijms13032590.

16. Belokon, Y.N. Chiral complexes of Ni(II), Cu(II), and Cu(I) as reagents, catalysts and receptors for asymmetric-synthesis and chiral recognition of amino-acids. Pure Appl. Chem. 1992, 64, 1917–1924, doi:10.1351/pac199264121917.

17. Belokon, Y.N.; Bespalova, N.B.; Churkina, T.D.; Cisarova, I.; Ezernitskaya, M.G.; Harutyunyan, S.R.; Hrdina, R.; Kagan, H.B.; Kocovsky, P.; Kochetkov, K.A.; et al. Synthesis of alpha-amino acids via asymmetric phase transfer-catalyzed alkylation of achiral nickel(II) complexes of glycine-derived Schiff bases. J. Am. Chem. Soc. 2003, 125, 12860–12871.

18. Belokon, Y.N.; Tararov, V.I.; Maleev, V.I.; Saveleva, T.F.; Ryzhov, M.G. Improved procedures for the synthesis of (S)-2-N-(N′-benzylprolyl)amino benzophenone (BPB) and Ni(II) complexes of Schiff's bases derived from BPB and amino acids. Tetrahedron Asymmetry 1998, 9, 4249–4252, doi:10.1016/S0957-4166(98)00449-2.

19. Hayashi, T.; Kishi, E.; Soloshonok, V.A.; Uozumi, Y. Erythro-selective aldol-type reaction of N-sulfonylaldimines with methyl isocyanoacetate catalyzed by gold(I). Tetrahedron Lett. 1996, 37, 4969–4972, doi:10.1016/0040-4039(96)00981-1.

20. Soloshonok, V.A.; Fokina, N.A.; Rybakova, A.V.; Shishkina, I.P.; Galushko, S.V.; Sorochinsky, A.E.; Kukhar, V.P.; Savchenko, M.V.; Svedas, V.K. Biocatalytic approach to enantiomerically pure beta-amino acids. Tetrahedron Asymmetry 1995, 6, 1601–1610, doi:10.1016/0957-4166(95)00204-3.

21. Soloshonok, V.A.; Soloshonok, I.V.; Kukhar, V.P.; Svedas, V.K. Biomimetic transamination of alpha-alkyl beta-keto carboxylic esters. Chemoenzymatic approach to the stereochemically defined alpha-alkyl beta-fluoroalkyl beta-amino acids. J. Org. Chem. 1998, 63, 1878–1884, doi:10.1021/jo971777m.

22. Deng, G.H.; Wang, J.; Zhou, Y.; Jiang, H.L.; Liu, H. One-pot, large-scale synthesis of Nickel(II) complexes derived from 2-N-(alpha-picolyl)amino benzophenone (PABP) and α- or β-amino acids. J. Org. Chem.2007, 72, 8932–8934, doi:10.1021/jo071011e.

23. Lin, D.Z.; Deng, G.H.; Wang, J.; Ding, X.; Jiang, H.L.; Liu, H. Efficient Synthesis of Symmetrical alpha,alpha-Disubstituted beta-Amino Acids and alpha,alpha-Disubstituted Aldehydes via Dialkylation of Nucleophilic beta-Alanine Equivalent. J. Org. Chem. 2010, 75, 1717–1722, doi:10.1021/jo902699t.

24. Wang, J.; Lin, D.Z.; Shi, J.M.; Ding, X.; Zhang, L.; Jiang, H.L.; Liu, H. Highly enantio- and diastereoselective mannich reaction of a chiral Nickel(II) glycinate with an alpha-imino ester for asymmetric synthesis of a 3-aminoaspartate. Synthesis 2010, 7, 1205–1208.

25. Wang, J.; Zhou, S.B.; Lin, D.Z.; Ding, X.; Jiang, H.L.; Liu, H. Highly diastereo- and enantioselective synthesis of syn-beta-substituted tryptophans via asymmetric Michael addition of a chiral equivalent of nucleophilic glycine and sulfonylindoles. Chem. Commun. 2011, 47, 8355–8257.

26. Wang, J.; Ji, X.; Shi, J.M.; Sun, H.F.; Jiang, H.L.; Liu, H. Diastereoselective Michael reaction of chiral nickel(II) glycinate with nitroalkenes for asymmetric synthesis of beta-substituted alpha,gamma-diaminobutyric acid derivatives in water. Amino Acids 2012, 42, 1685–1694, doi:10.1007/s00726-011-0870-x.

27. Wang, J.; Liu, H.; Acena, J.L.; Houck, D.; Takeda, R.; Moriwaki, H.; Sato, T.; Soloshonok, V.A. Synthesis of bis-α,α′-amino acids through diastereoselective bis-alkylations of chiral Ni(II)-complexes of glycine. Org. Biomol. Chem. 2013, 11, 4508–4515, doi:10.1039/c3ob40594j.

28. Smith, D.J.; Yap, G.P.A.; Kelley, J.A.; Schneider, J.P. Enhanced stereoselectivity of a Cu(II) complex chiral auxiliary in the synthesis of Fmoc-L-γ-carboxyglutamic acid. J. Org. Chem. 2011, 76, 1513–1520, doi:10.1021/jo101940k.

29. Xie, X.; Peng, C.; He, G.; Leng, H.-J.; Wang, B.; Huang, W.; Han, B. Asymmetric synthesis of a structurally and stereochemically complex spirooxindole pyran scaffold through an organocatalytic multicomponent cascade reaction. Chem. Commun. 2012, 48, 10487–10489.

30. He, J.; Ouyang, G.; Yuan, Z.; Tong, R.; Shi, J.; Ouyang, L. A facile synthesis of functionalized dispirooxindole derivatives via a three-component 1,3-dipolar cycloaddition reaction. Molecules 2013, 18, 5142–5154, doi:10.3390/molecules18055142.

31. Hou, X.; Luo, H.; Zhong, H.; Wu, F.; Zhou, M.; Zhang, W.; Han, X.; Yan, G.; Zhang, M.; Lu, L.; et al. Analysis of furo 3,2-c tetrahydroquinoline and pyrano 3,2-c tetrahydroquinoline derivatives as antitumor agents and their metabolites by liquid chromatography/electrospray ionization tandem mass spectrometry.Rapid Commun. Mass Spectr. 2013, 27, 1222–1230, doi:10.1002/rcm.6562.

32. Li, X.; Yang, L.; Peng, C.; Xie, X.; Leng, H.-J.; Wang, B.; Tang, Z.-W.; He, G.; Ouyang, L.; Huang, W.; et al. Organocatalytic tandem Morita-Baylis-Hillman-Michael reaction for asymmetric synthesis of a drug-like oxa-

spirocyclic indanone scaffold. Chem. Commun. 2013, 49, 8692–8694, doi:10.1039/c3cc44004d.

33. Wu, G.; Ouyang, L.; Liu, J.; Zeng, S.; Huang, W.; Han, B.; Wu, F.; He, G.; Xiang, M. Synthesis of novel spirooxindolo-pyrrolidines, pyrrolizidines, and pyrrolothiazoles via a regioselective three-component 3+2 cycloaddition and their preliminary antimicrobial evaluation. Mol. Diver. 2013, 17, 271–283, doi:10.1007/s11030-013-9432-3.

# Chapter 2

# SYNTHESIS, CHARACTERIZATION AND APPLICATION OF NOVEL BISAZO REACTIVE DYES ON VARIOUS FIBERS

Divyesh R. Patel, Nikul S. Patel, Hemant S. Patel and Keshav C. Patel*

Synthetic Organic Chemistry Research Laboratory, Department of Chemistry, Veer Narmad South Gujarat University, Surat (Gujarat), India-395 007

## ABSTRACT

Ten hot brand bisazo reactive dyes (D1 to D10) have been synthesized by coupling bis(diazotised), 4,4′-methylene bis(2,6-dichloroaniline) (A) with various 5-sulfo anthranilo cyanurated coupling components (R) and their dyeing performance as reactive dyes has been assessed on silk, wool and cotton fibres. The purity of dyes was checked by TLC. The IR spectra and 1H-NMR spectra prove the structure of newly reactive dyes. The percentage dye bath exhaustion on different fibres was reasonable good and acceptable. The dyes exhibited high levels of light, washing and rubbing fastness.

## INTRODUCTION

Reactive dyes are colored compounds which contain one or two groups capable of forming covalent bond between a carbon atom or phosphorus atom of the dyes ion or molecules and an oxygen atom, nitrogen atom or sulfur atom of a hydroxyl, an amino or a mercapto group respectively, of the substrate [1]. These dyes are generally used on higher value clothes, which are normally mercerized [2].

Reactive dyes though late entry in to the field of synthetic dyes, very soon attained a commercial status. Several new reactive systems have been introduced from time to time, which covers the subject of innumerable patents and publication [3]. It was for the first time that dyeing has been done by chemical reaction between the dye and the fibre, enabling one to get assortment of bright, attractive shades of adequate fastness with considerable ease of dyeing. It can also be easily understood that dyes with two reactive groups give a higher fixation yield than dyes with one reactive group for it one of the two dye-fibre bonds is hydrolyzed, one is still left for fixation [4, 5]. Reactive dyes are well known and applied for dyeing of different materials [6]. Among them triazine derivatives have an important place [7].

s-Triazine based chemicals have been applied variously in the manufacture of polymers, dyes, drugs, explosives, pesticides and commodity chemicals [8] as a consequence, theoretical and experimental studies on these chemicals have been widely carried out [9, 10] with the result that the s-triazine ring is known as an important conjugated heterocycle whose electronics properties are expected to show suitable differences from those of benzene due to the alternate replacement of -CH- group by nitrogen atoms [11].

s-Triazine plays an important role in synthesized dyes. The key compound of reactive dyes are a cyanuric chloride in synthesized dyestuff have two reactive groups in their structure which give high fixation yields, excellent wet fastness, brilliant shade and simple application techniques in textile printing. The advantage owing to the chloro triazine groups (s-triazine) is that due to the electrophilic property of the cyanuric group, a wide range of chromophores having good fastness to light, perspiration and chlorine.

Hot-Brand reactive dyes have been widely considered due to their fixation yield on various fibers [12].

Various bisazo reactive dyes (Figure 1 and Figure 2) have been reported earlier which shows good dyeing properties on silk, wool and cotton [13, 14].

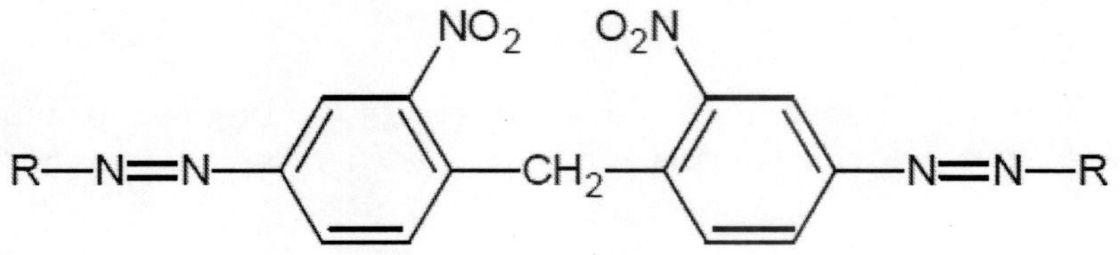

**Figure 1** Cold brand bisazo reactive dyes. Where R = Various cyanurated coupling components.

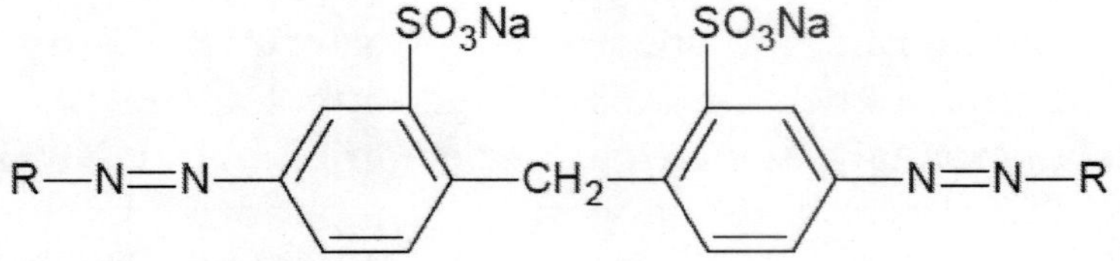

**Figure 2** Hot brand bisazo reactive dyes. Where R = Various m-nitro anilino cyanurated coupling components.

In a continuation of our work we report here the synthesis of some hot-brand reactive dyes with a higher degree of activity synthesis and study of the dyeing properties of the bisazo reactive dyes base on 4,4′-methylene bis(2,6-dichloroaniline). The reactive dyes of the following structure were prepared (Figure 3).

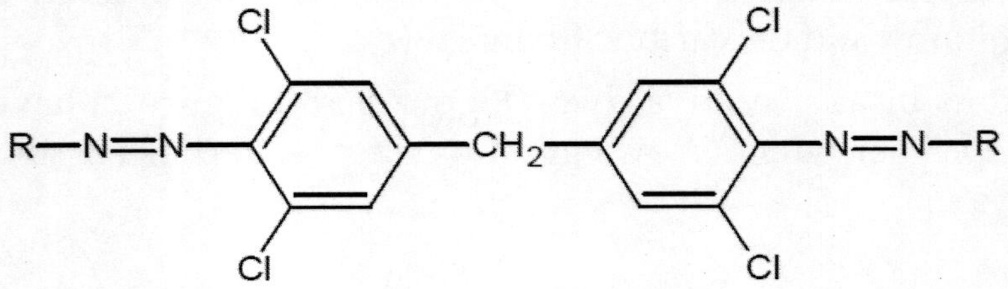

R—N=N—[3,5-dichloro-phenyl]—$CH_2$—[3,5-dichloro-phenyl]—N=N—R

**Figure 3** Hot brand bisazo reactive dyes. Where R = various 5-sulfo anthranilo cyanurated coupling components.

## MATERIAL AND METHODS

### General

All melting points are determined using a DSC 7, Perkin-Elmer (USA) Differential Scanning Caloriate (heating rate 5 °C/min, $N_2$ gas) digital melting apparatus and are uncorrected. IR spectra were recorded in KBr on a Perkin-Elmer model-377 spectrophotometer instrument. Elemental microanalyses were performed on a LECI CHN- 932 for C, H and N. Thin Layer chromatography was performed using silica-coated aluminum plates (60-F254, Merck) [15]. 1H NMR spectra were obtained with a Jeol JNM0- FX 200 at 300 MHz instrument, Using TMS as the internal standard and DMSO-*d*6 as a solvent. Chemical shifts are given in δ ppm. Absorption spectra were obtained with Becman DB-GT grafting spectrometer instrument. Fastness to light was assessed in accordance with BS 1006-1978 [16]. Rubbing fastness was carried out with an Atlas Crock meter in accordance with AATCC TM 8-1961 [17] and the wash fastness test in accordance with ISO: 765-1979 [18]. The entire regents were purchased from Merck and Renkem, which were of G. R. grade and used without further purification. All crude products were isolated as solids and purified by a combination of column chromatography and recrystallization.

### Chemistry

*Preparation Of 4-4′ Methylene Bis(2,6-Dichloroaniline) [19]:*

2,6-Dichloroaniline (14.8 G, 0.1 mole) was dissolved in water (125 mL) and 36.5% hydrochloric acid (25 mL) at 50 °C. The reaction mixture was then treated with 3% aqueous formaldehyde solution (35 mL). The temperature was maintained at 60 $^0$C and stirred for an hour and neutralized with 10% sodium hydroxide solution, yellow precipitates obtained were filtered, washed with hot water, dried and recrystallised from acetic acid. Yield 84%, m.p. 273 °C. IR (KBr) 3420 $cm^{-1}$, 3300 $cm^{-1}$ (N-H), 2850 $cm^{-1}$ (C-H). 1H NMR (DMSO): δ 8.2 (2H, s, $NH_2$), 3.45 (2H, s, $CH_2$), 7.05-7.15 (4H, m, Ar–H). Elemental analysis: Found C- 46.40%; H-2.94% N-8.32% $C_{13}H_{10}N_2Cl_4$ (MF require C-46.45%; H-2.99%; N-8.37%).

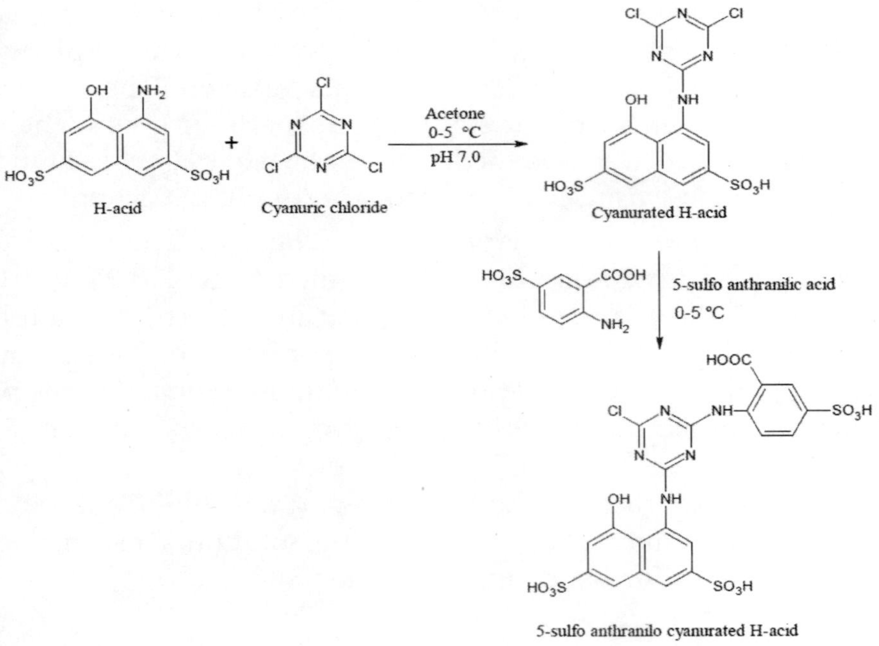

**Scheme 1** Preparation of 5-sulfo anthranilo cyanurated H-acid (R).

**Tetrazotisation Of 4-4′-Methylene Bis(2,6-Dichloroaniline): 4-4′-Methylene Bis(2,6-**dichloroaniline) (1.68 g, 0.005 mole) was suspended in $H_2O$ (60 mL). Hydrochloric acid (10 mL) was added drop wise to this well stirred suspension. The mixture was gradually heated up to 70 °C till clear solution obtained. The solution was cooled to 0-5 °C in an ice bath. A solution of $NaNO_2$ (1.38 g) in $H_2O$ (8 mL) previously cooled to 0 °C, was then added over a period of 5 minutes with stirring. The stirring was continued for an hour maintaining the same temperature, with positive test for nitrous acid with required amount of a solution of a sulphamic acid. The clear tetrazo solution (A) at 0-5 °C was used for subsequent coupling reaction.

## Preparation Of 5-Sulpho Anthranilo Cyanurated H-Acid (R)

- ***Cyanuration of h-acid:*** Cyanuric chloride (1.85 g, 0.01 mole) was stirred in acetone (50 mL) at a temperature below 5 °C for a period of an hour. A neutral solution of H-acid (3.19 g, 0.01 mole) aqueous sodium carbonate solution (10% w/v) was then added in small lots in about an hour, the neutral pH was maintained below 5 °C through this reaction. The reaction mass was then stirred at 0-5 °C for further four hours when a clear solution was obtained. The cyanurated H-acid solution thus formed was used for subsequent coupling reaction.
- ***Condensation with 5-sulpho anthranilic acid:*** The ice-cooled and well-stirred solution of cyanurated H-acid (4.67 g, 0.01 mole) was heated up to 40-50 °C for half an hour. To this 5-sulpho anthranilic acid (2.17 g, 0.01 mole) was added dropwise at same temperature during a period of 30 minutes maintaining the pH neutral, by simultaneous addition of sodium carbonate solution (1% w/v). After the addition was completed the stirring was continued for further 3 hours, to prepare a 5-sulpho anthranilo cyanurated H-acid (R). The resulting solution thus obtained was used for further coupling reaction.

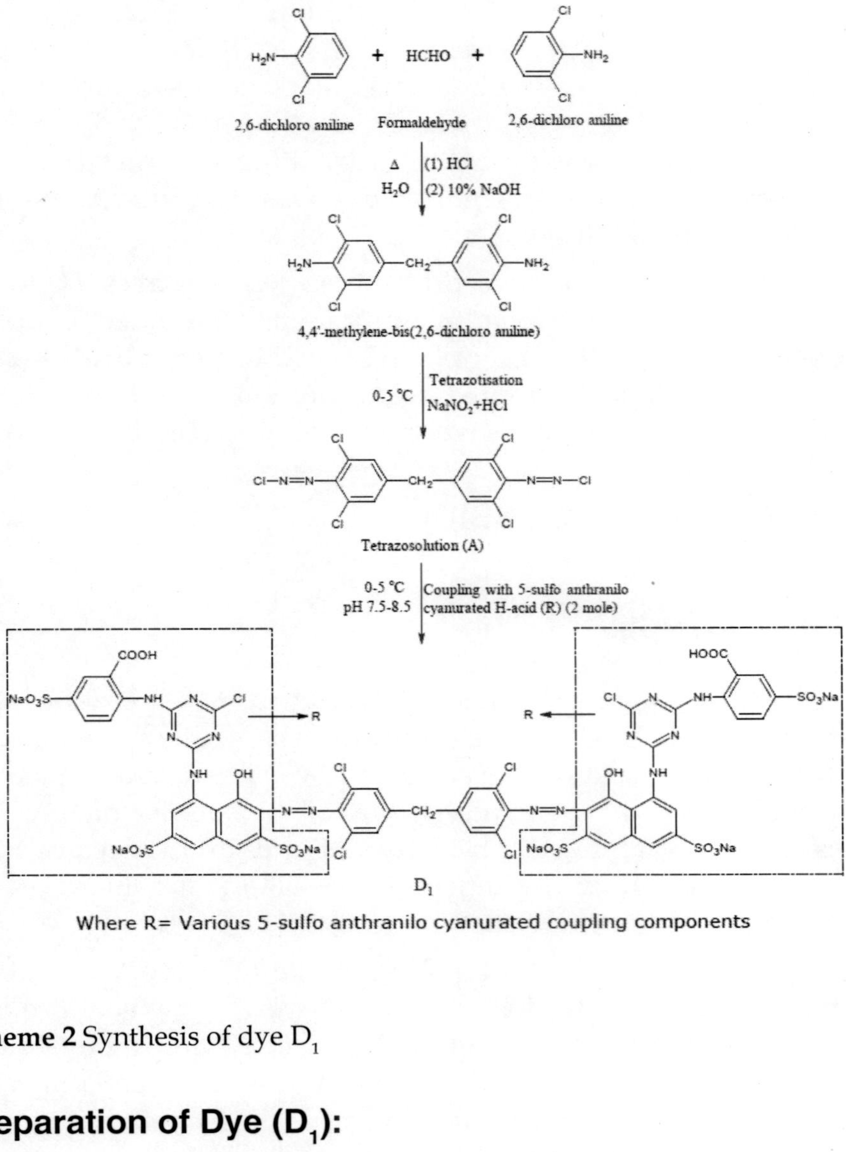

2,6-dichloro aniline + HCHO (Formaldehyde) + 2,6-dichloro aniline

Δ, $H_2O$; (1) HCl (2) 10% NaOH

4,4'-methylene-bis(2,6-dichloro aniline)

0-5 °C; Tetrazotisation $NaNO_2$+HCl

Tetrazosolution (A)

0-5 °C, pH 7.5-8.5; Coupling with 5-sulfo anthranilo cyanurated H-acid (R) (2 mole)

$D_1$

Where R= Various 5-sulfo anthranilo cyanurated coupling components

**Scheme 2** Synthesis of dye $D_1$

## Preparation of Dye ($D_1$):

To a well-stirred solution of 5-sulpho anthranilo cyanurated H-acid (R) (6.48 g, 0.01 mole), a freshly prepared solution of tetrazo solution (A) (2.155 g, 0.005 mole) was added drop wise over a period of 10-15 minutes. The pH was maintained at 7.5 to 8.5 by simultaneous addition of sodium carbonate solution (10% w/v). During coupling the purple solution was formed. Stirring was continued for 3-4 hours, maintaining the temperature below 5 °C. The reaction mixture was heated up to 60 °C and sodium chloride (15 g) added until the colouring material was precipitated. It was stirred for an hour, filtered and washed with a small amount of sodium chloride solution (5% w/v). The solid was dried at 80-90 °C and extracted with DMF. The dye was precipitated by diluting the DMF-extract with excess of chloroform. A purple dye was then filtered, washed with chloroform and dried at 60 °C.Yield 86%.

Following the above procedure other reactive dyes $D_2$ to $D_{10}$ were synthesized using 5-sulfo anthranilo cyanurated coupling components such as J-acid, N-phenyl-J-acid, Gamma-acid, K-acid, Chicago acid, Peri acid, Bronner acid, Tobias acid and Sulfo tobias acid. All the synthesized dyes were recorded in Table 1.

# RESULTS AND DISCUSSION

## Spectral Properties of Dyes

The absorption maxima ($\lambda_{max}$) of the dyes $D_1$ to $D_{10}$ were recorded in water and conc. H2SO4 and are shown in Table 4. The $\lambda_{max}$ values are directly proportional to the electronic power, nature and position of the substituents in the naphthy$_l$ ring of the coupler moiety. The values of log$_\varepsilon$ (molar extinction coefficient) are summarized in Table 3. All the values are in the range of 4.19-4.30, which indicates the dyes have high intensity of absorption.

Dye D2 have $\lambda_{max}$ is about 462 nm while D3 have $\lambda_{max}$ is about 482 nm. Here the introduction of phenyl ring which produce bathochromic effect and shifting the $\lambda_{max}$ value. Here 20 nm shifting in absorption is observed. Dye D4 have $\lambda_{max}$ is about 458 nm while D7 and D8 have $\lambda_{max}$ values are 435 nm and 440 nm respectively. Here the introduction of auxochrome like hydroxyl group which produce bathochromic effect and give rise to the 23 nm and 18 nm $\lambda_{max}$ value in D4 with respect to D7 and D8. Dyes D1, D5 and D6 have same groups but the positions of the groups are different so the oscillation of electron is fast in D5 and D6 as compare to D1. So dye D1 possesses higher $\lambda_{max}$ value as compare to D5 and D6. The shifting of 20 nm $\lambda_{max}$ value in D10 as compare to D9 due to the introduction of auxochrome like sulfonic acid group in D10, which increase the λmax value in D10 as compare to D9.

## Dyeing Properties of Dyes

All the dyes $D_1$ to $D_{10}$ were applied at 2% depth on silk, wool and cotton fibres according to usual procedure [21] in the dye bath containing materials as listed in Table 2.

**Table 1** Characterization data of dyes $D_1$ to $D_{10}$

| Dye No. | Various 5-sulfo anthranilo cyanurated coupling Components (R) | Molecular Formula | Mol. Wt. | Yield (%) | % C Found Req. | %H Found Req. | %N Found Req. | $^aR_f$ |
|---|---|---|---|---|---|---|---|---|
| $D_1$ | H-acid | $C_{53}H_{26}O_{24}N_{14}S_6Na_6Cl_6$ | 1788 | 86 | 35.57<br>35.60 | 1.46<br>1.51 | 10.96<br>10.99 | 0.43 |
| $D_2$ | J-acid | $C_{53}H_{24}O_{24}N_{14}S_4Na_4Cl_6$ | 1577 | 84 | 40.32<br>40.35 | 1.52<br>1.54 | 12.42<br>12.46 | 0.39 |
| $D_3$ | N-phenyl J-acid | $C_{65}H_{32}O_{24}N_{14}S_4Na_4Cl_6$ | 1839 | 84 | 42.41<br>42.45 | 1.74<br>1.78 | 10.65<br>10.68 | 0.40 |
| $D_4$ | Gamma-acid | $C_{53}H_{24}O_{24}N_{14}S_4Na_4Cl_6$ | 1577 | 85 | 40.33<br>40.38 | 1.52<br>1.57 | 12.42<br>12.45 | 0.44 |
| $D_5$ | K-acid | $C_{53}H_{26}O_{24}N_{14}S_6Na_6Cl_6$ | 1788 | 80 | 35.57<br>35.61 | 1.46<br>1.48 | 10.96<br>11.00 | 0.46 |
| $D_6$ | Chicago acid | $C_{53}H_{26}O_{24}N_{14}S_6Na_6Cl_6$ | 1788 | 85 | 35.57<br>35.62 | 1.46<br>1.49 | 10.96<br>11.01 | 0.42 |
| $D_7$ | Peri acid | $C_{53}H_{24}O_{24}N_{14}S_4Na_4Cl_6$ | 1545 | 87 | 41.17<br>41.20 | 1.55<br>1.59 | 12.68<br>12.71 | 0.45 |
| $D_8$ | Bronner acid | $C_{53}H_{24}O_{24}N_{14}S_4Na_4Cl_6$ | 1545 | 83 | 41.17<br>41.19 | 1.55<br>1.58 | 12.68<br>12.70 | 0.40 |
| $D_9$ | Tobias acid | $C_{53}H_{24}O_{24}N_{14}S_4Na_4Cl_6$ | 1545 | 80 | 41.17<br>41.22 | 1.55<br>1.59 | 12.68<br>12.71 | 0.40 |
| $D_{10}$ | Sulpho tobias acid | $C_{53}H_{22}O_{24}N_{14}S_6Na_6Cl_6$ | 1749 | 84 | 36.36<br>36.41 | 1.26<br>1.32 | 11.20<br>11.24 | 0.46 |

**Table 2** Dye-bath containing materials

| Materials | For silk | For wool | For cotton |
|---|---|---|---|
| Fabric | 2.0 g | 2.0 g | 2.0 g |
| Amount of dye | 40 mg | 40 mg | 40 mg |
| Glauber's salt (20% w/v) | 1.0 mL | 1.5 mL | 1.0 mL |
| Soda ash (10% w/v) | - | - | 1.0 mL |
| Acetic acid (10% w/v) | 1.0 mL | - | - |
| Formic acid (10% w/v) | - | 1.5 mL | - |
| pH | 3 | 3 | 8 |
| MLR | 1:40 | 1:40 | 1:40 |
| Dyeing time | 40 min | 60 min | 90 min |
| Dyeing temp. | 60-80 °C | 60-80 °C | 60-80 °C |
| Total volume | 80 mL | 80 mL | 80 mL |

## Infrared Spectra of Dyes

IR spectra [20] in general shows characteristic band at 3400-3430 $cm^{-1}$ indicates the N-H and O-H stretching vibrations. The band at 3050-3085 $cm^{-1}$ and 2870-2900 $cm^{-1}$ indicates the C-H stretching vibrations. The strong band observed at 1700-1720 $cm^{-1}$ indicates the C=O stretching vibration. The band at 1440-1445 $cm^{-1}$ shows the C-N stretching vibration. The bands at 1180-1192 $cm^{-1}$ and 1035-1050 $cm^{-1}$ shows S=O asymmetric and symmetric stretching vibrations. The azo and chloro groups are confirmed at the 1370-1385 $cm^{-1}$ and 760-780 $cm^{-1}$ respectively. (IR and 1H-NMR data are summarized in Table 3).

**Table 3** IR and 1H-NMR spectra of dyes $D_1$ to $D_{10}$

| Dye No. | IR (KBr): $v_{max}$ ($cm^{-1}$) | $^1H$-NMR (Chemical shift in δ ppm) |
|---|---|---|
| $D_1$ | $3400_{br}$ (O-H & N-H), $3050_m$, $2890_m$ (C-H), $1700_s$ (C=O), $1440_s$ (C-N), $1375_m$ (N=N) $1190_s$, $1045_s$ (S=O), $760_s$ (C-Cl). | 3.45 (s, 2H, $CH_2$), 5.07 (s, 2H, 2OH), 9.02 (s, 2H, 2COOH), 9.96 (s, 2H, 2NH), 10.75 (s, 2H, 2NH), 7.82-8.01 (m, 16H, Ar-H). |
| $D_2$ | $3410_{br}$ (O-H & N-H), $3060_m$, $2900_m$ (C-H), $1710_s$ (C=O), $1445_s$ (C-N), $1370_m$ (N=N), $1185_s$, $1042_s$ (S=O),$765_s$ (C-Cl). | 3.45 (s, 2H, $CH_2$), 5.06 (s, 2H, 2OH), 9.05 (s, 2H, 2COOH), 9.92 (s, 2H, 2NH),10.76 (s, 2H, 2NH), 7.86-8.32 (m, 18H, Ar-H). |
| $D_3$ | $3420_{br}$ (O-H & N-H), $3060_m$, $2900_m$ (C-H), $1720_s$ (C=O), $1440_s$ (C-N), $1375_m$ (N=N), $1192_s$, $1050_s$ (S=O), $760_s$ (C-Cl). | 3.46 (s, 2H, $CH_2$), 5.04 (s, 2H, 2OH), 9.02 (s, 2H, 2COOH), 10.75 (s, 2H, 2NH), 7.79-8.01 (m, 28H, Ar-H) |
| $D_4$ | $3420_{br}$ (O-H & N-H), $3070_m$, $2890_m$ (C-H), $1725_s$ (C=O), $1445_s$ (C-N), $1380_m$ (N=N), $1180_s$, $1035_s$ (S=O), $780_s$. | 3.44 (s, 2H, $CH_2$), 5.07 (s, 2H, 2OH), 9.02 (s, 2H, 2COOH), 10.02 (s, 2H, 2NH),10.75 (s, 2H, 2NH), 7.82-8.01 (m,18H, Ar-H) |
| $D_5$ | $3420_{br}$ (O-H & N-H), $3085_m$, $2890_m$ (C-H), $1700_s$ (C=O), $1445_s$ (C-N), $1385_m$ (N=N), $1185_s$, $1035_s$ (S=O), $780_s$ (C-Cl). | 3.45 (s, 2H, $CH_2$), 5.07 (s, 2H, 2OH), 9.01 (s, 2H, 2COOH), 9.96 (s, 2H, 2NH), 10.79 (s, 2H, 2NH), 7.82-8.01 (m,16H, Ar-H). |
| $D_6$ | $3400_{br}$ (O-H & N-H), $3060_m$, $2870_m$ (C-H), $1700_s$ (C=O), $1440_s$ (C-N), $1375_m$ (N=N), $1192_s$, $1048_s$ (S=O), $770_s$ (C-Cl). | 3.45 (s, 2H, $CH_2$), 5.08 (s, 2H, 2OH), 9.02 (s, 2H, 2COOH), 9.96 (s, 2H, 2NH), 10.78 (s, 2H, 2NH), 7.86-8.09 (m, 16H, Ar-H). |
| $D_7$ | $3420_{br}$ (N-H), 3055m, $2890_m$ (C-H), $1710_s$ (C=O), $1440_s$ (C-N), $1375_m$ (N=N), $1195_s$, $1030_s$ (S=O), $755_s$ (C-Cl). | 3.45 (s, 2H, $CH_2$), 9.02(s, 2H, 2COOH), 9.92 (s, 2H, 2NH), 10.62 (s, 2H, 2NH), 7.79-7.97 (m, 20H, Ar-H). |
| $D_8$ | $3420_{br}$ (N-H), 3050m, $2890_m$ (C-H), $1700_s$ (C=O), $1440_s$ (C-N), $1380_m$ (N=N), $1182_s$, $1045_s$ (S=O), $760_s$ (C-Cl). | 3.45 (s, 2H, $CH_2$), 9.03(s, 2H, 2COOH), 9.92 (s, 2H, 2NH), 10.64 (s, 2H, 2NH), 7.06-8.2 (m, 20H, Ar-H). |
| $D_9$ | $3400_{br}$ (N-H), 3080m, $2890_m$ (C-H), $1710_s$ (C=O), $1440_s$ (C-N), $1375_m$ (N=N), $1185_s$, $1035_s$ (S=O), $765_s$ (C-Cl). | 3.43 (s, 2H, $CH_2$), 9.01(s, 2H, 2COOH), 9.92 (s, 2H, 2NH), 10.81 (s, 2H, 2NH), 7.82-8.61 (m, 22H, Ar-H). |
| $D_{10}$ | $3430_{br}$ (N-H), $3050_m$, $2870_m$ (C-H), $1700_s$ (C=O), $1440_s$ (C-N), $1375_m$ (N=N), $1192_s$, $1048_s$ (S=O), $780_s$ C-Cl. | 3.42 (s, 2H, $CH_2$), 9.00(s, 2H, 2COOH), 9.96 (s, 2H, 2NH), 10.78 (s, 2H, 2NH), 7.88-8.63 (m, 20H, Ar-H) |

## Exhaustion and Fixation Study

The percentage exhaustion [22] of 2% dyeing on silk ranges from 67-75%, for wool ranges from 65-72% and for cotton ranges from 65-

73%. The percentage fixation [23] of 2% dyeing on silk fabric ranges from 84-92 %, for wool ranges from 85-93% and for cotton ranges from 85-92%.

All the dyes have good exhaustion value may be expected due to the diffusion of the dye molecule within the fabric proceed rapidly under dyeing condition. Also the introduction of triazine molecule in to the dye improves the exhaustion and fixation value (Table 4).

## Fastness Properties

All the dyes show generally fair to very good light fastness properties on cotton and wool and moderate to very good for silk. The washing and rubbing fastness for good to excellent fastness on silk, wool and cotton (Table 5).

**Table 4** Exhaustion and fixation data of the dyes $D_1$ to $D_{10}$

| Dye No. | Shade on dyed fibre | $\lambda_{max}$ nm ($H_2O$) | $\lambda_{max}$ nm ($H_2SO_4$) | Logε ($H_2O$) | % Exhaustion | | | % Fixation | | |
|---|---|---|---|---|---|---|---|---|---|---|
| | | | | | S | W | C | S | W | C |
| $D_1$ | Purple | 535 | 520 | 4.30 | 75.30 | 70.90 | 71.55 | 91.63 | 93.08 | 91.54 |
| $D_2$ | Yellow | 462 | 455 | 4.20 | 73.50 | 68.82 | 67.65 | 88.43 | 89.35 | 84.99 |
| $D_3$ | Orange | 482 | 472 | 4.27 | 70.60 | 70.47 | 68.72 | 85.69 | 91.52 | 86.57 |
| $D_4$ | Light yellow | 458 | 445 | 4.21 | 69.55 | 65.55 | 74.72 | 89.14 | 87.71 | 88.99 |
| $D_5$ | Light purple | 522 | 512 | 4.24 | 67.97 | 66.12 | 69.45 | 91.94 | 85.44 | 88.55 |
| $D_6$ | Light purple | 525 | 508 | 4.30 | 72.60 | 71.10 | 69.57 | 84.02 | 88.60 | 87.67 |
| $D_7$ | Greenish yellow | 435 | 410 | 4.30 | 69.35 | 68.00 | 71.82 | 85.80 | 84.55 | 90.49 |
| $D_8$ | Light yellow | 440 | 425 | 4.29 | 75.45 | 65.27 | 70.52 | 90.12 | 91.91 | 85.78 |
| $D_9$ | Light yellow | 432 | 418 | 4.26 | 69.65 | 68.40 | 65.17 | 89.73 | 86.94 | 84.38 |
| $D_{10}$ | Reddish yellow | 452 | 420 | 4.22 | 71.23 | 67.23 | 72.58 | 90.25 | 88.45 | 87.28 |

**Table 5** Fastness properties data of the dyes $D_1$ to $D_{10}$

| Dyes No. | Light fastness | | | Wash fastness | | | Rubbing fastness | | | | | |
|---|---|---|---|---|---|---|---|---|---|---|---|---|
| | | | | | | | Dry | | | Wet | | |
| | S | W | C | S | W | C | S | W | C | S | W | C |
| $D_1$ | 6 | 5-6 | 6 | 4 | 4 | 3 | 4-5 | 3 | 4 | 3-4 | 3 | 4-5 |
| $D_2$ | 3-4 | 5 | 4 | 4 | 4-5 | 4-5 | 4 | 4 | 3 | 5 | 5 | 4 |
| $D_3$ | 4-5 | 5 | 4-5 | 4-5 | 3 | 3 | 4 | 3-4 | 5 | 3 | 4-5 | 3 |
| $D_4$ | 4 | 4-5 | 5 | 4-5 | 4 | 3-4 | 5 | 5 | 3-4 | 4 | 5 | 4 |
| $D_5$ | 3 | 4 | 5-6 | 3-4 | 3 | 4 | 3 | 4 | 3 | 3-4 | 4-5 | 3-4 |
| $D_6$ | 5-6 | 6 | 4 | 5 | 4 | 4 | 4 | 3 | 4-5 | 3-4 | 5 | 3 |
| $D_7$ | 4 | 5 | 4 | 3 | 3-4 | 5 | 4-5 | 3-4 | 4-3 | 3 | 3 | 3-4 |
| $D_8$ | 4-5 | 4 | 4 | 4-5 | 5 | 3 | 3 | 3 | 3 | 4 | 4 | 5 |
| $D_9$ | 3 | 5 | 4-5 | 4 | 4-5 | 4-5 | 3-4 | 5 | 4 | 5 | 3-4 | 4-5 |
| $D_{10}$ | 5-6 | 6 | 6 | 3 | 3 | 4 | 4 | 3-4 | 3 | 3 | 4 | 3 |

## CONCLUSION

A series of bisazo reactive dyes containing 4,4′-methylene-bis(2,6-dichloro aniline) coupling moiety have been synthesized by conventional method and their colour properties examined by application on silk, wool and cotton fibres. These dyes give yellow to purple hues depending on the coupling components used. The exhaustion and fixation values of all the dyes are very good and show good fastness properties. These dyes are also used as a substitute for benzidine dyes which shows dangerous carcinogen properties.

## ACKNOWLEDGMENTS

The authors are thankful to Veer Narmad South Gujarat University, Surat, for research facilities. SAIF, Chandigarh for 1H NMR spectra and Atul Limited, Valsad for providing dyeing facilities, fastness tests and important chemicals.

## REFERENCES

1. Zollinger, H.; Color Chemistry, 2nd ed. Weinheim: VCH-Wiley, 1991.
2. Shah, K. M.; Hand Book of Synthetic Dyes and Pigments, Vol. I, 1st ed. Mumbai: Multi-Tech Publishing Co, 1994.
3. Siegel, E.; In: Venkataraman, K.; Editor. The Chemistry of Synthetic Dyes, Vol. I, New York: Academic Press: 1972.
4. Bredereck, K.; Schumacher, C. *Dyes and Pigments* 1993, 21, 23.
5. Masaki, M.; Ulrich, M.; Zollinger, H. *Journal of Society of Dyers and Colourists* 1988, 104, 425.
6. Warnig, M; Hallas, G.; Editor. The Chemistry and Application of Dyes, London: Plenum, 1990.
7. Konstantinova, T.; Petrova, P. *Dyes and Pigments* 2002, 52, 115.
8. Zhan, Z.; Mullner, M.; Lercher, J. A. *Catal. Today* 1996, 27, 167.
9. Korkin, A. A.; Bartlett, R. J. *J. Am. Chem. Soc.* 1996, *118*, 12244.
10. Stringfield, T. W.; Shephered, R. E. *Inorg. Chim. Acta.* 1999, *292*, 225.
11. Zheng, W.; Wong, N.; Zhon, G.; Liang, X.; Jinshan, L.; Tian, A. *New J. Chem.* 2004, *28*, 275.
12. Cassela, C. A.; G. P. 1,794,297 1959 (CA. 58:43253).
13. Mehta, J. R.; Patel, S. K.; Patel, K. C. *Colourage* 2008, *10*, 85.
14. Patel, D. R.; Patel, J. A.; Patel, K. C. *J. Saudi Chem. Soc.* 2009, *13*, 279.

15. Fried, B; Sherma, J.; Thin layer Chromatography Techniques and Application. New York and Basel: Marcel-Dekker. Inc., 1982.
16. Standard Test Method. BS 1006, (U.K.) 1978: ISO 105, (India) 1994.
17. AATCC Test Method 1961, 8.
18. Indian Standard. ISO: 1979, 765.
19. Nafziger, J. L.; US Pat. 4,554,378 1985; Cruros, Z.; Ger.Offen. 2,108,402 1971.
20. Colthup, N. B.; Daly, L. H.; Wiberley, S. E.; Introduction to Infrared and Raman Spectroscopy, 3rd ed. New York: Academic Press, 1991.
21. Shenai, V. A.; Chemistry of Dyes and Principles of Dyeing, Mumbai: Sevak Publication, 1973.
22. Iyer, V.; Gurjar, A. M.; Bhyrappa, K. S. *Indian J. Fibre and Text*. Res. 1993, 18, 120.
23. Venkatraman, K.; Ed. The Analytical Chemistry of Synthetic Dyes, New York: Wiley, 1977.

# Chapter 3

# ELEMENTARY REACTIONS AND THEIR ROLE IN GAS-PHASE PREBIOTIC CHEMISTRY

Nadia Balucani

Dipartimento di Chimica, Università degli Studi di Perugia, 06123 Perugia, Italy

## ABSTRACT

The formation of complex organic molecules in a reactor filled with gaseous mixtures possibly reproducing the primitive terrestrial atmosphere and ocean demonstrated more than 50 years ago that inorganic synthesis of prebiotic molecules is possible, provided that some form of energy is provided to the system. After that groundbreaking experiment, gas-phase prebiotic molecules have been observed in a wide variety of extraterrestrial objects (including interstellar clouds, comets and planetary atmospheres) where the physical conditions vary widely. A thorough characterization of the chemical evolution of those objects relies on a multi-disciplinary approach: 1) observations allow us to identify the molecules and their number densities as they are nowadays; 2) the chemistry which lies behind their formation starting from atoms and simple molecules is accounted for by complex reaction networks; 3) for a realistic modeling of such networks, a number of experimental parameters are needed and, therefore, the relevant molecular processes should be fully characterized in laboratory experiments. A survey of the available literature reveals, however, that much information is still lacking if it is true that only a small percentage of the elementary reactions considered in the models have been characterized in laboratory experiments. New experimental approaches to characterize the relevant elementary reactions in laboratory are presented and the implications of the results are discussed.

## INTRODUCTION

In the sequence of steps which are believed to have led from elementary particles to the emergence of life, an important one is certainly the formation of simple prebiotic molecules from parent species abundant in the Universe. The aggregation of H, O, C, N, S and other atoms into molecules and the subsequent chemical evolution are occurring also now in the Universe, as witnessed by the identification of more than one hundred molecules in the interstellar clouds (encompassing also prebiotic molecules such as hydrogen cyanide, glycol aldehyde, form amide and even, tentatively, glycine) and by the gas-phase chemical evolution of the atmospheres of several solar objects such as Titan. Certainly, these processes might seem relatively simple compared to the other unknown phenomena that have led to the first living organisms. Nevertheless, the formation mechanisms of many of the observed gaseous prebiotic molecules and radicals are far from being understood, while a comprehension of those processes can certainly help to set the stage for the emergence of life to occur.

If we compare the elemental composition of the Universe and that of a living cell, a remarkable difference is evident, especially as far as the abundances of carbon and nitrogen are concerned. And not only is the elemental composition different, but also the type of molecules: mostly large and complex in a cell, rare and simple in the Universe. Clearly a chemical differentiation has taken place, firstly in the formation and evolution of solar systems in star-forming regions and then in the passage from inanimate matter to living entities. A series of intermediate molecular species, characterized by some complexity and all the appropriate "ingredients", might be the link between these extremely different environments. In Table 1 are grouped a series of molecules which can be synthesized in abiotic processes (*e.g.* in the gas phase), but might be prebiotic in nature [1,2] and which have been identified in extraterrestrial gaseous environments (of course this does not mean that they are permanent gases under the typical conditions of Earth). For instance, gaseous molecules containing a C-N bond (such as HCN, $HC_3N$ and $CH_3CN$) are potential precursors of amino acids or nucleases in the presence of water, while molecules containing a C-O bond might be the building blocks for the synthesis of sugars and amino acids (notable examples are glycol-aldehyde and ethylene glycol).

**Table 1.** Simple gas-phase molecules and their possible relation to biologically relevant molecules.

| Gas-phase molecules | Potential precursor of | Examples |
|---|---|---|
| with C-N bonds | Aminoacids and Nucleobases | HCN, $CH_3CN$, $C_2N_2$, HCCCN, $CH_2NH$, $C_2H_3CN$ |
| with C-O bonds | Sugars and Aminoacids | $H_2CO$, $CH_3COH$, $(CH_2OH)_2$ |
| with C-C multiple bonds | Long carbon chain molecules and PAHs | From $C_2H_2$ up to polyynes |
| + other molecules such as $H_2O$, $NH_3$, $H_2S$, $NH_2CN$, HCOCN, $NH_2CH_2CN$, $HCONH_2$, $CH_3CONH_2$, $CH_2OHCHO$, $CH_3SH$ etc. | | |

Finally, gaseous molecules with multiple C-C bonds provide the skeleton for long carbon chain molecules and are possible precursors of polycyclic aromatic hydrocarbons (PAHs). Other molecules which might be synthesized in the gas phase are water, ammonia, form amide, sulfur and phosphorus compounds *etc.*

How these molecules, or even more complex ones, could have reached or been developed on our planet is a matter of debate. Essentially two possible scenarios have been suggested so far, the *endogenous* synthesis and the *exogenous* synthesis [3,4]. According to some theories dating back to the 'warm little pond' suggested by Darwin and the 'primordial soup' idea by Operon, the synthesis of organic molecules occurred directly in our planet starting from simple parent molecules (such as $N_2$ or $NH_3$, $H_2O$ or $H_2$, $CH_4$ or $CO_2$ and others) [5-9] which constituted a sort of secondary atmosphere of Earth, after the primordial one was swept away during the first period of the solar system life. There are some discussions on the composition of the atmosphere back then, which could be in a reduced or oxidized (or intermediate) state [7,8], and also on the role of the possible chemical processes, which could have been induced by various sources of energy, such as intense lightning, energetic solar photons, radioactivity, intense volcanic activity, and shock waves of different kinds [7-9]. This is the so-called endogenous synthesis vision, as the prebiotic species would have been synthesized directly on Earth from simple parent molecules. This vision gained support more than 50 years ago by the ground-breaking experiment of Stanley Miller [5], an experiment having also the merit of demonstrating that an abiotic synthesis of prebiotic molecules is possible starting from very simple gaseous parent molecules [7]. For completeness, it should be mentioned that there is an alternative endogenous synthesis theory, where gas-phase reactions are not involved, according to which some organic synthesis could have taken place in the proximity of oceanic hydrothermal vents because of the relative abundance of methane and ammonia [10].

The alternative scenario is the exogenous synthesis, according to which most of the organic molecules came from space, the carriers being comets, asteroids, meteorites or even interplanetary dust particles ([11-16] and references therein). The rationale behind this suggestion is that plenty of simple organic molecules have been identified in all these objects. The idea is that they were first synthesized in the nebula from which the formation of the solar system originated. Those molecules were then incorporated in comets and meteorites far from the central star and then reached our planet [12], where they could evolve in an aqueous medium after the terrestrial temperature decreased. In favor of this scenario there is the observation of many organic molecules in the interstellar clouds, including several star-forming regions. Amongst the more than 150 molecules identified in the interstellar medium [17], in fact, only 36 do not contain carbon atoms and most of the molecules listed in Table 1 (e.g. glycol aldehyde, ethylene glycol, HCN, $HC_3N$, form amide, methanimine) have been identified in the interstellar medium. In addition, complex, organic to some extent, molecules have been detected in comets, interplanetary dust particles and, especially, carbonaceous meteorites [11-16]. Following the observation, it remains to be demonstrated where and how these molecules are formed, whether they can be preserved during the early phases of solar-type star formation and in the solar nebula [11,12,18], how they can be incorporated into comets and meteorites and whether they can survive their entry into the terrestrial atmosphere [11,12]. The attractiveness of this theory is that, if it is proved that there is a link between prebiotic molecules synthesized in space and the appearance of life on Earth, a hint on the possibility that life is a widespread phenomenon in the Universe, rather than a local, fortuitous case, can be gained.

Whether in a planetary atmosphere or in the interstellar clouds (ISCs) that precede the formation of solar systems, the observation of prebiotic molecules in a wide variety of gaseous extraterrestrial objects, where physical conditions vary to a large extent, implies that gas-phase chemistry plays a role in their synthesis. Different types of molecular processes are believed to be involved, including radioactive association and recombination, surface-induced processes [19], photon- or particle-induced ionization and ion-molecule reactions [20], photon- or particle-induced dissociation and radical molecule reactions [20]. The concomitance of all these phenomena and the complexity of their environments require a modelling approach, where all the relevant molecular processes compatible with the boundary conditions should be considered with the appropriate parameters that describe them.

This strategy has been pursued by several research groups to model the chemical evolution of interstellar clouds, planetary atmospheres and commentary come (see for instance [21-30]). For the modeling of these complex networks of elementary reactions, a number of experimental parameters are needed and, therefore, the molecular processes necessary to construct a realistic model should be fully characterized in laboratory experiments. A survey of the available literature reveals, however, that much information is still lacking if it is true that only a small percentage of the elementary reactions included in the models have been characterized in laboratory experiments under conditions that simulate those prevailing in the various environments. To understand the reason for that, let us briefly examine two typical environments where this chemistry takes place, that is, ISCs and the atmospheres of planets and satellites of our solar system.

The ISCs are essentially formed by gaseous matter [21-26, 31, 32], with a small percentage (1% by mass) of dust particles. The average kinetic energy is typically confined to 0.8 kJ mol-1 (corresponding to a temperature of 100 K) in diffuse clouds and 0.08 kJ mol-1 (corresponding to a temperature of 10 K) in dark molecular clouds and, therefore, the formation of molecular species via gas-phase reactions can only involve reactions without appreciable activation energy. For this reason, since the first observation of molecules in ISCs, ion–molecule reactions have been considered to play a central role [29,32], as they are usually characterized by null activation energy and can, therefore, be fast at very low temperatures. Nevertheless, only a small percentage of atomic and molecular species are ionized in the clouds. Neutral-neutral gas-phase reactions have been correctly included in the modeling of ISC chemistry only after it has been possible to measure their rate constant down to very low temperatures (as low as 15 K) by means of the CRESU (the French acronym for *Cinétique de Reaction en Emolument Supersonique Uniformed*) technique [33]. From these studies experimental evidence has been obtained that some reactions involving atomic or radical species (such as atomic carbon or CN, $C_2H$ and C4H radicals) are very fast, with rate constants in the gas kinetics range, even at very low temperatures [33]. Also, the results from this technique have pointed out that the extrapolation at very low temperatures of the Arrhenius dependence of the rate constant outside the range of T investigated is not warranted and important deviations from it have been observed [34,35]. The inclusion of several fast neutral-neutral reactions in the complex reaction networks that model the chemical evolution of ISCs is a recent advance in the field [22,25]. In principle, however, all the possible reactions should be investigated at the low temperature of the various interstellar objects.

After numerous missions, airborne and ground-based observations, plenty of information is available on the structure and chemical composition of the atmospheres of the planets of our solar system. Similarly to the atmosphere of Earth, the atmospheres of the other planets (or satellites, such as Titan) can be described as giant photo reactors, where the energy deposited mainly by solar photons, but also by cosmic rays and other energetic particles, drives a complex gas-phase chemistry [27,36- 48]. In this case as well, to account for the chemical composition of the atmospheres, complex models including physical parameters, such as vertical transport, wind transport and temperature profiles, have been developed. These models are usually referred to as *photochemical* models, because the main source of energy that drives the various phenomena considered are solar photons. Among others, the atmosphere of Saturn's massive moon Titan has received considerable attention because it is considered to be somewhat reminiscent of the primeval atmosphere of Earth [49,50]. In an atmosphere mainly composed by molecular nitrogen with a small percentage of methane, the detection of gasphase nitriles (in trace amounts of a few parts per billion) has attracted a lot of interest, since these species are thought to be key intermediates towards the formation of biologically relevant molecules [1,251,52]. The surface temperature of Titan is ~94 K and even in the upper stratosphere the temperature does not exceed 187 K. Therefore, water is mainly in the solid state and the absence of liquid water prevents the evolution of life as we know it. On the other hand, exactly because of the absence of a biosphere, Titan provides us with the unique opportunity to investigate a chemical environment probably close enough to the primitive terrestrial atmosphere. In other words the study of the chemical evolution of Titan's atmosphere can help us to understand how biologically active molecules and their nitrile precursors can be synthesized in a mildly reducing atmosphere [49], possibly resembling that of our planet before the emergence of life drastically changed its composition.

Numerous photochemical models of increasing complexity have attempted to describe the atmospheric composition of Titan and important improvements have been achieved after the results of the Cassini- Huygens mission [41-48]. Nevertheless, even the latest models [45,48] include several elementary reactions (unimolecular, bimolecular and termolecular) never studied in the laboratory, with parameters estimated by analogy with 'similar' ones, whereas other elementary reactions have been included with the laboratory parameters measured under different conditions (particularly temperature). The extrapolation of the experimental data to temperatures different from those actually used in the experiments is not warranted [35], while the analogy with similar systems can lead to completely erroneous evaluations [34].

In the following sections, an overview will be given of the two experimental approaches that have allowed us to reach a better description of neutral-neutral gas-phase reactions of importance in the chemical modeling of ISCs and the atmosphere of Titan. This review paper is not meant to be a comprehensive survey of the very active and productive fields of prebiotic chemistry and iatrochemistry, where important contributions are also made by theoretical investigations (see, for instance, [53-55]), and the focus will be only on neutral-neutral bimolecular reactions. Firstly, the principles of the CRESU technique, that has finally allowed the measurements of several reaction rate constants at appropriately low temperatures, will be illustrated. This technique, as well as other typical kinetics techniques, usually provide us with the reactant disappearance rate constants, but more rarely are able to determine the nature of the reaction products and their branching ratio. This is, however, a very important piece of information, because in the reaction networks used in the models the products of one reaction are going to be the reagents of a subsequent one. An alternative approach is necessary and the best opportunity is furnished by free-collision experiments. This is the realm of *reaction dynamics*, a discipline which provides us with the most detailed knowledge of a gas-phase reaction and aims to verify whether a specific reaction pathway and its related products are really accessible by the system. In this respect, the crossed molecular beam (CMB) technique, especially when coupled to *universal* mass spectrometric (MS) detection, has proved to be an extraordinary tool [56]. Finally, the results on several reactive systems and their connection with the understanding of gas-phase prebiotic chemistry will be illustrated.

## The CRESU Technique

As pointed out in the Introduction, it is imperative to measure the rate constants of the potentially relevant reactions at the low temperatures typical of the ISCs and also of the atmospheres of the other objects of the solar system (with the only exception of Venus, all other atmospheres are characterized by an average temperature lower than that of our planet). There are two experimental approaches to cool gases to low temperatures and study their reactions: cryogenic cooling and expansion methods [33,57-59]. Cryogenic cooling is a simple approach that consists in cooling the entire reaction vessel at the desired temperature. This approach, however, is limited by the saturation vapour pressure of the reactant gases and there are practically no reactions that could be investigated at 10-20 K without condensation of the reactants on the walls of the refrigerated container. The alternative method to cool the reactant gases is the use of a supersonic expansion. The expansion from a high pressure zone to a low pressure one is an adiabatic, isentropic process which converts the thermal energy of the gas into kinetic energy. The use of a collimating axisymmetric, converging-diverging Laval nozzle, leads to the production of a cold and uniform supersonic flow of gas [33,57-59]. This is the approach used in the CRESU technique. Comprehensive descriptions of the principles of this technique can be found in several reviews [33,57-59]. An important characteristic of the expansion through a Laval nozzle is that a relatively dense medium (1016 – 1017 cm-3) is formed, in which the temperature and number density are constant along the axis of the flow. Different nozzles and carrier gases are used to produce a specific flow temperature and the supersonic flow regime is maintained as such also when 1-2% of additional gases are added. The chemically unstable reactants in the gas mixture (generally an atomic [60-62] or radical [63-70] species, but also unstable closed-shell species such as $C_2$ [71,72]) are generated via pulsed laser photolysis (PLP) of a molecular precursor. In a typical CRESU arrangement, the decay rate of the unstable reactant is followed by laser-induced fluorescence (LIF), see Figure 1. The photolysis and probe pulsed lasers co-propagate along the axis of the supersonic flow. On the one hand, the photolysis laser produces a homogeneous concentration of the desired radical starting from a suitable molecular precursor. On the other hand, the probe laser monitors the formed radicals by exciting them to an upper electronic state. The resulting fluorescence is collected at 10-50 cm downstream of the Laval nozzle exit. For a given concentration of the reaction partner, the decay rate of the radical is observed by varying the delay time between the pulse from the photolysis laser and the pulse from the probe laser. The fluorescence signal decreases with the delay time exactly because the radicals are consumed in the reaction during their transit along the flow. A severe limit of the CRESU technique is that it cannot be used for reactions with room temperature rate constants smaller than 10-12 cm3 molec-$^{1}$ s-$^{1}$. This order of magnitude is imposed by the fact that the supersonic flow is maintained uniform over a distance of a few tens of centimeters. This implies a time scale of 100 – 500 $\mu$s within which the change of concentration of the atom/radical reactant has to be detected.

In the classical CRESU technique, an extensive pumping system is required. To reduce the pumping system, a pulsed Laval nozzle technique was developed and used to study bimolecular reactions down to 53 K [73-83].

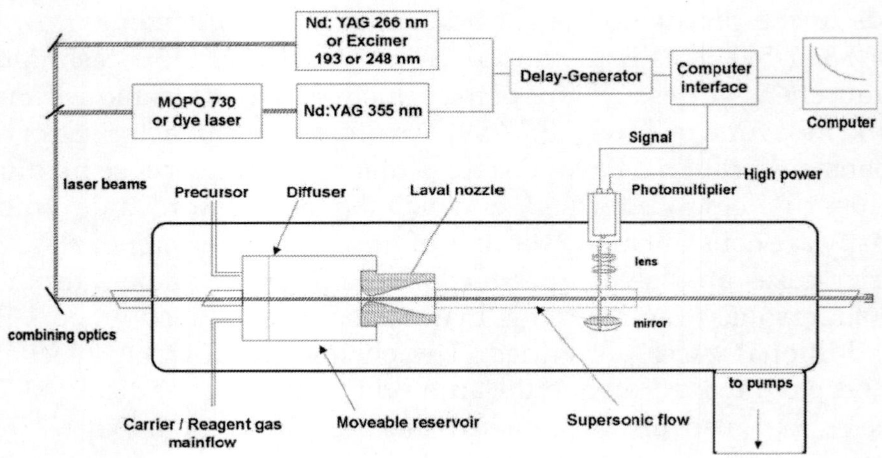


**Figure 1.** A schematic view of the PLP-LIF CRESU apparatus for the study of neutral neutral bimolecular reaction. Reprinted from Canosa, A.; Goulay, F.; Sims, I.R.; Rowe, B.R. Gas phase reactive collisions at very low temperature: recent experimental advances and perspectives. In *Low Temperatures and cold molecules,* Copyright 2008, published with permission from World Scientific Publishing Co. Pte. Ltd, Singapore.

## THE CROSSED MOLECULAR BEAM TECHNIQUE WITH MASS SPECTROMETRIC DETECTION

Amongst the possible experimental techniques to study a bimolecular reaction under collision-free conditions, the most versatile one is the crossed molecular beam (CMB) technique with mass spectrometric (MS) detection. This technique was first developed in the late '60s and used to address fundamental aspects of the reaction mechanisms at the microscopic level [56], while only recently, improvements in the production of beams of unstable species and vacuum technology have allowed the study of elementary reactions of interest in iatrochemistry [31,84-135]. The main advantage of CMB experiments is that the reactants are confined into distinct supersonic beams which cross each other at a specific angle; the species of each beam are characterized by a well-defined (both in magnitude and direction) velocity and are made to collide only with the molecules of the other beam, allowing us to observe the consequences of well-defined molecular collisions.

The products are formed at the collision center and then fly undisturbed towards the detector because of the large mean free path achieved by operating at a very low pressure ($10^{-5}$ Pa). In CMB experiments the product detection can be done by means of spectroscopic techniques or mass-spectrometry. Even though notable examples of CMB experiments coupled to spectroscopic detection on reactions of relevance in iatrochemistry are available in the literature [61,121-123,134], the use of MS detection is especially advantageous compared to spectroscopic techniques, the applicability of which requires the knowledge of the optical properties of the products - while, in many cases, their nature itself is unknown.

The coupling with MS detection, in fact, makes the method *universal*, that is, applicable to the study of (at least in principle) any reaction [56]. Every species can be ionized at the electron energy typically used in the ionizer ($\geq 70$ eV) which precedes the mass filter. Therefore, it is possible to determine the mass (and, in favorable cases, the gross formula) of all possible species produced from the reactions by ionizing the product and selecting different mass-to-charge ratios ($m/z$) in the mass selector. Some problems, such as dissociative ionization and background noise, have restricted the sensitivity of the method, but the recent implementation of the *soft* electron ionization (EI) approach has partly overcome this problem [113,114,127,128].

To be noted that an important advantage with respect to common MS flow reactors is the possibility to measure product angular and velocity distributions, which allows one to directly derive the amount of the total energy available to the products and, therefore, the energetics of the reaction. This is crucial when more isomers with the same gross formula can be produced [117]. In general, the CMB technique allows one to determine *(a)* the nature of the primary reaction products, *(b)* the branching ratios of competing reaction channels, *(c)* the microscopic reaction mechanisms *(d)* the product energy partitioning. More generally, we can say that it is possible to obtain information on the underlying potential energy surfaces (PES) governing the transformation from reactants to products.

The benefits of the CMB technique strongly motivate its extension to the study of reactions of interest in iatrochemistry. In the last few years the CMB method with MS detection has indeed been applied to the study of astronomically relevant reactions in different laboratories, especially in the research groups of Kaiser [31,84-112,135] and Casavecchia [113-133]. CMB machines equipped with MS detection can vary to some extent depending on the way used to generate the unstable species beams.

On the one hand, we have laser-based generation techniques, amongst which laser photolysis [135] or laser induced ablation from a refractory material rod and, eventually, subsequent reaction of the ablated species with another gas [136]. This method has been widely used in the laboratory of Kaiser to produce pulsed supersonic beams of atoms or radicals, such as C and B atoms, CN and $C_2H$ radicals, $C_2$ and $C_3$ [31,87,91,94]. Alternatively, beams of unstable species can be produced by flash pyrolysis or electrical discharge. A pulsed pyro lytic source has been used by Kaiser and coworkers to produce beams of $C_3H_5$ and $C_6H_5$ radicals [137,138], while a continuous pyro lytic source has been used by Casavecchia and coworkers to produce beams of $C_3$H5 and $CH_3$ [113,139].

A radiofrequency discharge beam source has been widely used by Casavecchia and coworkers to produce beams of C, O, N, S, Cl atoms, OH and CN radicals and $C_2$ [140]. An important, recent achievement in the crossed molecular beam machine with MS detection of Casavecchia is the set-up of a variable beam crossing angle configuration [113,114], see Figure 2. The relative collision energy, Ec, in a CMB experiment is given by 1/2 $\mu v_r{}^2$, where $\mu$ is the reduced mass of the system and vr is the relative velocity. In general $v_r$ is given by the equation $v_r{}^2 = v^1\ 2 + v^2\ 2\ \mu\ 2v^1v^2\cos\ \mu$, where $v^1$ and $v^2$ are the two reactant beam velocities in the laboratory frame and $\lambda$ is the crossing angle of the two beams.

Traditionally, CMB instruments with MS detection have featured a beam crossing angle $\lambda$ of 90 $\lambda$. In apparatuses with fixed $\lambda$ = 90°, the Ecs achievable are limited by the velocities with which the two beams can be produced. The relatively high Ecs normally achieved in CMB experiments might affect the results obtained by this technique on astronomically relevant reactions if there is a change of the reaction mechanism with varying Ec.

To overcome this problem, a variable beam crossing angle set-up, which allows crossing of the two reactant beams at three different angles of 45°, 90°, and 135° has been implemented. With this new setting, Ec could be varied over a much wider range than previously possible: with the beam velocities remaining constant, the $\lambda = 135°$ configuration allows reaching higher Ec while the $\lambda = 45°$ arrangement is intended for reaching very low Ec. In this way, the bimolecular reactions can be studied under collision free conditions at collision energies that resemble those of relevance in iatrochemistry with typical environments at low T.

Another important achievement obtained in the group of Casavecchia is the implementation of soft- EI by means of a tunable ionizer [113,114]. In this way, by tuning the electron energy below the threshold for dissociative ionization of interfering species, it is possible to eliminate or reduce the background or interfering signal of different sources [125,127,128] that, in several cases, would prohibit the reactive scattering experiments.

Remarkably, the use of a tunable electron impact ionizer in the CMB instrument permits measuring the EI efficiency curves of the reaction products as a function of electron energy down to their ionization thresholds and, from these, obtaining a direct estimate of the ionization energy of the product radicals [132]. In other words, it is possible to measure the unknown ionization energy of radical species by 'synthesizing' them in CMB reactive scattering experiments.

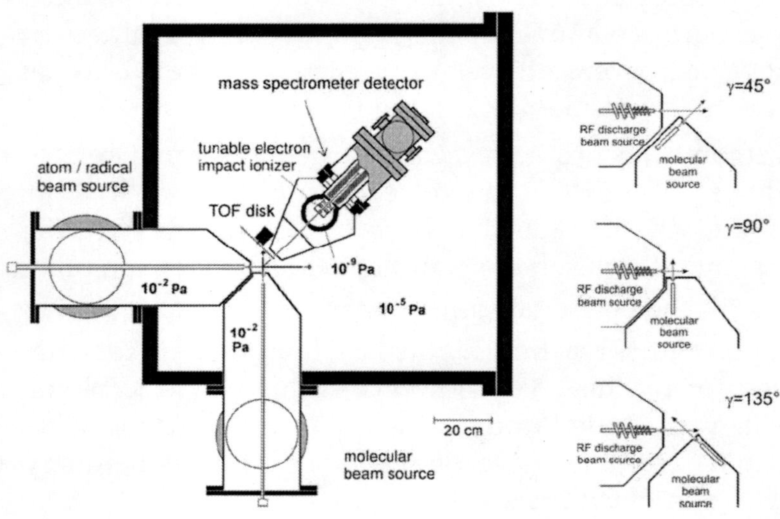


**Figure 2.** Schematic view of the Perugia crossed molecular beam apparatus (lhs). The three possible crossing beam geometries are also shown on the right-hand-side. Adapted with permission from reference [113].

## KEY RESULTS ON THE REACTIONS OF CN RADICALS

Cyanopolyynes, with the general chemical formula H-(C $\equiv$C)$_n$-CN, are ubiquitous in the ISCs and display high fractional abundances of up to 6 × 10-9 [141]. The simplest one, cyan acetylene HCCCN($X^1$ $\Sigma^+$), has been identified in the dark molecular clouds TMC-1 and OCM-1 as well as the carbon star IRC + 10,216 (CW Leo), towards high intensity IR sources (IRC + 10,216) and massive O-B stars within dense molecular clouds [142-144]. More complex cyanopolyynes up to $HC_{11}N$ have been assigned in TMC-1 [145,146]. Methylcyanoacetylene, $CH_3CCCN$ [147], vinyl cyanide, C2$H_3$CN, [148] and the two radicals $C_3N$ and $C_5N$ [149,150] have also been identified in various objects. Early models of ISC chemistry suggested that the dissociative recombination of the HCCCNH+ ion was the main formation mechanism of HCCCN and its isomer HCCNC. The ion HCCCNH+ was assumed to be formed via ion–

molecule reactions, amongst which N(4S) + $C_3H_3$ +. Nevertheless, this HCCCN formation route cannot explain either the observed [HCCCN]:[HCCNC]:[HNCCC] ratio or the HCCCN isotopomers ($H1_3CCCN$, $HC1_3CCN$, $HCCC1_3CN$) abundance towards TMC-1. Finally, a recent theoretical investigation of the possible ion-molecule reactions leading to HCCCNH+ showed that they are not feasible in molecular clouds [151].

Therefore, neutral-neutral reactions, based on the generalization of the reaction scheme

$$CN(X^2\Sigma^+) + H\text{-}(C{\equiv}C)_n\text{-}H \rightarrow H\text{-}(C{\equiv}C)_n\text{-}CN + H(^2S_{1/2}) \qquad (n = 1,2,..) \qquad (1)$$

Have been proposed (see [152-154] and references therein). Cyan acetylene has also been observed in the atmosphere of Titan, together with other CN containing molecules, such as HCN, $C_2N_2$ and $CH_3CN$ [155-157]. Even though they occur only in trace amounts of a few parts per billion, they are of particular importance because they are thought to be the key intermediates to the formation of biologically relevant molecules. Nitriles can be hydrolyzed and react via multistep synthesis ultimately to amino acids, thus providing one of the basic "ingredients" for life. In the upper layers of Titan's atmosphere, the energy deposition is mainly from high-energy electrons from Saturn's magnetosphere and short-wavelength solar ultraviolet photons ($\Sigma$ < 155 nm). Longer wavelength photons penetrate down to the stratosphere and photo dissociate HCN to cyan radicals, CN($X2\Sigma^+$). The cyan radical concentration profile overlaps with that of acetylene, so that their reaction is a plausible route of formation of cyan acetylene.

Certainly the possible importance of CN radical reactions with the hydrocarbons widely spread in ISCs and Titan's atmosphere became obvious after the CRESU results on these reactive systems. All the reactions with the abundant unsaturated hydrocarbons acetylene, methyl acetylene, allege and ethylene [67,158] have been found to be very fast, within the gas kinetic limit, down to very low temperatures (as low as 25 K). More surprisingly, the reaction CN + C2H6, the rate constant of which decreases with decreasing temperature up to ~ 100 K, was found to become faster at temperatures below 75 K and the rate constant reaches the value of 1.1 × 10-10 cm3 molec-1 s-1 at 25 K. The temperature dependence of the reaction CN + C2H6, see Figure 3, is probably the best example to demonstrate how the extrapolation at low temperatures of the rate constant outside the range of temperature investigated can be misleading. Recent theoretical calculations have been able to account for such a peculiar behavior [159]. After the CRESU experiments provided evidence that cyan radicals easily react also at the temperatures of Titan and ISM with relatively abundant hydrocarbons, only one parameter remained to be checked: the nature and the yield of the reaction products.

By combining CMB-MS experiments with the electronic structure calculations of the relevant stationary points along the minimum energy path, Kaiser and coworkers [84-87,98] have been able to demonstrate that HCCCN + H is the only open channel for the reaction CN + C2H2 (2), thus providing an experimental confirmation of the reaction scheme (1). Also, with the same technique the reactions CN + $C_2H_4$ (3) [97], CN + CH3CCH (4) [96,101], CN + $CH_2$=C=$CH_2$ (5) [96], CN + $CH_3CCCH_3$ (6) [100], CN + $C_6H_6$ (7) [99] and CN + $CH_3CH$=$CH_2$ (8) [95] have been investigated. All of them share a similar reaction mechanism: the electrophilic CN radical interacts with the $\pi$ electrons of the multiple bond of the unsaturated hydrocarbon without an entrance barrier, leading to the formation of a relatively stable addition intermediate. Because of its high energy content, the addition intermediate can directly undergo a C-H bond cleavage or isomerize to other intermediates before fragmenting into products. In this way, the identified molecular products were: cyan acetylene, HCCCN, from the reaction CN + $C_2H_2$, vinyl cyanide from the reaction CN + C2$H_4$, cyanomethylacetylene, CH3CCCN, and cyanoallene, CH2CCH(CN), from the reaction CN + CH3CCH, cyanoallene and 3-cyanomethylacetylene from the reaction CN + CH2=C=CH2, CH2CCCN(CH3) and $CH_3CCCN$ from the reaction CN + $CH_3CCCH_3$, cyan benzene from the reaction CN + $C_6H_6$ and $CH_2$=$CHCH_2CN$ and $CH_3CH$=CHCN from the reaction CN + CH3CH=CH2. The results obtained for reaction (2) have been confirmed by another CMB-MS study [117].

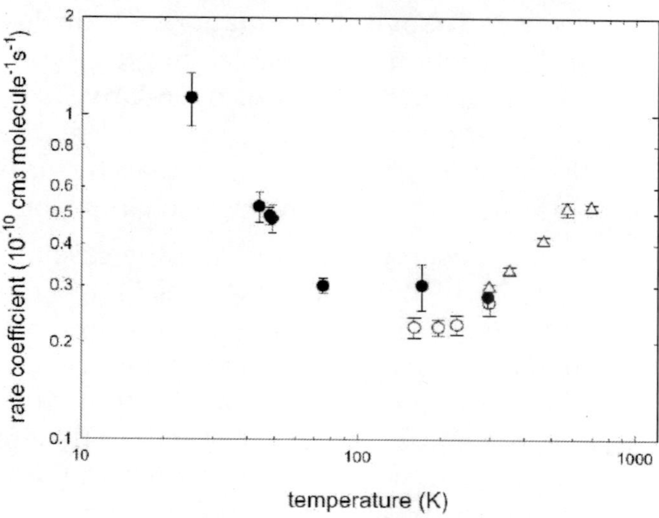


**Figure 3.** Rate coefficients for the reaction CN + C2H6 as a function of temperature, displayed on a log-log scale. The filled circles represent the results obtained in the CRESU experiment of Reference [67], open triangles are from Reference [163] while the open circles are from cooled cell experiments of Reference [67]. Adapted from Canosa, A.; Goulay, F.; Sims, I.R.; Rowe, B.R. Gas phase reactive collisions at very low temperature: recent experimental advances and perspectives. In *Low Temperatures and cold molecules,* Copyright 2008, published with permission from World Scientific Publishing Co. Pte. Ltd, Singapore

For the cases (2) and (3) the dominance of the H-displacement channel, leading to the molecular products listed above, has been confirmed by recent kinetic measurements where the H growth was monitored [160,161], while for the reaction of CN with propene (8) other reaction channels are open [161]. Finally, very recent experimental results obtained by the novel multiplexed time-resolved mass spectrometry with tunable synchrotron photoionization technique [162] have confirmed that $C_2H_3CN$ is the sole molecular product from reaction (3), while in the case of reaction (8) the channel leading to $C2H_3CN + CH_3$ is the dominant one (75 ± 15%), with the minor H-displacement channel producing the isomers 1-cyanopropene (57 ± 15%), 2-cyanopropene (43 ± 15%) and 3-cyanopropene (< 15%) [162].

In all reactions discussed above, the formation of is nitriles is never favored because all the reaction channels leading to these isomers are either endothermic by a few kJ mol-1 or the relative exit barriers are 5 - 30 kJmol-1 above the energy of the separated reactants [84]. In some cases, the theoretical calculations show that the formation of HCN and an alkyl radical is a possible reaction channel, but

HCN has never been observed, not even in the case of reaction (8) [162]. In summary, in all the reactions investigated the CN/H exchange channel is a main reaction pathway and products with a multiple C-C bond and a –CN group are formed. The structures of the products are reported in Figure 4. Furthermore, these reactions are also very fast at very low temperatures and therefore they represent a viable route for the formation of complex nitriles in the ISCs and cold planetary atmospheres, provided that the precursor molecules (HCN and unsaturated hydrocarbons) are present. In this respect, a nice example of how the results of CMB experiments can assist the astronomical search for hitherto unobserved molecules is furnished by the detection of cyanoallene towards the dark Taurus molecular cloud (TMC-1) [164,165]. Since both cyanomethylacetylene and cyanoallene are produced from the reaction between cyano radical and methyl acetylene, if the reaction (4) is the source of methyl acetylene in ISCs it is reasonable to expect the presence of cyanoallene as well. And this is in fact the case [164,165]

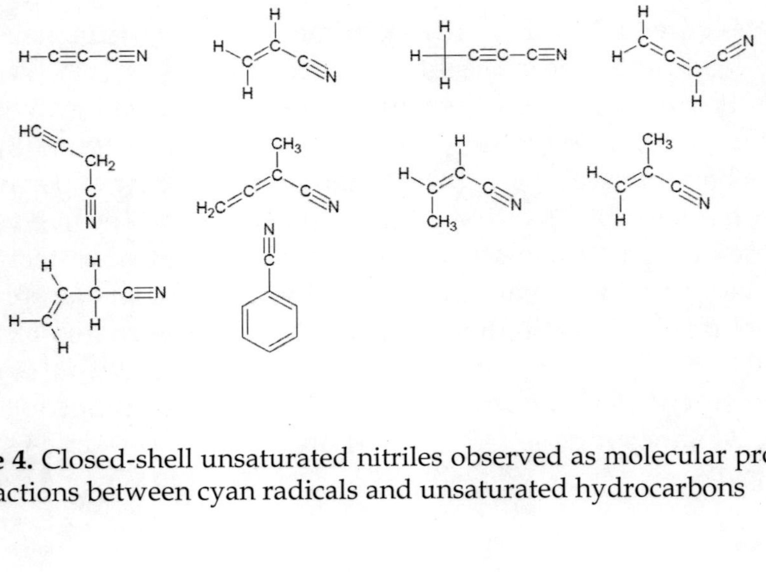

**Figure 4.** Closed-shell unsaturated nitriles observed as molecular products in the reactions between cyan radicals and unsaturated hydrocarbons

There is some interest also in the reaction of CN with saturated hydrocarbons. In particular, the reaction with methane, the most abundant hydrocarbon in the atmosphere of Titan, CN + CH4 (9), was invoked to account for the formation of acetonitrile in the atmosphere of Titan [43], according to the scheme CN + CH4 → $CH_3CN$ + H. This reaction is not as fast as those with unsaturated hydrocarbons and is characterized by a room temperature rate constant of 5.95 × 10-13 mole cm-3 s-1 [67]. An attempt to measure the H product yield was pursued by Gannon *et al.* [161], who were not able to observe any growth in H atom production matching the kinetics of CN removal. The main reaction channel is therefore the one leading to HCN + $CH_3$. An upper limit of 5% of yield for the channel leading to acetonitrile + H was given, but it was attributed to the background noise. Recent experiments by using the crossed-beam dc slice imaging [134] have been able to characterize the reaction mechanism of several reactions involving CN radicals and alkanes leading to alkyl radical + HCN.

## KEY RESULTS ON THE REACTIONS OF C2

C2 is one of the simplest diatomic molecules, but, in contrast to the similar $N_2$ or $O_2$ molecules, it is highly reactive. Another peculiarity of $C_2$ is that the first electronically excited metastable state ($a3\,\pi_u$) lies only 610 cm−1 (or 7.3 kJ mol-1) above the ground state ($X1\,\Sigma^+{}_g$). C2 has been identified in the interstellar medium (see for instance [166]) and cometary comae (see for instance [167-169]), while its detection in planetary atmospheres has never been reported in the literature. Nevertheless, formation and destruction processes of $C_2$ are included in the photochemical models of planetary atmospheres where $C_2H_2$ has been observed [27,37-45,47,48], since $C_2$ is expected to be produced (directly or in two steps) by acetylene photo dissociation. Because of the very low energy content of the a3 $\pi$u excited state, which can be produced by $C_2H$ photo dissociation [170], the presence of C2($a3\,\pi_u$) in various environments cannot be ruled out and its chemistry should be considered as well. Several experimental studies have been carried out to measure the rate coefficients of the removal of $C_2$(a3 $\pi_u$) and $C_2$($X1\,\Sigma_{+g}$) by hydrocarbons at room or higher temperature [171-174] and the results obtained for the reactions $C_2$($X1\,\Sigma_{+\,g}$) + $H_2$ and $C_2$($X1\,\Sigma_{+\,g}$) + $CH_4$ have been used in several photochemical models, after the extrapolation of the given rate coefficient temperature dependences to the conditions pertinent to planetary atmospheres. As already pointed out, such an extrapolation is not warranted. In the most recent model on the atmosphere of Titan by Lavvas *et al.* [47,48] the reactions of $C_2(X1\Sigma^+{}_g)$ with several hydrocarbons (including acetylene) have been included with the rate constants measured at 300 K. Recent CRESU kinetic experiments in the temperature range between 24 and 300 K [71,72] on several reactions of 1C2 have now furnished the correct temperature dependence for the reactions $C_2(X1\Sigma + g) + H2$ and $C2(X1\Sigma^+{}_g) + CH_4$, and also for other reactions with simple hydrocarbons.

Remarkably, those studies have demonstrated that the reactivity of $C2(X1\Sigma + g)$ with simple saturated (C2H6 and C3H8) and unsaturated (C2H2 and C2H4) hydrocarbons is very high with rate constants in the gas kinetics range. According to these new experimental results it has been established that [71]: *(a)* the reaction $C_2(X1\Sigma_{+\ g}) + CH_4$ is more efficient at low temperatures than what could be estimated by extrapolating the high temperature rate coefficients of Pitts *et al.* [173] *(2)* since the rate constants for the reactions $C2(X1\Sigma^{+\ g}) + C_2H_2$ and $C_2(X1\Sigma + g) + C_2H_4$ are larger than estimated, they should be included in the photochemical models of Titan, whereas the reaction of $C_2(X1\Sigma + g)$ with $H_2$ (presently included) actually has a negligible impact; *(3)* in the atmospheres of Saturn and Uranus, the reactions $C_2(X1\Sigma + g) + C_2H_2$ and $C_2(X1\Sigma + g) + C2H6$ should be included in the models. More recent results on $C_2(X1\Sigma + g) + C_2H_2$ obtained by Daugey *et al.* [175] using another CRESU apparatus at temperatures as low as 77 K confirm the results by Canosa *et al.* [71]. Also the kinetics involving $C_2$ in the triplet *a*3 $\pi_u$ state have been characterized with the CRESU technique [72]. Even though the trend with the temperature was quite different with respect to that of the singlet reaction, the rate constants for the reactions of $C_2(a3\Sigma u)$ with $C_2H_2$ and other unsaturated hydrocarbons are relatively large. In conclusion, the CRESU experiments point to an efficient removal of both C2(X1□+g) and C2(*a*3□u) by acetylene (and other hydrocarbons) at the temperatures of relevance to Titan, Giant Planets and ISCs.

As far as the nature of the reaction products is concerned, a systematic investigation by the CMBMStechnique has been pursued by Kaiser and co-workers [88-94] for these reactions as well and quite interesting results have been obtained. For instance, it has been demonstrated that the products of the $C_2$ + $C_2H_4$ reaction are not $C_2H$ + $C_2H_3$ or $C_2H_2$ + $C_2H_2$, as previously assumed, but *n*-C4H3 + H [92], that is, the $C_2$ + $C_2H_4$ reaction leads to a carbon chain elongation. Similarly, C4H has been identified as the primary product of the $C_2$ + $C_2H_2$ reaction, leading again to a carbon chain elongation [89,90]. Also, the 2,4-pentadiynyl-1 radical, $HCCCCCH_2$, is the main product of the reaction $C_2$ + CH3CCH [93], while the 1,3,5-hexatriynyl radical, $C_6H$, is the main product of the reaction $C_2$ + $C_4H_2$ (diacetylene) [88]. In all cases a carbon chain elongation was observed.

The reaction $C_2$ + $C_2H_2$ has also been investigated by Casavecchia and co-workers [125,126]. In this more recent study, the $C_2$ internal state population has been characterized by laser induced fluorescence to assist the interpretation of the reactive scattering results [125]. In spite of some differences in the characterization of the reaction mechanism, the results by Casavecchia and coworkers confirmed that $C_4H$ + H is the main reaction channel [125,126] and that $C_2$ reactions with unsaturated hydrocarbons can be a viable route to synthesize polymers in the interstellar medium.

## KEY RESULTS ON THE REACTIONS OF ATOMIC NITROGEN, N(2D)

The atmosphere of Titan is a gaseous environment dominated by molecular nitrogen and methane. In this condition, it is intuitive that the formation of nitriles is initiated by the reactions of active forms of nitrogen - such as nitrogen atoms or ions, which can be formed in the upper atmosphere from the abundant parent molecule $N_2$. In particular, atomic nitrogen can be produced by $N_2$ dissociation induced by electron impact, extreme ultra-violet photolysis and dissociative photoionization, galactic cosmic ray absorption, and $N_2$ + dissociative recombination. These processes lead to atomic nitrogen in the ground, 4S3/2, and first electronically excited, $2D_3$/2,5/2, states with comparable yields. The radioactive lifetimes of the metastable states $2D^3/2$ and $2D^5/2$ are quite long (12.3 and 48 hours, respectively),because the transition from a doublet to a quartet state is strongly forbidden. In addition, collisional deactivation of $N(^2D)$ by N2 is a slow process [176] and, therefore, the main fate of $N(^2D)$ above 800 km is chemical reaction with other constituents of Titan's atmosphere. The production of N atoms in the 2D state is an important fact, because N(4S) atoms exhibit very low reactivity with closed-shell molecules and the probability of collision with an open-shell radical is small. On the contrary the reactions of $N(^2D)$ with several molecules identified in the atmosphere of Titan (including the relatively abundant $CH_4$ and $H_2$) can make an important contribution to the chemical evolution of the atmosphere. Yung *et al.* [41] suggested that the main products of the $N(^2D) + CH_4$ (10) reaction are NH and $CH_3$ and, on this assumption, drew some reaction cycles, which ultimately lead to the formation of HCN. Also, Yung [177] suggested that the reaction $N(^2D) + C_2H_2$ (11) can be responsible for cyanogen, $C_2N_2$, and $C_4N_2$ formation in the case it produces HCCN + H. More recent photochemical models for calculating the vertical distribution of Titan's neutral atmosphere compounds rely on a similar suggestion [45,48]

Accurate laboratory experiments on the $N(^2D)$ reactions have become available only in the late 1990s, because of the experimental difficulties in studying those systems. The room temperature rate constants for reactions (10) and (11) have been found to be slightly larger than those used in the models [176]. As far as reaction (10) is concerned, however, the results of recent reaction dynamics studies do not confirm the assumptions of Yung about the reaction mechanism and the nature of the products. According to the *ab initio* calculations of the $NCH^4$ PES [178], the possible products are (10a) $CH_3N$ + H, (10b) NH + $CH_3$, (10c) $CHNH^2$ + H, and (10d) $CH_2NH$ + H. A recent spectroscopic study under collision-free conditions [179] derived an absolute yield of 0.3 ± 0.1 for NH (from channel 10b) and 0.8±0.2 for H (from channels 10a, 10c, 10d) production. The nature of the $CH^3N$ isomer(s) produced in conjunction with H has been established in a CMB-MS study of the $N(^2D) + CH_4$ reaction as a function of collision energy (from ~ 20 kJ mol-1 to ~ 60 kJ mol-1) [117-119,180]. Interestingly, the channels leading to CH2NH (methanimine) and $CH_3N$ (methylnitrene) were found to be both open and the relative branching ratio, σ (CH2NH + H) / σ ($CH_3N$ + H), was found to vary considerably with the amount of energy available [119]. In Figure 5 we report the results obtained at Ec = 37.2 kJ mol-1 where significant contributions to the reaction come from both channels 10a and 10d (the channel 10c is negligible). At Ec = 37.2 kJ $mol^{-1}$ σ ($CH_2NH$ + H) / σ ($CH_3N$ + H) is 0.25. A large yield of CH3N is not surprising at such a relatively large Ec and corresponds to a well-known dynamical effect: the energy which is released after the insertion of N(2D) into the C-H bond (corresponding to the exoergicity of 427 kJ mol-1 plus Ec) is initially 'localized' into the two new N-H and C-N bonds of the $CH_3NH$ intermediate. If one of the two newly formed (and very stressed) bonds breaks apart, we have the formation of either NH + $CH_3$ (corresponding to the fission of the new C-N bond) or $CH_3N$ + H (corresponding to the fission of the new N-H bond). The fission of one of the pre-existing C-H bonds, not directly involved in the insertion process, can only occur if some energy redistribution takes place.

For this to happen, the $CH_3NH$ should live long enough to allow for energy rearrangement. The trend of relative branching ratio with Ec (it varies from 1.3 at Ec σ 20 kJ mol-1 to 0.03 at σ 60 kJ mol-1 [119]), confirms the above explanation: the larger is the total energy available to the system, the shorter is the intermediate lifetime and smaller the chance that the energy redistribution takes place efficiently. Notably, the trend with Ec for the two channels (10a) and (10b) is expected to be similar, because in both cases the fission of one of the newly formed bonds is involved. Several implications for the atmospheric chemistry of Titan follow. The assumption that only NH and $CH_3$ are the products of reaction (10) is wrong since that reactive channel is minor in the laboratory experiments and, by analyzing the trend of the branching ratio as a function of the available energy, we can presume that it will be even minor under the low temperature conditions of Titan's atmosphere. Therefore, the nitrogen chemistry that relies on the dominance of the (10b) channel in the photochemical models of the atmosphere of Titan should be reconsidered. Furthermore, our results suggest that the reaction (10) is an active route of formation of methanimine, $CH_2NH$, a closed-shell molecule. $CH_2NH$ has been recently identified in the upper atmosphere of Titan [46] and in bulk experiments that mimic the atmosphere of Titan [181]. Interestingly, $CH_2NH$ contains a novel C-N bond, thus demonstrating that such a bond can be generated directly by a reaction involving an active form of $N_2$, the main constituent of the atmosphere of Titan, and $CH_4$, the second most abundant species. The presence of an unsaturated CN bond renders $CH_2NH$ a reactive molecule and in a reducing environment like the atmosphere of Titan, methanimine could be converted to the fully saturated analogue methylamine

Alternatively, it could photo dissociate to HCN/HNC in one or two steps. Some laboratory information is available on the UV absorption cross sections of $CH_2NH$ [182] and on the fragmentation of internally excited methanimine [183]. It is, therefore, possible that an important route of HCN formation in the atmosphere of Titan is:

$$N(^2D) + CH_4 \rightarrow CH_2NH + H \tag{10d}$$

$$CH_2NH + h\nu(UV) \rightarrow HCN + H_2 \text{ (or 2H)} \tag{12}$$

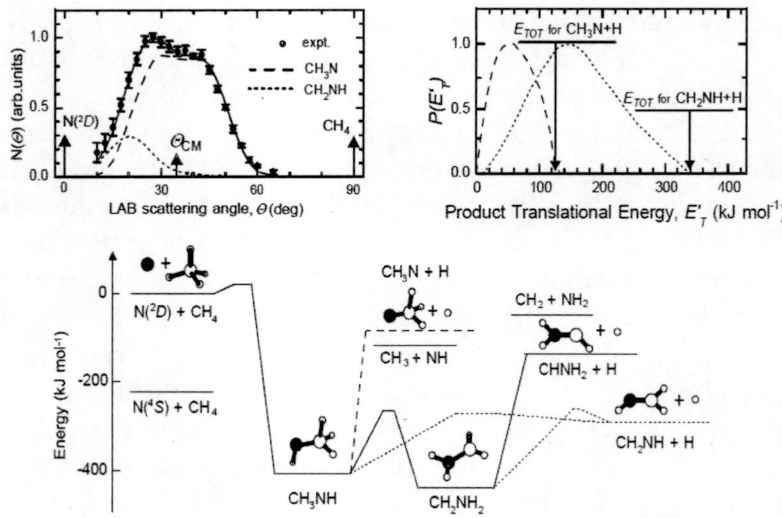


**Figure 5.** Top, left: Laboratory angular distribution of the $CH_3N/CH_2NH$ products from the reaction $N(^2D) + CH_4$ at Ec = 37.2 kJ mol-1. The separate contributions to the total LAB angular distribution (solid line) from the $CH_3N$ channel (dashed line) and the $CH_2NH$ channel (dotted line) are shown; top, right: best-fit CM product translational energy distributions for the methylnitrene (dashed line) and the methanimine (dotted line) forming channels. The arrows indicate the total available energy for the various isomer channels. Bottom: schematic potential energy surface for the $CH_4N$ system. Reprinted with permission from reference [180].

The reaction $N(^2D) + C_2H_2$ (11) was also investigated in CMB experiments with MS detection and the experimental results were consistent with both HCCN + H and cyclic-HC(N)C + H formation [115] and essentially confirm the potential role of the $N(^2D)$ + C2H2 reaction as the first step in the formation of $C_2N_2$ and $C_2N_4$. Probably the case which better represents the potentiality of our experimental technique is the reaction $N(^2D) + C_2H_4$ (13) [116]. This reaction was not considered in the early models, even though ethylene is almost as abundant as acetylene, or was considered erroneously [43]. The CMB-MS results clearly indicate that also this reaction proceeds through the formation of an addition intermediate and that products of general formula $C_2H_3N$ are formed through a N/H exchange channel [116]. The experimental findings gained support from the PES calculations by Takayanagi *et al.* [184] and are consistent with the formation of ketenimine and $^2$*H*-azirine. It is very interesting to note, however, that according to the energy release – a piece of information which can be obtained only from a reaction dynamics experiment – a large fraction of the 2*H*-azirine and ketenimine molecules are formed with enough internal energy to spontaneously tautomerize to the most stable isomer acetonitrile [116], even in a collision free environment. Therefore, reaction *(13)* can be considered an effective route which ultimately leads to the formation of $CH_3CN$. Our suggestion has been accepted and included in the models recently developed by Wilson and Atreya [45] and Lavvas *et al.* [48]. Very recent experimental results have become available on the reaction of N(2D) with another abundant hydrocarbon in the atmosphere of Titan, that is, C2H6. According to the preliminary analysis of the experimental distributions, $C_2H_5N$ isomers have been identified as a result of an H-displacement channel [185].

Quite interestingly, all the reactions between $N(_2D)$ and hydrocarbons investigated so far are characterized by the formation of active radicals (such as $CH_3N$, HCCN or $C_2H_5N$), cyclic nitriles (such as $^2$*H*-azirine) and unsaturated nitriles (such as ketenimine, methanimine and ethanimine) (see Figure 6). These species can obviously react further with the other components of the relatively dense atmosphere of Titan. In particular, $CH_2NH$ can polymerize (in a similar manner as $H_2CO$ does) or copolymerize with the other unsaturated nitriles to form the Titan aerosol, which has been recently analyzed by the Aerosol Collector and Pyrolysis during the Huygens probe descent and found to be quite rich in nitrogen [186]

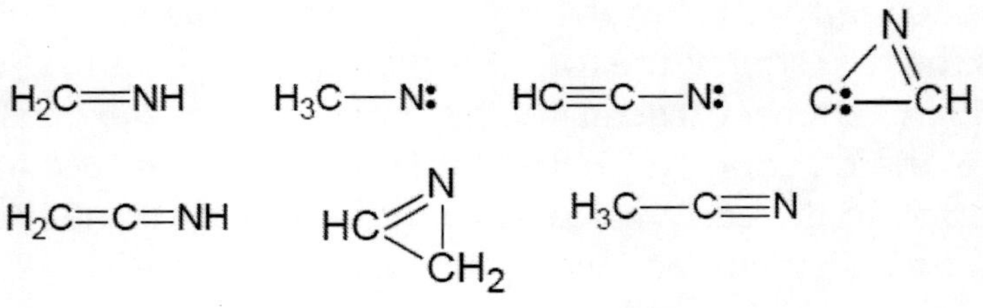

$H_2C{=}NH$ $H_3C{-}N{:}$ $HC{\equiv}C{-}N{:}$ N C: CH

$H_2C{=}C{=}NH$ N HC $CH_2$ $H_3C{-}C{\equiv}N$

**Figure 6.** Nitriles and nitrile radicals observed as molecular products in the reactions of $N(^2D)$ with $CH_4$, $C_2H_2$ and $C_2H_4$.

## KEY RESULTS ON THE REACTIONS OF ATOMIC SULFUR, $S(^1D)$

Simple organ sulfur (C-S containing) compounds have been observed in ISCs, commentary come and also in planetary atmospheres, raising the question of how they are formed under such different conditions [187-194]. Amongst those observations, probably the most striking one is the detection of CS, $CS_2$ and $S_2$ after the impact of the comet Shoemaker-Levy 9 on Jupiter, where those species were observed after the impact for several days [190]. Some possible mechanisms of C-S bond formation were suggested [191] and several investigated in crossed beam experiments [192,193] and electronic structure calculations [194]. A substantial lack of experimental data, both at the level of kinetic investigation and of reaction dynamics, has so far prevented the role of the reactions of atomic sulfur with hydrocarbons or hydrocarbon radicals from being established in all of the above mentioned environments. In the light of the general interest of sulfur atom reactions and the paucity of experimental information on their reaction kinetics and especially dynamics, a systematic investigation of sulfur atom reaction kinetics and dynamics by means of the CRESU and CMB-MS techniques has been recently started [132,133].

The first results have just appeared in the literature [132,133] and revealed interesting information concerning the chemical behavior of the sulfur atoms in the electronically excited state, $^1D$. S($^1D$) can be produced in planetary atmospheres or commentary come by the UV photo dissociation of several precursor molecules, such as OCS [198] and CS2 [199] and even the photo dissociation of $H_2S$ at the Lyman-α wavelength can occur via a three body dissociation with S($^1D$) formation [200]. Also the UV photo dissociation of radicals, such as SH, leads to the formation of atomic sulfur in the first electronically excited state [201]. Since the lifetime of the metastable excited state is long enough (28 s), sulfur atoms in the excited $^1D$ state may well react with other gaseous species provided that the density is not too low. To be noted that the UV photo dissociation of H2S in the primeval terrestrial atmosphere has been suggested as a possible initiating step of gas-phase prebiotic synthesis [7].

The reaction mechanisms inferred for the two reactions S($^1D$) + $C_2H$= and S($^1D$) + $C_2H_4$ share similar features [132,133]: the electrophilic S($^1D$) atom adds, without any barrier, to the multiple bond of acetylene or ethylene, forming an internally excited cyclic intermediate (thiirene and thiirane in the two cases). Once formed, thiirene can directly rearrange to H2CCS (thioketene, the most stable isomer along the singlet $C_2H_2S$ PES), which, in turn, dissociates into HCCS + H. In an alternative pathway, the PES minimum configuration of thioketene is never reached and the succession of rearrangements after thiirene formation is isomerization to CC(H)SH followed by isomerization to HCCSH which, in turn, can undergo an S-H bond fission to HCCS + H. In both pathways, HCCS + H are the only possible products. Tirana, the cyclic intermediate formed by the reaction of S($^1D$) with ethylene, can directly undergo ring-opening and three-center H2 elimination to thioketene ($CH_2CS$) + $H_2$ (the least abundant channel, with a yield of 7%) or, more readily, isomerize to vinylthiol (CH2CHSH) which can undergo C-H bond rupture to thiovinoxy ($CH_2CH_S$) + H (the second channel in importance, with a yield of 0.15). Vinylthiol can also isomerize to the slightly more stable thio-acetaldehyde ($CH_3CHS$), which can undergo C-C bond cleavage to $CH_3$ + HCS (the main reaction channel, with a yield of 78%). The reaction S($^1D$) + $C_2H_4$ was also investigated by the CRESU technique and was found to remain rapid down to the very low temperature of 23 K, occurring without any activation energy. This is corroborated by the theoretical calculations of the singlet C2H4S PES which did not find any appreciable reaction barrier to addition of S($^1D$) to the ethane molecule forming an initial tirade stable intermediate. These first results on S($^1$D) reactions with hydrocarbons have shown that simple organo-sulfur compounds can be formed in relatively dense gaseous environments, provided that simple precursor molecules of sulfur atoms and UV light are available in the presence of simple hydrocarbons. These experimental findings might have several implications for the formation of prebiotic important organ sulfur compounds.

## CONCLUSIONS

In this paper, two powerful experimental techniques devoted to the study of bimolecular, neutral neutral reactions have been described and their contributions to the understanding of several gas-phase reactions leading to prebiotic molecules illustrated. It should be stressed that these experimental approaches are such that they can only focus on one elementary reaction at a time, while the complex reaction networks that describe the chemistry of interstellar clouds and planetary atmospheres comprise hundreds or thousands elementary reactions. It is, therefore, important that a collaboration is established with modelers and astronomers, so that the possibly relevant reactions and molecular species are recognized and the experimental effort is directed towards them. In this respect, positive initiatives have been two multidisciplinary EU networks, *Molecular Universe* and *Euro planet*, where physical chemists and astronomers/ planetologists have had the chance to communicate and exchange ideas. Also, sensitivity methods applied to the models of ISCs and planetary atmospheres are of great help in identifying the critical reactions that should be better characterized in the laboratory to improve the reliability of the models [202-205].

## ACKNOWLEDGEMENTS

I wish to thank Andre Canoas for providing the figures on the CRESU experimental set-up and results. Special thanks go to my previous and current collaborators, in particular Ralf Kaiser and Piero Casavecchia. I am grateful to Dimitris Skouteris for comments on the manuscript. Finally, I wish to thank C. Ceccarelli, S. Viti,

A. Lazcano, J.R. Brucato, E. Gallori, E. De Mauro, J.I. Lunine and Y. Keheyan for stimulating discussions on astrochemistry and prebiotic chemistry. I acknowledge financial support from the Italian MIUR (Ministero Istruzione Università Ricerca) under projects PRIN (2007H9S8SW_004) and European Commission through the Coordination Action 001637 (Europlanet). This work is also supported by the European Union Marie-Curie human resources and mobility programmer under contract MCRTN-CT-2004-512302 (Molecular Universe).

## REFERENCES AND NOTES

1. Eschenmoser, A.; Loewenthal, E. Chemistry of potentially prebiological natural products. *Chem. Soc. Rev.* 1992, *21*, 1-16.
2. Eschenmoser, A. Chemistry of potentially prebiological natural products. *Orig. Life Evol. Biosph.* 1994, *24*, 389-423.
3. Bernstein, M. Prebiotic materials from on and off the early Earth. *Phil. Trans. R. Soc. B* 2006, *361*, 1689-1702.
4. Chyba, C.; Sagan, C. Endogenous production, exogenous delivery and impact-shock synthesis of organic molecules: an inventory for the origins of life. *Nature* 1992, *355*, 125-132.
5. Miller, S.L. Production of amino acids under possible primitive Earth conditions. *Science* 1953,*117*, 528-529
6. Bada, J.J.; Lazcano, A. Prebiotic soup - revisiting the miller experiment. *Science* 2003, *300*,745-746.
7. Miller, S.L. Current status of the prebiotic synthesis of small molecules. *Chem. Scripta* 1986,*26B*, 5-11.
8. Cleaves, H.J.; Chalmers, J.H.; Lazcano, A.; Miller, S.L.; Bada J.L. A reassessment of prebiotic organic synthesis in neutral planetary atmospheres. *Orig. Life Evol. Biosph.* 2008, *38*, 105-115.
9. Johnson, A.P.; Cleaves, H.J.; Dworkin, J.P.; Glavin, D.P.; Lazcano, A.; Bada, J.L. The Miller volcanic spark discharge experiment. *Science* 2008, *322*, 404-404.
10. Holm, N.G.; Andersson, E. Hydrothermal simulation experiments as a tool for studies of the origin of life on earth and other terrestrial planets: a review. *Astrobiology* 2005, *5*, 444-460; And references therein.
11. Ehrenfreund, P.; Irvine, W.; Becker, L.; Blank, J.; Brucato, J.R.; Colangeli, L.; Derenne, S.; Despois, D.; Dutrey, A.; Fraaije, H.; Lazcano, A.; Owen, T.; Robert, F. Astrophysical and astrochemical insights into

the origin of life. *Rep. Prog. Phys.* 2002, *65*, 1427-1487.

12. Ehrenfreund, P.; Charnley, S.B. Organic molecules in the interstellar medium, comets, and meteorites: A voyage from dark clouds to the early earth. *Ann. Rev. Astron. Astrophys.* 2000, *38*, 427-483.
13. Orò, J.; Mills, T.; Lazcano, A. Comets and the formation of biochemical compounds on the primitive Earth - a review. *Orig. Life Evol. Biosph.* 1991, *21*, 267-277.8.
14. Pizzarello, S. The chemistry of life's origin: a carbonaceous meteorite perspective. *Acc. Chem. Res.* 2006, *39*, 231-237.
15. Martins, Z.; Botta, O.; Fogel, M.L.; Sephton, M.A.; Glavin, D.P.; Watson, J.S.; Dworkin, J.P.; Schwartz, A.W.; Ehrenfreund, P. Extraterrestrial nucleobases in the Murchison meteorite. *Earth Planet. Sci. Lett.* 2008, *270*, 130-136.
16. Flynn, G.J.; Keller, L.P.; Jacobsen, C.; Wirick, S. An assessment of the amount and types of organic matter contributed to the Earth by interplanetary dust. *Adv. Space Res.* 2004, *33*, 57-66.
17. A continuously updated list of the observed molecules is available online: http://www.astrochymist.org.
18. Ehrenfreund, P.; Bernstein, M.P.; Dworkin, J.P.; Sandford, S.A.; Allamandola, L.J. The photostability of amino acids in space. *Astrophys. J.* 2001, *550*, L95-L99.
19. See, for instance, Stantcheva, T.; Herbst, E. Models of gas-grain chemistry in interstellar cloud cores with a stochastic approach to surface chemistry. *Astron. Astrophys.* 2004, *423*, 241-251.
20. Roberge, W.G.; Jones, D.; Lepp, S.; Dalgarno A. Interstellar photodissociation and photoionization rates. *Astrophys. J. Suppl. Ser.* 1991, *77*, 287-297.
21. Tielens, A.G.G.M. *The physics and chemistry of the interstellar medium*; Cambridge University Press: Cambridge, UK, 2005.
22. Herbst, E.; Millar, T.I. The chemistry of cold interstellar cloud cores. In *Low Temperatures and cold molecules*; Smith, I.W.M., Ed.; Imperial College Press, World Scientific Publishing Co. Pte. Ltd.: Singapore, 2008; pp. 1-54.
23. Mitchell, G.F. Effects of shocks on the molecular composition of a dense interstellar cloud. *Astrophys. J. Suppl. Ser.* 1984, *54*, 81-101.
24. Herbst, E. Chemistry of Star-Forming Regions. *J. Phys. Chem. A* 2005, *109*, 4017-4029
25. Le Teu, Y.H.; Millar, T.J.; Markwick, A.J. The UMIST database for astrochemistry 1999. *Astron. Astrophys. Suppl. Ser.* 2000, *146*, 157-168.

26. Turner, B.E.; Herbst, E.; Terzieva, R. The physics and chemistry of small translucent molecular clouds. XIII. The basic hydrocarbon chemistry. *Astrophys. J. Suppl. Ser.* 2000, *26*, 427-460.

27. Yung, Y.L.; DeMore, W.B. *Photochemistry of Planetary Atmospheres;* Oxford University Press: New York, USA, 1998.

28. Mitchell, G.F.; Prasad, S.S.; Huntress, W.T. Chemical model calculations of C2, C3, CH, CN, OH, and NH2 abundances in cometary comae. *Astrophys. J.* 1981, *244*, 1087-1093.

29. Anicich, V.G. A survey of bimolecular ion-molecule reactions for use in modeling the chemistry of planetary-atmospheres, cometary comae, and interstellar clouds. *Astrophys. J. Suppl. Ser.* 1993, *84*, 215-315.

30. Cottin, H.; Gazeau, M.C.; Raulin, F. Cometary organic chemistry: a review from observations, numerical and experimental simulations. *Planet. Space Sci.* 1999, *47*, 1141-1162.

31. Kaiser, R.I. Experimental investigation on the formation of carbon-bearing molecules in the interstellar medium via neutral-neutral reactions. *Chem. Rev.* 2002, *102*, 1309-1358.

32. Smith, D. The ion chemistry of interstellar clouds. *Chem. Rev.* 1992, 92, 1473-1485.

33. Canosa, A.; Goulay, F.; Sims, I.R.; Rowe, B.R. Gas phase reactive collisions at very low temperature: recent experimental advances and perspectives. In *Low Temperatures and cold molecules*; Smith, I.W.M., Ed.; Imperial College Press, World Scientific Publishing Co. Pte. Ltd.: Singapore, 2008; pp. 55-120.

34. Smith, I.W.M.; Herbst, E.; Chang, Q. Rapid neutral-neutral reactions at low temperatures: a new network and first results for TMC-1. *Mon. Not. R. Astronom. Soc.* 2004, *350*, 323-330.

35. Smith, I.W.M. The temperature-dependence of elementary reaction rates: beyond Arrhenius. *Chem. Soc. Rev.* 2008, *37*, 812-826.

36. Strobel, D.F. Photochemistry in outer Solar System atmospheres. *Space Sci. Rev.* 2005, *116*, 155-170.

37. Moses, J.I.; Bezard, B.; Lellouch, E.; Gladstone, G.R.; Feuchtgruber, H.; Allen, M. Photochemistry of Saturn's atmosphere. I. Hydrocarbon chemistry and comparisons with ISO observations. *Icarus* 2000, *143*, 244–298.

38. Atreya, S.K. Uranus photochemistry and prospects for Voyager 2 at Neptune. *Adv. Space Res.* 1990, *10*, 113-119.

39. Bishop, J.; Romani, P.N.; Atreya, S. Voyager 2 ultraviolet spectrometer solar occultations at Neptune: photochemical modeling of the 125-165

nm light curves. *Planet. Space Sci.* 1998, *46*, 1-20.

40. Romani, P.N.; Bishop, J.; Bezard, B.; Atreya, S. Methane photochemistry on Neptune: ethane and acetilene mixing ratios and haze production. *Icarus* 1993, *106*, 442-463.
41. Yung, Y.; Allen, M.; Pinto, J. Photochemistry of the atmosphere of Titan - comparison between model and observations. *Astrophys. J. Suppl. Ser.* 1984, *55*, 465-506.
42. Toublanc, D.; Parisot, J.; Brillet, J.; Gautier, D.; Raulin, F.; McKay, C. Photochemical modeling of Titan's atmosphere. *Icarus* 1995, *113*, 2-26.
43. Lara, L.M.; Lellouch, E.; López-Moreno, J.J.; Rodrigo, R. Vertical distribution of Titan's atmospheric neutral constituents. *J. Geophys. Res.* 1996, *101*, 23261-23283.
44. Lebonnois, S.; Toublanc, D.; Hourdin, F.; Rannou, P. Seasonal variations of Titan's atmospheric composition. *Icarus* 2001, *152*, 384-406.
45. Wilson, E.H.; Atreya, S. Current state of modeling the photochemistry of Titan's mutually dependent atmosphere and ionosphere. *J. Geophys. Res.* 2004, *109*, E06002.
46. Vuitton, V.; Yelle, R.V.; Anicich, V.G. The nitrogen chemistry of Titan's upper atmosphere revealed. *Astrophys. J.* 2006, *647*, L175-L178.
47. Lavvas, P.P.; Coustenis, A.; Vardavas, I.M. Coupling photochemistry with haze formation in Titan's atmosphere, part I: Model description. *Planet. Space Sci.* 2008, *56*, 27-66.
48. Lavvas, P.P.; Coustenis, A.; Vardavas, I.M. Coupling photochemistry with haze formation in Titan's atmosphere. Part II: results and validation with Cassini/Huygens data. *Planet. Space Sci.* 2008, *56*, 67-99.
49. Sagan, C., Thompson, W.R., Khare, B.N. Titan: a laboratory for prebiological organic chemistry *Acc. Chem. Res.* 1992, 25, 286-292.
50. Coustenis, A.; Taylor, F.W. *Titan: the Earth-like Moon*; World Scientific: Singapore, 1999.
51. Owen, T.; Raulin, F.; McKay, C.P.; Lunine, J.I.; Lebreton, J.P.; Matson, D.L. The Relevance of Titan and Cassini/Huygens to Pre-Biotic Chemistry and the Origin of Life on Earth. In *Huygens: Science, Payload and Mission*; Lebreton, J.P., Wilson, A., Eds.; ESA-ESTEC: Noordwijk, the Netherlands, August, 1997; pp. 231-233, ESA SP-1177.
52. Raulin, F. Astrobiology and habitability of Titan. *Space Sci. Rev.* 2008, *135*, 37-8.
53. Roy, D.; Najafian, K.; Schleyer, P.v.R. Chemical evolution: the mechanism of the formation of adenine under prebiotic conditions.

*Proc. Natl. Acad. Sci. U.S.A.* 2007, *104*, 17272-17277.

54. Glaser, R.; Hodgen, B.; Farrelly, D.; McKee, E. Adenine synthesis in interstnellar space: mechanisms of prebiotic pyrimidine-ring formation of monocyclic HCN-pentamers. *Astrobiology* 2007, *7*, 455-470.

55. Woon, D.E. Pathways to glycine and other amino acids in ultraviolet-irradiated astrophysical ices determined via quantum chemical modeling. *Astrophys. J.* 2002, *571*, L177-L180.

56. Lee, Y.T. Molecular beam studies of elementary chemical processes. In *Nobel Lectures in Chemistry* 1981-1990; Fraegsmyr, T., Malstrom, B.G., Eds.; World Scientific: Singapore, 1992; pp. 320-357.

57. Smith, I.W.M.; Rowe, B.R. Reaction kinetics at very low temperatures: laboratory studies and interstellar chemistry. *Acc. Chem. Res.* 2000, *33*, 261-268.

58. Smith, I.W.M. The Liversidge Lecture 2001 – 02. Chemistry amongst the stars: reaction kinetics at a new frontier. *Chem. Soc. Rev.* 2002, *31*, 137-146.

59. Smith, I.W.M. Laboratory studies of atmospheric reactions at low temperature. *Chem. Rev.* 2003, *103*, 4549-4564.

60. Chastaing, D.; Le Picard, S.D.; Sims, I.R.; Smith I.W.M. Rate coefficients for the reactions of C(3PJ ) atoms with $C_2H_2$, $C_2H_4$, $CH_3$≡CCH and $H_2CCCH_2$ at temperatures down 15 K. *Astron. Astrophys.* 2001, *365*, 241-247.

61. Chastaing, D.; Le Picard, S.D.; Sims, I.R.; Smith I.W.M.; Geppert, W.D.; Naulin, C.; Costes, M. Rate coefficients and cross-sections for the reactions of C(3PJ) atoms with methylacetylene and allene. *Chem. Phys. Lett.* 2000, *331*, 170-176.

62. Canosa, A.; Le Picard, S.D.; Gougeon, S.; Rebrion-Rowe, C.; Travers, D.; Rowe, B.R. Rate coefficients for the reactions of Si(3PJ) with C2H2 and C2H4: Experimental results down to 15 K. *J. Chem. Phys.* 2001, *115*, 6495-6503.

63. Sims, I.R.; Queffelec, J.L.; Defrance, A.; Rebrion-Rowe, C.; Travers, D.; Bocherel, P.; Rowe, B.R.; Smith I.W.M. Ultralow temperature kinetics of neutral-neutral reactions - the technique and results for the reactions CN + O2 down to 13 K and CN + NH3 down to 25 K. *J. Chem. Phys.* 1994, *100*, 4229-4241.

64. Goulay, F.; Rebrion-Rowe, C.; Biennier, L.; Le Picard, S.D.; Canosa, A.; Rowe, B.R. The reaction of anthracene with CH radicals: An experimental study of the kinetics between 58 K and 470 K. *J. Phys. Chem. A.* 2006, *110*, 3132-3137.

65. Goulay, F.; Rebrion-Rowe, C.; Le Garrec, J.L.; Le Picard, S.D.; Canosa, A.; Rowe, B.R. The reaction of anthracene with OH radicals: an experimental study of the kinetics between 58 K and 470 K. *J. Chem. Phys.* 2005, 122, 104308. 66.

66. Brownsword, R.A.; Canosa, A.; Rowe, B.R.; Sims, I.R.; Smith, I.W.M.; Stewart, D.W.A.; Symonds, A.C.; Travers, D. Kinetics over a wide range of temperature (13 – 744 K): rate constants for the reactions CH(v = 0) with H2 and D2 and for the removal of CH(v = 1) by H2 and D2 *J. Chem. Phys.* 1997, 106, 7662-7677.

67. Sims, I.R.; Queffelec, J.L.; Travers, D.; Rowe, B.R.; Herbert, L.B.; Karthauser, J., Smith, I.W.M. Rate constants for the reactions of CN with hydrocarbons at low and ultra-low temperature. *Chem. Phys. Lett.* 1993, *211*, 461-468.

68. Canosa, A.; Sims, I.R.; Travers, D.; Smith, I.W.M.; Rowe, B.R. Reactions of the methylidine radical with $CH_4$, $C_2H_2$, $C_2H_4$, $C_2H_6$ and but-1-ene studied between 23 and 295 K with a CRESU apparatus. *Astron. Astrophys.* 1997, *323*, 644-651.

69. Chastaing, D.; James, P.L.; Sims, I.R.; Smith, I.W.M. Neutral-neutral reactions at the temperatures of interstellar clouds - Rate coefficients for reactions of $C_2H$ radicals with O2, C2H2, C2H4 and C3H6 down to 15 K. *Faraday Discuss.* 1998, *109*, 165-181.

70. Berteloite, C.; Le Picard, S.D.; Birza, P.; Gazeau, M.C.; Canosa, A.; Benilan, Y.; Sims, I.R. Low temperature (39 K – 298 K) kinetic study of the reactions of C4H radical with various hydrocarbons observed in Titan's atmosphere: CH4, C2H2, C2H4, C2H6, C3H4 and C3H8. *Icarus* 2008, *194*, 746-757.

71. Canosa, A.; Paramo, A.; Le Picard, S.D.; Sims, I.R. An experimental study of the reaction kinetics of $C_2$(X1Σ+ g) with hydrocarbons ($CH_4$, $C_2H_2$, $C_2H_4$,$C_2H_6$ and $C_3H_8$) over the temperature range 24 – 300 K. Implications for the atmospheres of Titan and the Giant Planets. *Icarus* 2007, *187*, 558-568.

72. Paramo, A.; Canosa, A.; Le Picard, S.D.; Sims, I.R. Rate coefficients for the reactions of $C_2$(a3 $\pi_u$) and $C_2$(X 1Σ+ g) with various hydrocarbons ($CH_4$, $C_2H_2$, $C_2H_4$, $C_2H_6$, and $C_3H_8$): A gasphase experimental study over the temperature range 24-300 K. *J. Phys. Chem. A* 2008, *112*, 9591-9600.

73. Atkinson, D.B.; Smith, M.A. Design and characterization of pulsed uniform supersonic expansions for chemical applications. *Rev. Sci. Instrum.* 1995, *66*, 4434-4446.

74. Lee, S.; Hoobler, R.J.; Leone, S.R. A pulsed Laval nozzle apparatus with laser ionization mass spectrometry for direct measurements of

rate coefficients at low temperatures with condensable gases. *Rev. Sci. Instrum.* 2000, *71*, 1816-1823.

75. Spangenberg, T.; Kohler, S.; Hansmann, B.; Wachsmuth, U.; Abel, B.; Smith, M.A. Lowtemperature reactions of OH radicals with propene and isoprene in pulsed Laval nozzle expansions. *J. Phys. Chem. A* 2004, *108*, 7527-7534.

76. Mullen, C.; Smith, M.A. Low temperature NH(X3Σ−) radical reactions with NO, saturated, and unsaturated hydrocarbons studied in a pulsed supersonic Laval nozzle flow reactor between 53 and 188 K. *J. Phys. Chem. A* 2005, *109*, 1391-1399.

77. Vakhtin, A.B.; Heard, D.E.; Smith, I.W.M.: Leone, S.R. Kinetics of C2H radical reactions with ethene, propene and 1-butene measured in a pulsed Laval nozzle apparatus at T-103 and 296 K. *Chem. Phys. Lett.* 2001, *348*, 21-26.

78. Goulay, F.; Leone, S.R. Low-temperature rate coefficients for the reaction of ethynyl radical (C2H) with benzene. *J Phys. Chem. A* 2006, *110*, 1875-1880.

79. Nizamov, B.; Leone, S.R. Kinetics of C2H reactions with hydrocarbons and nitriles in the 104 – 296 K temperature range. *J Phys. Chem. A* 2004, *108*, 1746-1752.

80. Vakhtin, A.B.; Murphy, J.E.; Leone, S.R. Low-temperature kinetics of reactions of OH radical with ethene, propene, and 1-butene. *J Phys. Chem. A* 2003, *107*, 10055-10062.

81. Murphy, J.E.; Vakhtin, A.B.; Leone, S.R. Laboratory kinetics of C2H radical reactions with ethane, propane, and n-butane at T = 96 – 296 K: implications for Titan. *Icarus* 2003, *163*, 175-181.

82. Vakhtin, A.B.; Heard, D.E.; Smith, I.W.M.; Leone, S.R. Kinetics of reactions of C2H radical with acetylene, O2, methylacetylene, and allene in a pulsed Laval nozzle apparatus at T = 103 K. *Chem. Phys. Lett.* 2001, *344*, 317-324.

83. Hansmann, B.; Abel B. Kinetics in cold Laval nozzle expansions: From atmospheric chemistry to oxidation of biomolecules in the gas phase. *Chem. Phys. Chem.* 2007, *8*, 343-356.

84. Balucani, N.; Asvany, O.; Huang, L.C.L.; Lee, Y.T.; Kaiser, R.I.; Osamura, Y.; Bettinger, H.F. Formation of nitriles in the interstellar medium via reactions of cyano radicals, CN(X2Σ+), with unsaturated hydrocarbons. *Astrophys. J.* 2000, *545*, 892-906.

85. Balucani, N.; Asvany, O.; Osamura, Y.; Huang, L.C.L.; Lee, Y.T.; Kaiser, R.I. Laboratory investigation on the formation of unsaturated nitriles in Titan's atmosphere. *Planet. Space Sci.* 2000, *48*, 447-462.

86. Kaiser, R.I.; Balucani, N. Astrobiology - the final frontier in chemical reaction dynamics. *Int. J. Astrobiology* 2002, *1*, 15-23.

87. Kaiser, R.I.; Balucani, N. The formation of nitriles in hydrocarbon-rich atmospheres of planets and their satellites: Laboratory investigations by the crossed molecular beam technique. *Acc. Chem. Res.* 2001, *34*, 699-706.

88. Zhang, F.T.; Kim, S.; Kaiser, R.I.; Mebel, A.M. Formation of the 1,3,5-hexatriynyl radical (C6H(X2)) via the crossed beams reaction of dicarbon (C2(X1+ g/a3IIu)), with diacetylene (C4H2(X1+ g)). *J Phys. Chem. A* 2009, *113*, 1210-1217.

89. Gu, X.B.; Guo, Y.; Mebel, A.M.; Kaiser, R.I. Chemical dynamics of the formation of the 1,3- butadiynyl radical (C4H(X2+)) and its isotopomers. *J Phys. Chem. A* 2006, *110*, 11265-11278

90. Kaiser, R.I.; Balucani, N.; Charkin, D.O.; Mebel, A.M. A crossed beam and ab initio study of the C2(X1+ g/a3u) + C2H2(X1+ g) reactions. *Chem. Phys. Lett.* 2003, *382*, 112-119.

91. Kaiser, R.I.; Le, T.N.; Nguyen, T.L.; Mebel, A.M.; Balucani, N.; Lee, Y.T.; Stahl, F.; Schleyer, P.V.; Schaefer, H.F. A combined crossed molecular beam and ab initio investigation of C2 and C3 elementary reactions with unsaturated hydrocarbons - pathways to hydrogen deficient hydrocarbon radicals in combustion flames. *Faraday Discuss.* 2001, *119*, 51-66.

92. Balucani, N.; Mebel, A.M.; Lee, Y.T.; Kaiser, R.I. A combined crossed molecular beam and ab initio study of the reactions C2(X1+ g/a3IIu) + C2H4 n-C4H3(X2A') + H(2S1/2). *J Phys. Chem. A* 2001, *105*, 9813-9818.

93. Guo, Y.; Gu, X.; Balucani, N.; Kaiser, R.I. Formation of the 2,4-pentadiynyl-1 radical (H2CCCCCH, X2B1) in the crossed beams reaction of dicarbon molecules with methylacetylene. *J Phys. Chem. A* 2006, *110*, 6245-6249.

94. Gu, X.B.; Guo, Y.; Zhang, F.T.; Mebel, A.M.; Kaiser, R.I. Reaction dynamics of carbon-bearing radicals in circumstellar envelopes of carbon stars. *Faraday Discuss.* 2006, *133*, 245-275.

95. Gu, X.B.; Zhang, F.T.; Kaiser, R.I. Reaction dynamics on the formation of 1-and 3- cyanopropylene in the crossed beams reaction of ground-state cyano radicals (CN) with propylene (C3H6) and its deuterated isotopologues. *J Phys. Chem. A* 2008, *112*, 9607-9612.

96. Balucani, N.; Asvany, O.; Kaiser, R.I.; Osamura, Y. Formation of three C4H3N isomers from the reaction of CN(X2+) with allene, H2CCCH2 (X1A1), and methylacetylene, CH3CCH (X1A1): a combined crossed beam and ab initio study. *J Phys. Chem. A* 2002, *106*, 4301-4311.

97. Balucani, N.; Asvany, O.; Chang, A.H.H.; Lin, S.H.; Lee, Y.T.; Kaiser, R.I.; Osamura, Y. Crossed beam reaction of cyano radicals with hydrocarbon molecules. III. Chemical dynamics of vinylcyanide (C2H3CN; X1A') formation from reaction of CN(X2+) with ethylene, C2H4(X 1Ag). *J. Chem. Phys.* 2000, *113*, 8643-8655.

98. Huang, L.C.L.; Asvany, O.; Chang, A.H.H.; Balucani, N.; Lin, S.H.; Lee, Y.T.; Kaiser, R.I.; Osamura, Y. Crossed beam reaction of cyano radicals with hydrocarbon molecules. IV. Chemical dynamics of cyanoacetylene (HCCCN; X1+) formation from reaction of CN(X2+) with acetylene, C2H2(X1+ g). *J. Chem. Phys.* 2000, *113*, 8656-8666.

99. Balucani, N.; Asvany, O.; Chang, A.H.H.; Lin, S.H.; Lee, Y.T.; Kaiser, R.I.; Bettinger, H.F.; Schleyer, P.V.; Schaefer, H.F. Crossed beam reaction of cyano radicals with hydrocarbon molecules. I. Chemical dynamics of cyanobenzene (C6H5CN; X1A1) and perdeutero cyanobenzene (C6D5CN; X1A1) formation from reaction of CN(X2+) with benzene C6H6(X 1A1g)), and d6-benzene C6D6(X1A1g). *J. Chem. Phys.* 1999, *111*, 7457-7471.

100. Balucani, N.; Asvany, O.; Chang, A.H.H.; Lin, S.H.; Lee, Y.T.; Kaiser, R.I.; Bettinger, H.F.; Schleyer, P.V.; Schaefer, H.F. Crossed beam reaction of cyano radicals with hydrocarbon molecules. II. Chemical dynamics of 1-cyano-1-methylallene (CNCH3CCCH2; X 1A') formation from reaction of CN(X2+) with dimethylacetylene CH3CCCH3 (X 1A1'). *J. Chem. Phys.* 1999, *111*, 7472-7479.

101. Huang, L.C.L.; Balucani, N.; Lee, Y.T.; Kaiser, R.I.; Osamura, Y. Crossed beam reaction of the cyano radical, CN(X2+), with methylacetylene, CH3CCH (X1A1): Observation of cyanopropyne, CH3CCCN (X1A1), and cyanoallene, H2CCCHCN (X1A'). *J. Chem. Phys.* 1999, *111*, 2857-2860

102. Kaiser, R.I.; Ochsenfeld, C.; Head-Gordon, M.; Lee, Y.T. Neutral-neutral reactions in the interstellar medium. II. Isotope effects in the formation of linear and cyclic C3H and C3D radicals in interstellar environments. *Astrophys. J.* 1999, *510*, 784-788.

103. Kaiser, R.I.; Ochsenfeld, C.; Stranges, D.; Head-Gordon, M.; Lee, Y.T. Combined crossed molecular beams and ab initio investigation of the formation of carbon-bearing molecules in the interstellar medium via neutral-neutral reactions. *Faraday Disc.* 1998, *109*, 183-204.

104. Kaiser, R.I.; Stranges, D.; Lee, Y.T.; Suits, A.G. Neutral-neutral reactions in the interstellar medium. I. Formation of carbon hydride radicals via reaction of carbon atoms with unsaturated hydrocarbons. *Astrophys. J.* 1997, *477*, 982-989.

105. Kaiser, R.I.; Ochsenfeld, C.; Head Gordon, M.; Lee, Y.T.; Suits, A.G.

A combined experimental and theoretical study on the formation of interstellar C3H isomers. *Science* 1996, *274*, 1508-1511.

106. Kaiser, R.I.; Lee, Y.T.; Suits, A.G. Crossed-beam reaction of carbon atoms with hydrocarbon molecules .I. Chemical dynamics of the propargyl radical formation C3H3 (X2B2), from reaction of C(3Pj) with ethylene, C2H4(X1Ag) *J. Chem. Phys.* 1996, *105*, 8705-8720.

107. Zhang, F.T.; Kim, Y.S.; Zhou, L.; Chang, A.H.H.; Kaiser, R.I. A crossed molecular beam study on the synthesis of the interstellar 2,4-pentadiynylidyne radical (HCCCCC). *J. Chem. Phys.* 2008, *129*, 134313.

108. Guo, Y.; Gu, X.; Zhang, F.; Sun, B.J.; Tsai, M.F.; Chang, A.H.H.; Kaiser, R.I. Unraveling the formation of HCPH(X2A') molecules in extraterrestrial environments: Crossed molecular beam study of the reaction of carbon atoms, C(3Pj), with phosphine, PH3(X1A1) . *J Phys. Chem. A* 2007, *111*, 3241-3247.

109. Gu, X.B.; Guo, Y.; Zhang, F.T.; Kaiser, R.I. Investigating the chemical dynamics of the reaction of ground-state carbon atoms with acetylene and its isotopomers. *J Phys. Chem. A* 2007, *111*, 2980-2992.

110. Kaiser, R.I.; Stahl, F.; Schleyer, P.V.; Schaefer, H.F. Atomic and molecular hydrogen elimination in the crossed beam reaction of d1-ethinyl radicals C2D(X2+) with acetylene,C2H2(X1+ g): dynamics of d1-diacetylene (HCCCCD) and d1-butadiynyl (DCCCC) formation. Phys. Chem. Chem. Phys. 2002, 4, 2950-2958.

111. Stahl, F.; Schleyer, P.V.; Bettinger, H.F.; Kaiser, R.I.; Lee, Y.T.; Schaefer, H.F. Reaction of the ethynyl radical, C2H, with methylacetylene, CH3CCH, under single collision conditions:Implications for astrochemistry. *J. Chem. Phys.* 2001, *114*, 3476-3487.

112. Kaiser, R.I.; Chiong, C.C.; Asvany, O.; Lee, Y.T.; Stahl, F.; Schleyer, P.V.; Schaefer, H.F.Chemical dynamics of d1-methyldiacetylene (CH3CCCCD; X1A1) and d1-ethynylallene (H2CCCH(C2D); X1A') formation from reaction of C2D(X2+) with methylacetylene, CH3CCH(X1A1) *J. Chem. Phys.* 2001, *114*, 3488-3496.

113. Casavecchia, P.; Leonori, F.; Balucani, N.; Petrucci, R.; Capozza, G.; Segoloni, E. Probing the dynamics of polyatomic multichannel elementary reactions by crossed molecular beam experiments with soft electron-ionization mass spectrometric detection. *Phys. Chem. Chem. Phys.* 2009, *11*, 46–65.

114. Balucani, N.; Capozza, G.; Leonori, F.; Segoloni, E.; Casavecchia, P. Crossed molecular beam reactive scattering: from simple triatomic to multichannel polyatomic reactions. *Int. Rev. Phys. Chem.* 2006, *25*, 109-163

115. Balucani, N.; Alagia, M.; Cartechini, L.; Casavecchia, P.; Volpi, G.G.; Sato, K.; Takayanagi, T.; Kurosaki, Y. Cyanomethylene formation from the reaction of excited nitrogen atoms with acetylene: a crossed beam and ab initio study. *J. Am. Chem. Soc.* 2000, *122*, 4443-4450.

116. Balucani, N.; Cartechini, L.; Alagia, M.; Casavecchia, P.; Volpi, G.G. Observation of nitrogenbearing organic molecules from reactions of nitrogen atoms with hydrocarbons: a crossed beam study of N(2D) + ethylene. *J. Phys. Chem. A* 2000, *104*, 5655-5659.

117. Casavecchia, P.; Balucani, N.; Cartechini, L.; Capozza, G.; Bergeat, A.; Volpi, G.G. Crossed beam studies of elementary reactions of N and C atoms and CN radicals of importance in combustion. *Faraday Discuss.* 2001, *119*, 27-49.

118. Balucani, N.; Casavecchia, P. Gas-phase reactions in extraterrestrial environments: laboratory investigations by crossed molecular beams. *Orig. Life Evol. Biosph.* 2006, 36, 443-450.

119. Balucani, N.; Bergeat, A.; Cartechini, L.; Volpi, G.G.; Casavecchia, P.; Skouteris, D.; Rosi, M. Combined crossed molecular beam and theoretical studies of the N(2D) + CH4 reaction and implications for atmospheric models of Titan. *J. Phys. Chem. A,* submitted.

120. 120. Balucani, N.; Casavecchia, P.; Banares, L.; Aoiz, F.J.; Gonzalez-Lezana, T.; Honvault, P.; Launay, J.M. Experimental and theoretical differential cross sections for the N(2D) + H2 reaction, *J. Phys. Chem. A* 2006, *110*, 817-829.

121. Cartechini, L.; Bergeat, A.; Capozza, G.; Casavecchia, P.; Volpi, G.G.; Geppert, W.D.; Naulin, C.; Costes, M. Dynamics of the C + C2H2 reaction from differential and integral cross-section measurements in crossed-beam experiments. *J. Chem. Phys.* 2002, *116*, 5603-5611.

122. Geppert, W.D.; Naulin, C.; Costes, M.; Capozza, G.; Cartechini, L.; Casavecchia, P.; Volpi, G. G. Combined crossed-beam studies of C(3PJ) + C2H4 $\rightarrow$ C3H3 + H reaction dynamics between 0.49 and 30.8 kJ mol-1. *J. Chem. Phys.* 2003, *119*, 10607-10617.

123. Costes, M.; Daugey, N.; Naulin, C.; Bergeat, A.; Leonori, F.; Segoloni, E.; Petrucci, R.; Balucani, N.; Casavecchia, P. Crossed-beam studies on the dynamics of the C + C2H2 interstellar reaction leading to linear and cyclic C3H + H and C3 + H2. *Faraday Discuss.* 2006, *133*, 157–176.

124. Leonori, F.; Petrucci, R.; Segoloni, E.; Bergeat, A.; Hickson, K.M.; Balucani, N.; Casavecchia, P. Unraveling the dynamics of the C(3P,1D) + C2H2 reactions by the crossed molecular beam scattering technique. *J. Phys. Chem. A* 2008, *112*, 1363-1379.

125. Leonori, F.; Petrucci, R.; Hickson, K.M.; Segoloni, E.; Balucani, N.; Le Picard, S.D.; Foggi, P.; Casavecchia, P. Crossed molecular beam

study of gas phase reactions relevant to the chemistry of planetary atmospheres: the case of C2 + C2H2. *Planet. Space Sci.* 2008, *56*, 1658-1673.

126. Balucani, N.; Leonori, F.; Petrucci, R.; Hickson, K.M.; Casavecchia, P. Crossed molecular beam studies of C(3P,1D) and C2(X1Σ+ g, a3u) reactions with acetylene. *Phys. Scr.* 2008, 78, 058117.

127. Capozza, G.; Segoloni, E.; Leonori, F.; Volpi, G.G.; Casavecchia, P. Soft electron impact ionization in crossed molecular beam reactive scattering: The dynamics of the O(3P) + C2H2 reaction. *J. Chem. Phys.* 2004, *120*, 4557-4560.

128. Casavecchia, P.; Capozza, G.; Segoloni, E.; Leonori, F.; Balucani, N.; Volpi, G.G. dynamics of the O(3P) + C2H4 reaction: identification of five primary product channels (vinoxy, acetyl, methyl, methylene, and ketene) and branching ratios by the crossed molecular beam technique with soft electron ionization. *J. Phys. Chem. A* 2005, *109*, 3527-3530

129. Balucani, N.; Stranges, D.; Casavecchia, P.; Volpi, G.G. Crossed beam studies of the reactions of atomic oxygen in the ground 3P and first electronically excited 1D states with hydrogen sulfide. *J. Chem. Phys.* 2004, *120*, 9571-9582.

130. Alagia, M.; Balucani, N.; Casavecchia, P.; Stranges, D.; Volpi, G.G. Crossed beam studies of four-atom reactions: the dynamics of OH + D2. *J. Chem. Phys.* 1993, *98*, 2459.

131. Alagia, M.; Balucani, N.; Casavecchia, P.; Stranges, D.; Volpi, G.G. Crossed beam studies of four-atom reactions: the dynamics of OH + CO. *J. Chem. Phys.* 1993, *98*, 8341.

132. Leonori, F.; Petrucci, R.; Balucani, N.; Hickson, K.M.; Hamberg, M.; Geppert, W.D.; Casavecchia, P.; Rosi, M. Combined crossed beam and theoretical studies of the $S(^1D)$ + C2H2 reaction. *J. Phys. Chem. A* 2009, *113*, 4330-4339.

133. Leonori, F.; Petrucci, R.; Balucani, N.; Casavecchia, P.; Rosi, M.; Berteloite, C.; Le Picard, S.D.; Canosa, A.; Sims, I.R. Observation of organosulfur products (thiovinoxy, thioketene and thioformyl) in crossed-beam experiments and low temperature rate coefficients for the reaction S(1D) + $C_2H_4$. *Phys. Chem. Chem. Phys.* 2009, in press.

134. Huang, C.; Li, W.; Estillore, A.D.; Suits, A.G. Dynamics of CN plus alkane reactions by crossedbeam dc slice imaging. *J. Chem. Phys.* 2008, *129*, 074301.

135. Zhang, F.; Kim, S.; Kaiser, R.I. A crossed molecular beams study of the reaction of the ethynyl radical ($C_2H(X_2\Sigma+)$) with allene ($H_2CCCH_2(X_1A_1)$). *Phys. Chem. Chem. Phys.* 2009, in press. DOI:

10.1039/b822366a.

136. Kaiser, R.I.; Ting, J.; Huang, L.C.L.; Balucani, N.; Asvany, O.; Lee, Y.T.; Chan, O.; Stranges, D.; Gee, D. A versatile source to produce high intensity, pulsed supersonic radical beams for crossed beam experiments – the cyano radical CN($X_2\Sigma$ +), as a case study. *Rev. Sci. Instrum.*1999, *70*, 4185-4184.

137. 137. Guo, Y.; Mebel, A.M.; Zhang, F.; Gu, X.; Kaiser, R.I. Crossed molecular beam studies of the reactions of allyl radicals, $C_3H_5(X_2A_2)$, with methylacetylene ($CH_3$CCH(X1A1)), allene

138. (H=CC$CH_2(X_1A_1)$), and their isotopomers. *J. Phys. Chem. A* 2007, *111*, 4914-4921. Gu, X.; Kaiser, R.I. Reaction dynamics of phenyl radicals in extreme environments: a crossed molecular beam study. *Acc. Chem. Res.* 2009, *42*, 290-302.

139. Leonori, F.; Balucani, N.; Capozza, G.; Segoloni, E.; Stranges, D.; Casavecchia, P. Crossed beam studies of radical-radical reactions: O(3P) + C3H5 (allyl). *Phys. Chem. Chem. Phys.* 2007, *9*, 1307-1311.

140. Alagia, M.; Aquilanti, V.; Ascenzi, D.; Balucani, N.; Cappelletti, D.; Cartechini, L.; Casavecchia, P.; Pirani, F.; Sanchini, G.; Volpi, G.G. Magnetic analysis of supersonic beams of atomic oxygen, nitrogen, and chlorine generated from a radio-frequency discharge. *Israel J. Chem.* 1997, *37*, 329-342.

141. Bettens, R.P.A.; Lee, H.H.; Herbst, E. The importance of classes of neutral-neutral reactions in the production of complex interstellar-molecules. *Astrophys. J.* 1995, *443*, 664-674.

142. Turner, B.E. Detection of Interstellar Cyanoacetylene. *Astrophys. J.* 1971, *163*, L35-L41.

143. Cernicharo, J.; Guelin, M.; Kahane, X. A lambda 2 mm molecular line survey of the C-star envelope IRC + 10216. *Astronom. Astrophys. Suppl. Ser.* 2000, *142*, 181-215.

144. Wyrowski, F.; Schilke, P.; Walmsley, C.M. Vibrationally excited $HC_3N$ toward hot cores. *Astronom. Astrophys.* 1999, *341*, 882-895.

145. Bell, M.B.; Feldman, P.A.; Travers, M.J.; McCarthy, M.C.; Gottlieb C.A.; Thaddeus, P. Detection of HC11N in the Cold Dust Cloud TMC-1 *Astrophys. J.* 1997, *483*, L61-L64.

146. Bell, M.B.; Watson, J.K.G.; Feldman, P.A.; Travers, M.J. The excitation temperatures of HC9N and other long cyanopolyynes in TMC-1. *Astrophys. J.* 1998, *508*, 286-290.

147. Broten, N.W; MacLeod, J.M.; Avery, L.W.; Friberg, P.; Hjalmarson, A.; Hoglund, B.; Irvine, W.M.The detection of interstellar methylcyanoacetylene. *Astrophys. J.* 1984, *276*, L25-L29.

148. Nummelin, A.; Bergman, P. Vibrationally excited vinyl cyanide in Sgr $B_2$(N). *Astron. Astrophys.* 1999, *341*, L59-L62.

149. Guelin, M.; Neininger, N.; Cernicharo, J. Astronomical detection of the cyanobutadiynyl radical $C_5N$. *Astron. Astrophys.* 1998, *335*, L1-L4.

150. Guelin, M.; Mezaoui, A.; Friberg, P. Astronomical study of the $C_3N$ and $C_4H$ radicals - Hyperfine interactions and Rho-type doubling. *Astron. Astrophys.* 1982, *109*, 23-31.

151. Takagi, N.; Fukuzawa, K.; Osamura, Y.; Schaefer III, H.F. Ion-molecule reactions producing $HC_3NH$+ in interstellar space: forbiddenness the reaction between cyclic $C_3H_3$\ + and the N atom. *Astrophys. J.* 1999, *525*, 791-798.

152. Smith, I.W.M.; Sims, I.R. General Discussion. *J. Chem. Soc. Faraday Trans.* 1993, 89, 2165.

153. Fukuzawa, K.; Osamura, Y. Molecular orbital study of neutral-neutral reactions concerning HC3N formation in interstellar space. *Astrophys. J.* 1997, *489*, 113-121; and references therein.

154. Winstanley, N.; Nejad, L.A.M. Cyanopolyyne Chemistry in TMC-1. *Astrophys. Space Sci.* 1996, *240*, 13-37.

155. 155. Kunde, V.G., Aikin, A.C., Hand, R.A., Jennings, D.E., Maguire, W.C., Samuelson, R.E. $C_4H_2$, $HC_3N$, and $C_2N_2$ in Titan's atmosphere. *Nature* 1981, *292*, 686-688.

156. Teanby, N.A.; Irwin, P.G.J.; de Koka, R.; Vinatier, S.; Bézard, B.; Nixon, C.A.; Flasar, F.M.; Calcutt, S.B.; Bowles, N.E.; Fletcher, L.; Howett, C.; Taylor, F.W. Vertical profiles of HCN, $HC_3N$, and $C_2H_2$ in Titan's atmosphere derived from Cassini/CIRS data. *Icarus* 2007, *186*, 364-384.

157. Bezard, B.; Marten, A.; Paubert, G. Detection of acetonitrile on Titan. *Bull. Am. Astron. Soc.* 1993, *25*, 1100-1104.

158. Carty, D.; Le Page, V.; Sims, I.R.; Smith, I.W.M. Low temperature rate coefficients for the reactions of CN and $C_2H$ radicals with allene ($CH_2$=C=$CH_2$) and methyl acetylene ($CH_3C$ equivalent to CH). *Chem. Phys. Lett.* 2001, *344*, 310-316.

159. Georgievskii, Y.; Klippenstein, S.J. Strange kinetics of the $C_2H_6$ + CN reaction explained. *J. Phys. Chem. A* 2007, *111*, 3802-3811.

160. Choi, N.; Blitz, M.A.; McKee, K.; Pilling, M.J.; Seakins, P.W. H atom branching ratios from the reactions of CN radicals with $C_2H_2$ and $C_2H_4$. *Chem. Phys. Lett.* 2004, *384*, 68-72.

161. Gannon, K.L.; Glowacki, D.R.; Blitz, M.A.; Hughes, K.J.; Pilling, M.J.; Seakins, P.W. H atom yields from the reactions of CN radicals with $C_2H_2$, $C_2H_4$, $C_3H_6$, trans-2-$C_4H_8$, and iso-$C_4H_8$. *J. Phys. Chem. A* 2007,

*111*, 6679-6692.

162. Trevitt, A.J.; Goulay, F.; Meloni, G.; Osborn, D.L.; Taatjes, C.A.; Leone, S.R. Isomer-specific product detection of CN radical reactions with ethene and propene by tunable VUV photoionization mass spectrometry. *Int. J. Mass Spectr.* 2009, *280*, 113-118

163. Herbert, L.; Smith, I.W.M.; Rowland, D.; Spencer-Smith, R.D. Rate constants for the elementary reactions between CN radicals and $CH_4$, $C_2H_6$, $C2H_4$, $C_3H_6$, and $C_2H_2$ in the range - 295 < T/K < 700K. *Int. J. Chem. Kinet.* 1992, *24*, 791-802.

164. Chin, Y.; Kaiser, R.I.; Lemme, C.; Henkel, C. Detection of interstellar cyanoallene and its implications for astrochemistry. *AIP Conf. Proc.* 2006, *855*, 149-153.

165. Lovas, F.J.; Remijan, A.J.; Hollis, J.M.; Jewell, P.R.; Snyder, L.E. Hyperfine structure identification of interstellar cyanoallene toward TMC-1. *Astrophys. J.* 2006, *637*, L37-L40.

166. van Dishoeck, E.F.; Black, J.H. The excitation of interstellar $C_2$. *Astrophys. J.* 1982, *258*, 533-547.

167. Altwegg, K.; Balsinger, H.; Geiss, J. Composition of the volatile material in Halley's coma from in situ measurements. *Space Sci. Rev.* 1999, *90*, 3-18.

168. Geiss, J.; Altwegg, K.; Balsinger, H.; Graf, S. Rare atoms, molecules and radicals in the coma of P/Halley. *Space Sci. Rev.* 1999, *90*, 253-268.

169. Sorkhabi, O.; Blunt, V.M.; Lin, H.; A'Hearn, M.F.; Weaver, H.A.; Arpigny, C.; Jackson, W.M. Using photochemistry to explain the formation and observation of $C_2$ in comets. *Planet. Space Sci.* 1997, *45*, 721-730.

170. Mebel, A. M.; Hayashi, M.; Jackson, W. M.; Wrobel, J.; Green, M.; Xu, D.; Lin, S. H. Branching ratios of $C_2$ products in the photodissociation of $C_2H$ at 193 nm. *J. Chem. Phys.* 2001, *114*, 9821-9831.

171. Becker, K.H.; Donner, B.; Dinis, C.M.F.; Geiger, H.; Schmidt, F., Wiesen, P. Kinetics of the $C_2(a3\pi_u)$ radical reacting with selected molecules and atoms. *Z. Phys. Chem.* 2000, *214*, 503-517.

172. Pasternack, L.; McDonald, J.R. Reactions of $C_2(X1\Sigma + g)$ produced by multi-photon UV-excimer laser photolysis. *Chem. Phys.* 1979, *43*, 173-182.

173. Pitts, W.M.; Pasternack, L.; McDonald, J.R. Temperature-dependence of the $C_2(X1\Sigma^+ g)$ reaction with $H_2$ and $CH_4$ and $C_2(X1\Sigma_+ g$ and $a3\pi_u$ equilibrated states) with $O_2$. *Chem. Phys.* 1982, *68*, 417-422.

174. Reisler, H.; Mangir, M.S.: Wittig, C. Kinetics of free-radicals generated

by IR laser photolysis. 4. Intersystem crossings and reactions of C2(X1 $\Sigma$ + g) and C2(a3 $\pi$ u) in the gaseous-phase. *J. Chem. Phys.* 1980, *73*, 2280-2286.

175. Daugey, N.; Caubet, P.; Bergeat, A.; Costes, M.; Hickson, K. M. Reaction kinetics to low temperatures. Dicarbon + acetylene, methylacetylene, allene and propene from 77 $\leq$ T $\leq$ 296 K. *Phys. Chem. Chem. Phys.* 2008, *10*, 729-737.
176. Herron, J.T. 1999. Evaluated chemical kinetics data for reactions of $N(^2D)$, $N(^2P)$ and N2(A3u +) in the gas phase. *J. Phys. Chem. Ref. Data* 1999, *28*, 1453-1483.
177. Yung,Y. L. An update of nitrile photochemistry in Titan, *Icarus* 1987, *72*, 468-472.
178. Kurosaki, Y.; Takayanagi, T.; Sato, K.; Tsunashima, S. Ab initio molecular orbital calculations of the potential energy surfaces for the N(2D) + CH4 reaction. *J. Phys. Chem. A* 1998, *102*,\ 254-259.
179. Umemoto, H.; Nakae, T.; Hashimoto, H.; Kongo, K.; Kawasaki M. Reactions of N(2D) with methane and deuterated methanes, *J. Chem. Phys.* 1998, *109*, 5844-5848
180. Balucani, N.; Casavecchia, P. Neutral-neutral gas-phase reactions in extraterrestrial environments: laboratory investigations by crossed molecular beams. *AIP Conference Proceedings* 2006, *855*, 17-30.
181. Fujii, T.; Arai, N. Analysis of N-containing hydrocarbon species produced by a CH4/N2 microwave discharge: Simulation of Titan's atmosphere. *Astrophys J.* 1999, *519*, 858-863.
182. Teslja, A.; Nizamov, B.; Dagdigian, P.J. The electronic spectrum of methyleneimine. *J. Phys. Chem. A* 2004, *108*, 4433-4439.
183. Larson, C.; Ji, Y.Y.; Samartzis, P., Wodtke, A.M.; Lee, S.H.; Lin, J.J.M.; Chaudhuri, C.; Ching T.T. Collision-free photochemistry of methylazide: Observation of unimolecular decomposition of singlet methylnitrene. *J. Chem. Phys.* 2006, *125*, 133302.
184. Takayanagi, T.; Kurosaki, Y.; Sato, K.; Tsunashima, S. Ab initio molecular orbital study of the $N(^2D)$ + ethylene reaction. *J. Phys. Chem. A* 1998, *102*, 10391-10398.
185. Leonori, F.; Petrucci, R.; Stazi, M.; Balucani, N.; Casavecchia, P.; Skouteris, D.; Rosi, M. Combined crossed-beam and theoretical studies of the $N(^2D)+C_2H_6$ reaction. In preparation.
186. Israel, G.; Szopa, C.; Raulin, F.; Cabane, M.; Niemann, H.B.; Atreya, S.K.; Bauer, S.J.; Brun, J.F.; Chassefiere, E.; Coll, P.; Conde, E.; Coscia, D.; Hauchecorne, A.; Millian, P.; Nguyen, M.J.; Owen, T.; Riedler, W.; Samuelson, R.E.; Siguier, J.M.; Steller, M.; Sternberg, R.; Vidal-Madjar,

C. Complex organic matter in Titan's atmospheric aerosols from in situ pyrolysis and analysis. *Nature* 2005, *438*, 796-799.

187. Yamada, M.; Osamura, Y.; Kaiser, R.I. A comprehensive investigation on the formation of organo-sulfur molecules in dark clouds via neutral-neutral reactions. *Astronom. Astrophys.* 2002, *395*, 1031-1044.

188. Petrie, S. Formation of interstellar CCS and CCCS: A case for radical/neutral chemistry? *Mon. Not.R. Astronom. Soc.* 1996, *281*, 666-672.

189. Wakelam, V.; Ceccarelli, C.; Castets, A.; Lefloch, B.; Loinard, L.; Faure, A.; Schneider, N.; Benayoun, J.J. Sulphur chemistry and molecular shocks: The case of NGC 1333-IRAS 2. *Astron. Asrophys.* 2005, *437*, 149-158.

190. Wakelam, V.; Caselli, R.; Ceccarelli, C.; Herbst, E.; Castets, A. Resetting chemical clocks of hot cores based on S-bearing molecules. *Astron. Asrophys.* 2004, *422*, 159-169.

191. Wakelam, V.; Castets, A.; Ceccarelli, C.; Lefloch, B.; Caux, E.; Pagani, L. Sulphur-bearing species in the star forming region L1689N. *Astron. Asrophys.* 2004, *413*, 609-622.

192. Dello Russo, N.; DiSanti, M.A.; Mumma, M.J.; Magee-Sauer, K.; Rettig, T.W. Carbonyl sulfide in comets C/1996 B2 (Hyakutake) and C/1995 O1 (Hale-Bopp): Evidence for an extended source in Hale-Bopp. *Icarus*, 1998, *135*, 377-388.

193. Noll, K.S.; McGrath, M. A.; Trafton, L. M.; Atreya, S. K.; Caldwell, J.J.; Weaver, H. A.; Yelle, R. V.; Barnet, C.; Edgington, S. HST spectroscopic observations of Jupiter after the collision of comet Shoemaker-Levy 9. *Science* 1995, *267*, 1307-1313.

194. Atreya, S.K.; Edginton, S.G.; Trafton, L. M.; Caldwell, J.J.; Noll, K. S.; Weaver, H. A. Abundances of ammonia and carbon-disulfide in the jovian stratosphere following the impact of comet Shoemaker-Levy 9. *Geophys. Res. Lett.* 1995, *22*, 1625-1628

195. Kaiser, R. I.; Ochsenfeld, C.; Head-Gordon, M.; Lee, Y. T. The formation of HCS and HCSH molecules and their role in the collision of comet Shoemaker-Levy 9 with Jupiter. *Science* 1998,\ *279*, 1181.

196. Kaiser, R. I.; Yamada, M.; Osamura, Y. A crossed beam and ab initio investigation of the reaction of hydrogen sulfide, $H_2S$(X1A1), with dicarbon molecules, C2(X1Σ+ g). *J. Phys. Chem. A* 2002, *106*, 4825.

197. Woon, D.E. Quantum chemical evaluation of the astrochemical significance of reactions between S atom and acetylene or ethylene. *J. Phys. Chem. A* 2007, *111*, 11249-11253.

198. Kim, M. H.; Li, W.; Lee, S.K.; Suits, A.G. Probing of the hot-band excitations in the photodissociation of OCS at 288 nm by DC slice

imaging. *Can. J. Chem.* 2004, *82*, 880-884.

199. Black, G.; Jusinski, L.E. Branching ratio for S(33PJ) AND S(31D2) atom production in the photodissociation of $CS_2$ at 193 nm. *Chem. Phys. Lett.*, 1986, 124, 90-92.

200. Cook, P.A.; Langford, S.R.; Dixon, R.N.; Ashfold M.N.R. An experimental and ab initio reinvestigation of the Lyman-$\alpha$ photodissociation of H2S and D2S. *J. Chem. Phys.* 2001, *114*, 1672-1684.

201. Janssen, L.M.C.; van der Loo, M.P.J.; Groenenboom, G.C.; Wu, S.-M.; Radenović, D.C.; van Roij, A.J.A.; Garcia, I.A.; Parker, D.H. Photodissociation of vibrationally excited SH and SD radicals at 288 and 291 nm: The S(1D2) channel. *J. Chem. Phys.* 2007, *126*, 094304.

202. Vasyunin, A.I.; Semenov, D.; Wakelam, V.; Herbst, E.; Sobolev, A.M. Chemistry in protoplanetary disks: A sensitivity analysis. *Astrophys. J.* 2008, *672*, 629-641.

203. Wakelam, V.; Loison, J.C.; Herbst, E.; Talbi, D.; Quan, D.; Caralp, F. A sensitivity study if the neutral-neutral reactions C + C3 and C + C5 in cold dense interstellar clouds. *Astron. Asrophys.* 2009, *495*, 513-521.

204. Carrasco, N.; Alcaraz, C.; Dutuit, O.; Plessis, S.; Thissen, R.; Vuitton, V.; Yelle, R.; Pernot, P. Sensitivity of a Titan ionospheric model to the ion-molecule reaction parameters. *Planet. Space Sci.* 2008, *56*, 1644-1657.

205. Dobrijevic, M.; Carrasco, N.; Hebrard, E.; Pernot, P. Epistemic bimodality and kinetic hypersensitivity in photochemical models of Titan's atmosphere *Planet. Space Sci.* 2008, *56*, 1630-1643. 

# Chapter 4

# ENHANCEMENT OF ESTERIFICATION OF PROPIONIC ACID WITH ISOPROPYL ALCOHOL BY PERVAPORATION REACTOR

Ajit P. Rathod,[1] Kailas L. Wasewar,[1,2] and Chang Kyoo Yoo[2]

[1]Advance Separation and Analytical Laboratory (ASAL), Department of Chemical Engineering, Visvesvaraya National Institute of Technology (VNIT), Nagpur 440010, India

[2]Environmental Management & Systems Engineering Lab (EMSEL), Department of Environmental Science and Engineering, College of Engineering, Kyung Hee University, Seocheon-dong 1, Giheung-gu, Yongin-si, Gyeonggi-do 446-701, Republic of Korea

## ABSTRACT

With increasing cost of raw materials and energy, there is an increasing inclination of chemical process industries toward new processes that result in lesser waste generation, greater efficiency, and substantial yield of the desired products. Esterification is a chemical reaction in which two reactants carboxylic acid and alcohol react to form an ester and water. This reaction is a reversible reaction and the equilibrium conversion can be altered by varying the process parameters. Pervaporation reactor can enhance the conversion by shifting the equilibrium of reversible esterification reactions. Polyvinyl alcohol-polyether sulfone composite hydrophilic membrane was used for pervaporation-assisted esterification of propionic acid with isopropyl alcohol. The experiments were carried out in the presence of sulphuric acid as a catalyst at 50°C to 80°C with various reactants ratios. The esterification was carried out for catalyst loadings of 0.089 kmol/m$^3$ to 0.447 kmol/m$^3$. The molar ratios of isopropyl to propionic acid used for the experiment were 1 to 1.5. Maximum conversion was obtained for the ratio of 1.4. Also effect of other parameters such as process temperature and catalyst concentration was discussed. It was found that the use of pervaporation reactor increased the conversion of the propionic acid considerably.

## INTRODUCTION

Esterification reaction is a classic example of an equilibrium-limited reaction. The conversion is generally low due to limits imposed by thermodynamic equilibrium. An ester and water as byproducts were obtained from this reaction. Esters are widely used in various process industries. It is necessary to shift the position of the equilibrium to the ester side by either using a large excess of one of the reactants (generally the alcohol). The use of a large excess of reactant increased the cost for separation [1]. Pervaporation reactor shows potential alternative due to rate-controlled separation process and low-energy utilization as compared to distillation. Further, pervaporation can be operated at a temperature that matches the optimal temperature for reaction with an appropriate membrane. The last characteristic is mainly important for esterifications due to temperature constraints [2].

Pervaporation membrane reactors have been studied for esterification of acetic acid and ethanol [1], acetic acid and isopropanol [3], oleic acid and ethanol [4, 5], tartaric acid and ethanol [6], oleic acid and butanol [7], and valeric acid and ethanol [8] with various acids or lipases as catalysts. In some cases, the membrane itself is catalytically active [9].

There is increasing attention towards the use of membrane processes to increase the yield of equilibrium-limited reactions by selective elimination of a reaction product [10].The review on PV separation of water-acetic acid mixtures and its important results are detailed [11]. Novel poly(vinyl alcohol)-titanium dioxide (PVA-$TiO_2$) mixed matrix membranes prepared by incorporating nano-sized titanium dioxide (21 nm) and titanium dioxide surface modified with polyaniline (PANI) into PVA and cross-linked with glutaraldehyde

[12]. Poly(vinyl alcohol) (PVA)-based nanocomposite membranes were prepared by coprecipitation of different amounts of Fe(II) and Fe(III) taken in an alkaline medium and their pervaporation (PV) performances were investigated to dehydrate isopropanol, 1,4-dioxane and tetrahydrofuran (THF) from aqueous feeds containing 10–20 wt.% of water in isopropanol and 1,4-dioxane, 5–15 wt.% of water in THF [13]. Mixed matrix membranes of poly(vinyl alcohol) (PVA), loaded with phosphomolybdic heteropoly acid (HPA) and cross-linked with glutaraldehyde prepared by the solution casting technique. Pervaporation (PV) experiments were performed to separate water–isopropanol feed mixtures [14]. Mixed matrix membranes of sodium alginate (NaAlg) and poly(vinyl alcohol) (PVA) containing 5 and 10 wt.% silicalite-1 particles were fabricated by solution casting method and the cured membranes were cross-linked with glutaraldehyde. These membranes were used in pervaporation (PV) dehydration of isopropanol [15]. Mixed matrix membranes of sodium alginate (NaAlg) were prepared by solution casting incorporating 2.5, 5, 7.5, and 10 wt.% of zeolite beta particles. The membranes thus prepared were cross-linked with glutaraldehyde and tested for the pervaporation (PV) dehydration of ethanol and acetic acid [16]. Filled mixed matrix membranes (MMM) of sodium alginate (NaAlg) were prepared by solution casting method and were filled with 4A zeolite particles in varying compositions from 0 to 10 wt% with respect to weight of NaAlg polymer. Membranes were cross-linked with glutaraldehyde and tested for pervaporation (PV) dehydration of acetic acid and ethanol [17]. A very few work has been available on pervaporation reactor for esterification reactions of propionic acid with isopropanol. The reaction studied for the application of pervaporation reactor is an esterification of propionic acid with isopropanol to give isopropyl propionate and water:

In the present paper, experiments were conducted for esterification of propionic acid with isopropanol coupled with pervaporation and without pervaporation. Also effect of various parameters such as initial mole ratio of isopropyl alcohol over propionic acid, process temperature, and catalyst concentration on the performance of pervaporation reactor was discussed.

$$C_3H_6O_2 + C_3H_8O \longleftrightarrow C_6H_{12}O_2 + H_2O. \qquad (1)$$

# EXPERIMENT

## Membrane

Polyvinyl alcohol-polyether sulfone (PVA-PES) composite kind of hydrophilic membrane was used in flat sheet module of pervaporation reactor (supplied by Permionics Membranes Pvt. Ltd., India, size of test cell: 240 mm × 180 mm × 25 mm, and MOC of test cell SS-316) with active membrane area 0.0155 $m^2$.

## Thermogravimetric Analysis (TGA)

Thermal stability of PVA-PES membrane was determined with SII Co. Exstar TG/DTA (Japan). Thermogravimetric analyzer was used in order to know their upper temperature limit. The sample (6 mg) was placed in an aluminum pan and heated over a temperature range of 30 to 400°C at a heating rate of 10°C $min^{-1}$. The thermal stability was determined over temperature range from 30–400°C. In PVA-PES membrane, the onset of thermal decomposition started at 250.2°C and decomposition ends around 261.6°C with melting point 256.4°C as shown in Figure 1 observed total weight loss was 30.3%. This shows a high thermal stability of the PVA-PES membrane.

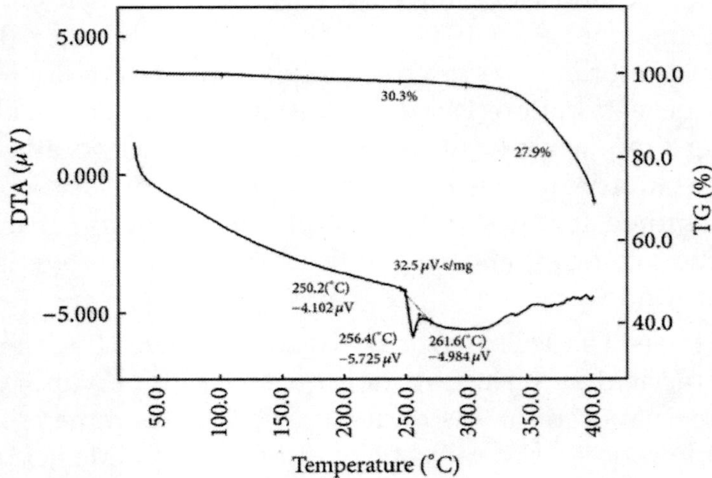


**Figure** 1: Thermogravimetric analyzer (TGA) analysis of membrane of PVA-PES.

## Apparatus

Esterification reaction and separation were carried out in a laboratory-scale semibatch pervaporation reactor supplied by Permionics Membranes Pvt. Ltd., India (Figure 2). A vacuum of less than 8 mbar was generated with a vacuum pump (Make: ILMVAC GmbH; Model: Diaphragm pump chem. Resistant, MPC 301Z). Cold water was supplied from a chiller to condense the permeate water vapor. For pervaporation experiments, the flat sheet membrane test cell was used. Recirculation pump was used (Make: Ravel Hiteks pvt Ltd; Model: RH-P-120L) to circulate the retentate solution.

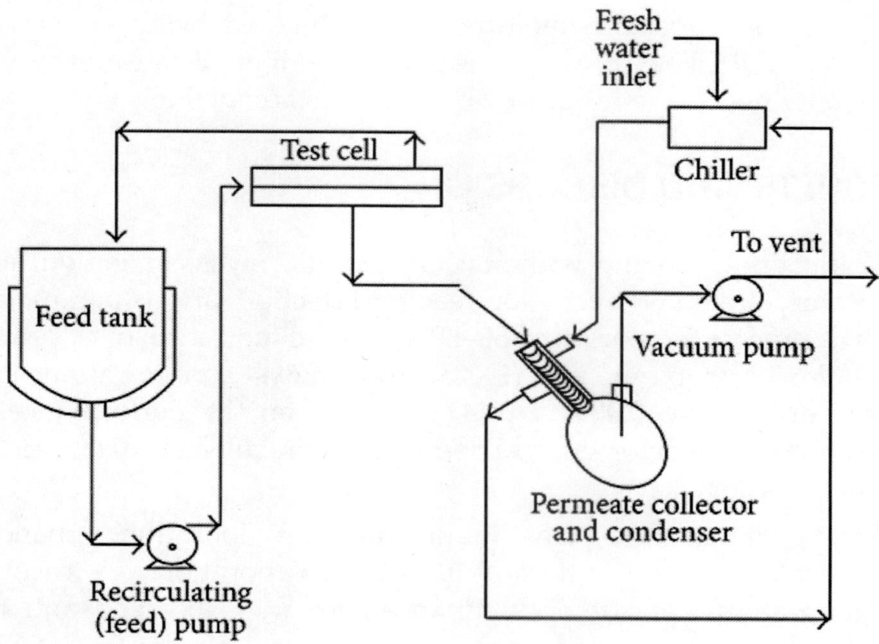


Figure 2: Typical diagram of pervaporation reactor (PVR).

## Esterification

Analytical grade chemicals were used in the esterification reaction. The CAS number, source, and grades of the chemicals used are as follows: propionic acid (CAS 79-09-4, Qualigens 99%), isopropyl alcohol (CAS 67-63-0, Qualigens 99%), and Sulphuric acid (CAS 7664-93-9, Qualigens 99.9%). All the chemicals were used without any further treatment. Esterification experiments utilized the equivalent mole of propionic acid (500 mL) and isopropyl alcohol (510 mL), giving a mole ratio of isopropyl alcohol to propionic acid as 1 : 1. Experiments were carried out with sulphuric acid as a catalyst. The water was removed by applying vacuum on the permeate side. Permeate was condensed and collected in round bottom flask.

### Analysis

The permeate composition and the amount of product (isopropyl propionate) in reaction mixtures were obtained by titration with 0.125 N NaOH. Few titration was performed in triplicate and results were obtained in the range of ±3%, which is acceptable.

## RESULTS AND DISCUSSION

Esterification with and without pervaporation was carried out in a laboratory-scale pervaporation reactor. Effect of various parameters such as process temperature (50–80°C), initial mole ratio of isopropyl alcohol over propionic acid (1–1.5), and catalyst concentrations that were varied from 0.089 to 0.447 kmol/$m^3$ on the performance of pervaporation reactor was carried out. The results are discussed in further section.

From Figure 3, it was found that pervaporation enhanced the conversion more than without pervaporation. In case of pervaporation, considerable enhancement was observed and the conversion was increased from its equilibrium value of 66% to 87% using a PVA-PES membrane. This membrane was the most selective one for water transport of membranes used in this work [18], if the removal of water is indeed the major mechanism for the enhancement of yield.

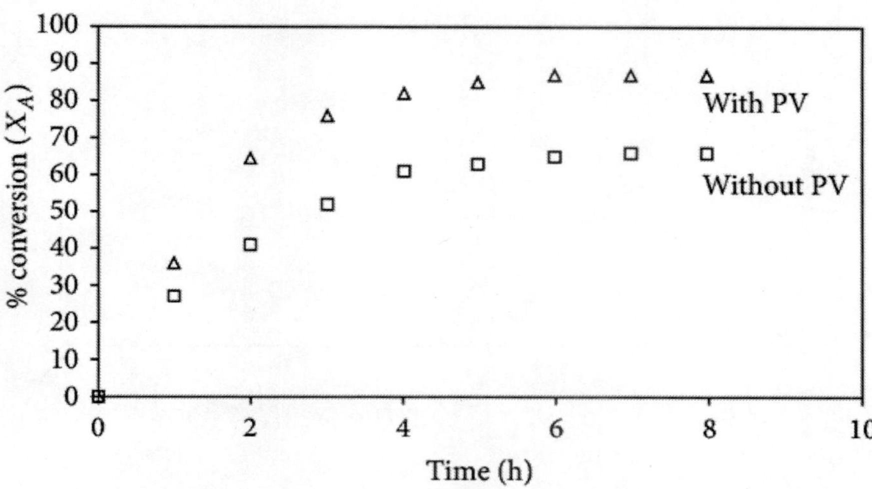


Figure 3: Comparison of experimental results for propionic acid conversion (; kmol/m$^3$; ; m$^2$/m$^3$): with PV (Δ) and without PV (□).

Catalyst concentration may be an alternative way to accelerate ester production [19]. In view of this, the catalyst concentration was varied from 0.089 kmol/m$^3$ to 0.447 kmol/m$^3$.The experimental results for conversion of propionic acid during pervaporation process over various catalyst concentrations were presented in Figure 4. It can be seen that the conversion is increasing with catalyst concentration. At higher concentration of catalyst, the percentage in conversion is also increasing but not with much significance (<5%). Hence, catalyst concentration of = 0.358 kmol/m$^3$ is optimal.

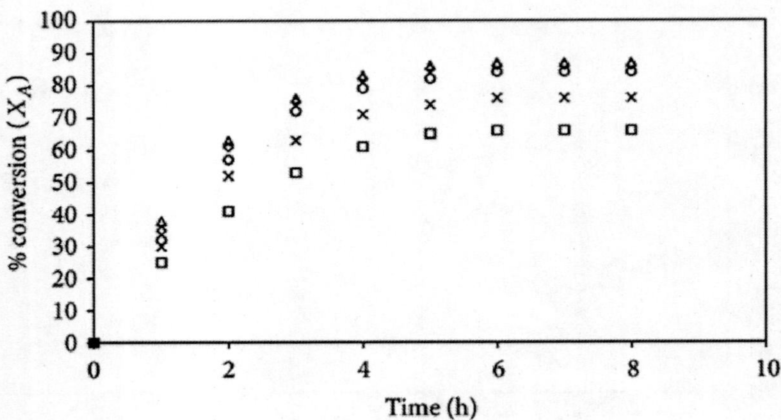


Figure 4: Effect of change in catalyst concentration on conversion of propionic acid (; ; m$^2$/m$^3$): kmol/m$^3$ (□), kmol/m$^3$ (), kmol/m$^3$ (), kmol/m$^3$ (◇), kmol/m$^3$ (Δ).

Figure 5 depicts the effect of initial molar reactant ratio on the propionic acid conversion. The reactant ratio was varied from 1 to 1.5 for fixed values of the other parameters. The higher conversion was observed for higher ratios. It can be found that  played a part in reaction rate but exerted no effect on kinetics of PV [20].

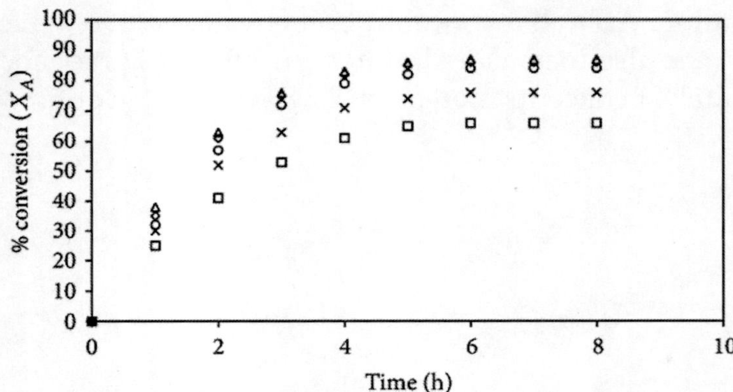


Figure 5: Effect of change in ratio of initial reactants on conversion of propionic acid (; kmol/m$^3$; m$^2$/m$^3$): (□), (), (), (◇), (Δ).

## CONCLUSION

Experiments for esterification of propionic acid with isopropyl alcohol coupled with and without pervaporation were carried out. The performance of pervaporation reactor was analyzed by studying effect of various parameters such as temperature, catalyst concentration, and reactant ratio. Using pervaporation reactor considerable enhancement was observed and the conversion of propionic acid was increased from its equilibrium value of 66% to 87% using a PVA-PES membrane. The presented data can be extended for study and design of pervaporation reactor for similar kind of reactions.

## REFERENCES

1. Ó. de la Iglesia, R. Mallada, M. Menéndez, and J. Coronas, "Continuous zeolite membrane reactor for esterification of ethanol and acetic acid," Chemical Engineering Journal, vol. 131, no. 1–3, pp. 35–39, 2007. View at Publisher · View at Google Scholar · View at Scopus
2. X. Feng and R. Y. M. Huang, "Studies of a membrane reactor: esterification facilitated by pervaporation," Chemical Engineering Science, vol. 51, no. 20, pp. 4673–4679, 1996. View at Publisher · View at Google Scholar · View at Scopus
3. M. T. Sanz and J. Gmehling, "Esterification of acetic acid with isopropanol coupled with pervaporation. Part II. Study of a pervaporation reactor," Chemical Engineering Journal, vol. 123, no. 1-2, pp. 9–14, 2006. View at Publisher · View at Google Scholar · View at Scopus
4. H. Kita, S. Sasaki, K. Tanaka, K. Okamoto, and M. Yamamoto, "Esterification of carboxylic acid with ethanol accompanied by pervaporation," Chemistry Letters, vol. 17, no. 12, pp. 2025–2028, 1988.View at Publisher · View at Google Scholar
5. K. Okamoto, M. Yamamoto, Y. Otoshi et al., "Pervaporation-aided esterification of oleic acid,"Journal of Chemical Engineering of Japan, vol. 26, no. 5, pp. 475–481, 1993. View at Publisher · View at Google Scholar · View at Scopus
6. J. T. F. Keurentjes, G. H. R. Janssen, and J. J. Gorissen, "The esterification of tartaric acid with ethanol: kinetics and shifting the equilibrium by means of pervaporation," Chemical Engineering Science, vol. 49, no. 24, pp. 4681–4689, 1994. View at Scopus
7. S. J. Kwon, K. M. Song, W. H. Hong, and J. S. Rhee, "Removal of water produced from lipase-catalyzed esterification in organic solvent by pervaporation," Biotechnology and Bioengineering, vol. 46, no. 4, pp. 393–

395, 1995. View at Scopus

8. X. Ni, Z. Xu, Y. Shi, and Y. Hu, "Modified aromatic polyimide membrane preparation and pervaporation results for esterification system," Water Treatment, vol. 10, no. 2, pp. 115–120, 1995.View at Scopus
9. L. Bagnell, K. Cavell, A. M. Hodges, A. W.-H. Mau, and A. J. Seen, "The use of catalytically active pervaporation membranes in esterification reactions to simultaneously increase product yield, membrane permselectivity and flux," Journal of Membrane Science, vol. 85, no. 3, pp. 291–299, 1993.View at Publisher · View at Google Scholar · View at Scopus
10. J. J. Jafar, P. M. Budd, and R. Hughes, "Enhancement of esterification reaction yield using zeolite A vapour permeation membrane," Journal of Membrane Science, vol. 199, no. 1, pp. 117–123, 2002.View at Scopus
11. T. M. Aminabhavi and U. S. Toti, "Pervaporation separation of water-acetic acid mixtures using polymeric membranes," Designed Monomers and Polymers, vol. 6, no. 3, pp. 211–236, 2003. View at Publisher · View at Google Scholar · View at Scopus
12. M. Sairam, M. B. Patil, R. S. Veerapur, S. A. Patil, and T. M. Aminabhavi, "Novel dense poly(vinyl alcohol)-$TiO_2$ mixed matrix membranes for pervaporation separation of water-isopropanol mixtures at 30°C," Journal of Membrane Science, vol. 281, no. 1-2, pp. 95–102, 2006. View at Publisher · View at Google Scholar · View at Scopus
13. M. Sairam, B. V. K. Naidu, S. K. Nataraj, B. Sreedhar, and T. M. Aminabhavi, "Poly(vinyl alcohol)-iron oxide nanocomposite membranes for pervaporation dehydration of isopropanol, 1,4-dioxane and tetrahydrofuran," Journal of Membrane Science, vol. 283, no. 1-2, pp. 65–73, 2006. View at Publisher · View at Google Scholar · View at Scopus
14. S. B. Teli, G. S. Gokavi, M. Sairam, and T. M. Aminabhavi, "Mixed matrix membranes of poly(vinyl alcohol) loaded with phosphomolybdic heteropolyacid for the pervaporation separation of water-isopropanol mixtures," Colloids and Surfaces A, vol. 301, no. 1–3, pp. 55–62, 2007. View at Publisher · View at Google Scholar · View at Scopus
15. S. G. Adoor, B. Prathab, L. S. Manjeshwar, and T. M. Aminabhavi, "Mixed matrix membranes of sodium alginate and poly(vinyl alcohol) for pervaporation dehydration of isopropanol at different temperatures," Polymer, vol. 48, no. 18, pp. 5417–5430, 2007. View at Publisher · View at Google Scholar · View at Scopus
16. S. G. Adoor, L. S. Manjeshwar, S. D. Bhat, and T. M. Aminabhavi, "Aluminum-rich zeolite beta incorporated sodium alginate mixed matrix membranes for pervaporation dehydration and esterification of ethanol and acetic acid," Journal of Membrane Science, vol. 318, no. 1-2, pp. 233–246, 2008. View at Publisher · View at Google Scholar · View at Scopus
17. S. D. Bhat and T. M. Aminabhavi, "Pervaporation-aided dehydration and esterification of acetic acid with ethanol using 4A zeolite-filled cross-linked sodium alginate-mixed matrix membranes,"Journal of Applied Polymer

Science, vol. 113, no. 1, pp. 157–168, 2009. View at Publisher · View at Google Scholar · View at Scopus

18. F. R. Chen and H. F. Chen, "Pervaporation separation of ethylene glycol-water mixtures using crosslinked PVA-PES composite membranes. Part I. Effects of membrane preparation conditions on pervaporation performances," Journal of Membrane Science, vol. 109, no. 2, pp. 247–256, 1996.View at Publisher · View at Google Scholar · View at Scopus

19. Y. Zhu and H. Chen, "Pervaporation separation and pervaporation-esterification coupling using crosslinked PVA composite catalytic membranes on porous ceramic plate," Journal of Membrane Science, vol. 138, no. 1, pp. 123–134, 1998. View at Publisher · View at Google Scholar · View at Scopus

20. Q. L. Liu and H. F. Chen, "Modeling of esterification of acetic acid with n-butanol in the presence of $Zr(SO_4)_2{\cdot}4H_2O$ coupled pervaporation," Journal of Membrane Science, vol. 196, no. 2, pp. 171–178, 2002. View at Publisher · View at Google Scholar · View at Scopus

# Chapter 5

# 2-METHOXYETHANOL: A REMARKABLY EFFICIENT AND ALTERNATIVE REACTION MEDIUM FOR IODINATION OF REACTIVE AROMATICS USING IODINE AND IODIC ACID

Arvind M. Patil, Sainath B. Zangade, Yeshwant B.Vibhute*,and Sarala N. Kalyankar

Laboratory of Organic Synthesis, Department of studies in Chemistry, Yeshwant Mahavidyalaya, Nanded 431602 (M.S) India.

## ABSTRACT

Remarkably effective iodination of reactive aromatics carried out using iodine and iodic acid in 2- methoxyethanol as an efficient and alternative reaction medium. The comparison has been made by carrying out iodination reaction in acetic acid and ethanol. The 2-methoxyethanol is found to be excellent reaction solvent in terms of clean reaction conditions, short reaction time giving quantitative yields of product and no need of further purification.

## INTRODUCTION

Aromatic iodocompounds are valuable and versatile synthetic intermediates inorganic chemistry [1]. They react with nucleophiles such as amines or alkoxides to give the corresponding substituted products and can be lithiated to introduce electrophiles via halogen lithium exchange reaction [2]. They are also important and most reactive intermediate for various cross-coupling reactions and especially useful for formation of carbon-carbon and carbon-heteroatom bonds[3].

The moderate reactivity of iodine with aromatic substrates requires the addition of activating agents for its utilization. Generally, aromatic compounds are iodinated using iodine in presence of Lewis acid or an oxidizing agent [4]

Synthesis of iodoaromatics using various reaction medium and iodinating reagent involves I2/petroleum ether [5], KI/H2SO4 in H2SO4 [6], NaI/Chloramine-T in methanol [7], NaOCl/NaI in aqueous alcohol medium [8], KI/H2O2 in CH3COOH [9], KClO3 /KI/HCl in aqueous medium [10], and I2/NaBO3.H2O in ionic liquid [11]. The utility of alternative reaction solvents such as water [12], ionic liquid [13], flourous [14], supercritical media [15] and polyethylene glycol [PEG] [16] is rapidly growing. These solvents have attracted the attention of organic chemists due to their solvating ability and aptitude to act as a phase transfer catalyst, negligible vapour pressure, non-hazardous, easy work- up and economical cost. However many of these reported procedures have one or more disadvantages such as use of expensive catalysts, long reaction time, low selectivity, requirements of special apparatus and side reaction.

In continuation of earlier research program on iodination of reactive aromatics using iodine and iodic acid [17], herein we wish to report the use of 2-methoxyethanol as a reaction solvent.

Material and Methods

Melting points were determined in an open capillary tube and are uncorrected. IR spectra were recorded in KBr on a Perkin-Elmer spectrometer. 1H NMR spectra were recorded on a Gemini 300-MHZ instrument in CDCl3 as solvent and TMS as an internal standard. The mass spectra were recorded on EI-Shimadzu-GC-MS spectrometer. Elemental analyses were performed on a Carlo Erba 106 Perkin-Elmer model 240 analyzer.

General procedures for iodination of hydroxyl aromatic aldehydes, hydroxyl acetophenones, substituted anilines and phenols in ethanol, acetic acid and 2-methoxyethanol:

Mixture of different aromatic compounds (50 mmol), iodine (20 mmol) dissolved in 5 mL of 2-methoxyethanol and iodic acid (10 mmol) dissolved in water (1 mL) was added with shaking and refluxed for 3-8 minutes (tables 1-4). On cooling reaction mixture, crystalline solid product separated out (reaction monitored on TLC).

Obtained solid product was filtered through Buchner funnel. Physical data is given in Tables 1-4. For synthesis of di-iodo product 40 mmol of iodine and 20 mmol of iodic acid with 50 mmol of substrate was used.

Similar procedure was carried out by using 25 mL of ethanol or 20 mL of acetic acid. Results obtained by these procedures are shown in tables 1-4.

## Spectral data of some selected compounds

### *4-Hydroxy-3, 5-diiodo-benzaldehyde*

IR ($cm^{-1}$): 2852 (C-H stretch of CHO),

**Table 1**.Physical data of iodole hydroxyl benzaldehydes

| Entry No. | Substrate | Product | M.P. (°C) Found (Reported) | Effect of solvent on iodination of hydroxy aldehydes | | | | | |
|---|---|---|---|---|---|---|---|---|---|
| | | | | Ethanol | | Acetic acid | | 2-Methoxyethanol | |
| | | | | Time (min) | Yield (%) | Time (min) | Yield (%) | Time (min) | Yield (%) |
| 1 | 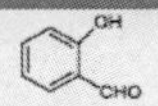 OH, CHO | 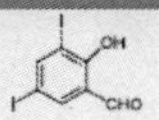 I, OH, I, CHO | 110 (110) [19] | 05 | 82 | 18 | 58 | 03 | 96 |
| 2 | 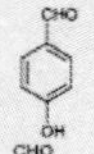 CHO, OH | 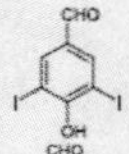 CHO, I, I, OH | 194 (195) [18] | 07 | 80 | 20 | 62 | 05 | 88 |
| 3 | 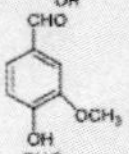 CHO, $OCH_3$, OH | 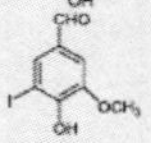 CHO, I, $OCH_3$, OH | 184 (185) [18] | 09 | 85 | 22 | 50 | 05 | 90 |
| 4 | 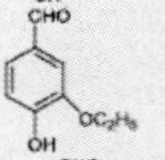 CHO, $OC_2H_5$, OH | 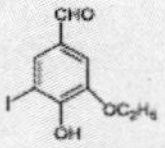 CHO, I, $OC_2H_5$, OH | 185 (185) [19] | 05 | 80 | 17 | 63 | 04 | 86 |
| 5 | 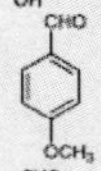 CHO, $OCH_3$ | 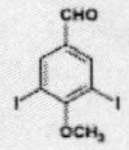 CHO, I, I, $OCH_3$ | 105 (107) [19] | 08 | 80 | 21 | 55 | 05 | 92 |
| 6 | 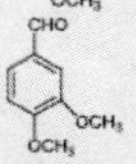 CHO, $OCH_3$, $OCH_3$ | 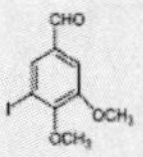 CHO, I, $OCH_3$, $OCH_3$ | 70 (69-72) [19] | 10 | 80 | 23 | 68 | 05 | 85 |

**Table 2.**Physical data of iodole hydroxyl acetophenones

| Entry No. | Substrate | Product | M.P.(°C) Found (Reported) | Effect of solvent on iodination of hydroxy acetophenones | | | | | |
|---|---|---|---|---|---|---|---|---|---|
| | | | | Ethanol | | Acetic acid | | 2-Methoxyethanol | |
| | | | | Time (min) | Yield (%) | Time (min) | Yield (%) | Time (min) | Yield (%) |
| 1 | 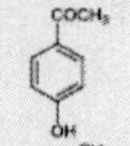 | 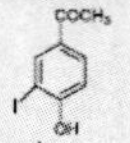 | 164 (162) [19] | 05 | 80 | 18 | 58 | 05 | 87 |
| 2 | 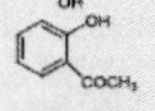 | 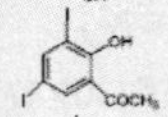 | 132 (132) [18] | 08 | 85 | 22 | 64 | 06 | 92 |
| 3 | 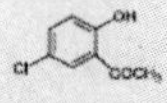 | 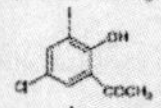 | 89 (90) [19] | 05 | 84 | 17 | 60 | 04 | 95 |
| 4 | 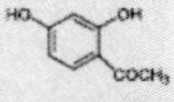 | 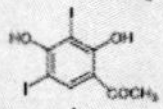 | 177 (178) [18 | 09 | 85 | 24 | 62 | 05 | 90 |
| 5 | 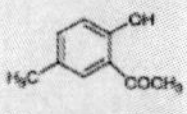 | 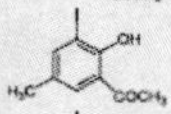 | 90 (90) [19] | 05 | 82 | 18 | 50 | 03 | 85 |
| 6 | 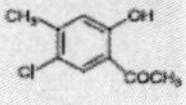 | 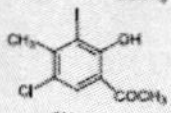 | 78 (76) [19] | 06 | 85 | 25 | 70 | 05 | 88 |
| 7 | 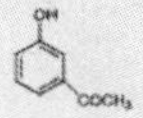 | 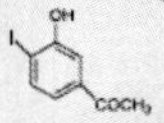 | 156 (155) [19] | 07 | 82 | 19 | 62 | 04 | 90 |
| 8 | 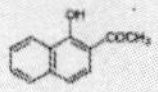 | 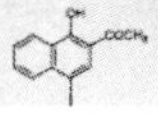 | 156 (156) [19] | 05 | 79 | 16 | 55 | 05 | 96 |

**Table 3.**Physical data of iodole anilines

| Entry No. | Substrate | Product | M.P.(°C) Found (Reported) | Effect of solvent on iodination of anilines | | | | | |
|---|---|---|---|---|---|---|---|---|---|
| | | | | Ethanol | | Acetic acid | | 2-Methoxyethanol | |
| | | | | Time (min) | Yield (%) | Time (min) | Yield (%) | Time (min) | Yield (%) |
| 1 | 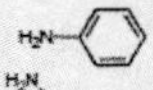 | 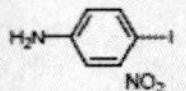 | 62 (62-63) [20] | 10 | 80 | 25 | 62 | 05 | 88 |
| 2 | 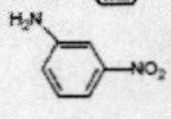 | 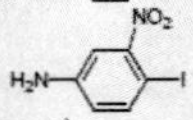 | 120 (122) [20] | 05 | 75 | 17 | 54 | 03 | 84 |
| 3 | 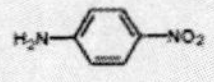 | 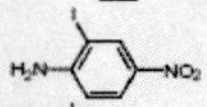 | 115 (116) [19] | 06 | 80 | 19 | 60 | 07 | 82 |
| 4 | 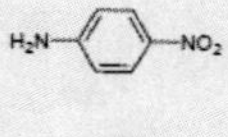 | 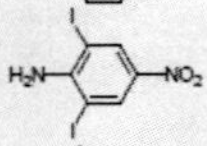 | 252 (251-253) [19] | 09 | 75 | 23 | 50 | 08 | 90 |
| 5 | 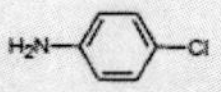 | 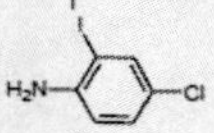 | 40 (39-41) [19] | 07 | 75 | 21 | 58 | 05 | 94 |
| 6 | 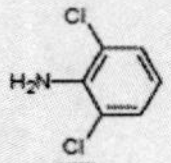 | 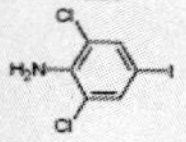 | 85 (80) [19] | 05 | 80 | 17 | 60 | 08 | 86 |
| 7 | 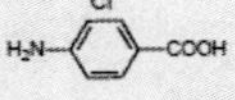 | 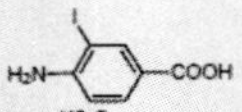 | 255 (258) [19] | 10 | 70 | 28 | 48 | 06 | 82 |
| 8 | 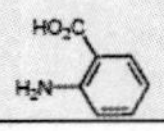 | 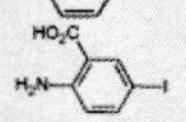 | 222 (220-225) [19] | 06 | 60 | 25 | 42 | 08 | 85 |

**Table 4.**Physical data of iodole phenols

| Entry No. | Substrate | Product | M.P (°C) Found (Reported) | Effect of solvent on iodination of Phenols | | | | | |
|---|---|---|---|---|---|---|---|---|---|
| | | | | Ethanol | | Acetic acid | | 2-Methoxyethanol | |
| | | | | Time (min) | Yield (%) | Time (min) | Yield (%) | Time (min) | Yield (%) |
| 1 |  | 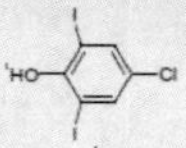 | 105 (107) [21] | 10 | 82 | 27 | 52 | 08 | 88 |
| 2 | 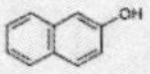 | 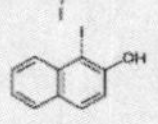 | 108 (108-110) [19] | 06 | 80 | 22 | 60 | 05 | 92 |
| 3 | 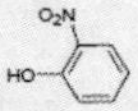 | 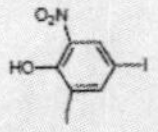 | 95 (93) [22] | 12 | 85 | 30 | 66 | 04 | 90 |
| 4 | 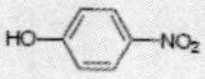 | 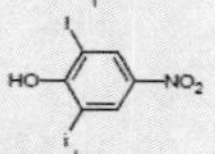 | 150 (152) [19] | 13 | 80 | 28 | 52 | 07 | 82 |
| 5 | 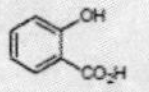 | 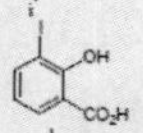 | 185 (190) [19] | 09 | 80 | 25 | 60 | 05 | 85 |
| 6 | 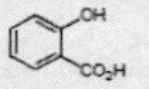 | 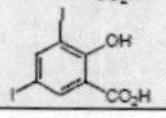 | 220 (220) [19] | 12 | 80 | 27 | 52 | 08 | 86 |

1660 (C=O), 1558 (C=C stretch). $^1H$ NMR ($CDCl_3$): δ 7.72 (s, 2H, Ar-H), 8.42 (s, 1H, OH), 10.02 (s, 1H, CHO). $^{13}C$ NMR (75 MHz, DMSO*d6*, δ, ppm):: 91.12 (C *of two* Ar-I), 131.64 (C of Ar-C), 139.42 (C *of two* Ar-H), 182.21 (C *of* Ar-OH), 191.37 (C *of* CHO). MS m/z: 374 ($M^+$). Anal. calcd. for: $C_7H_4O_2I_2$: C, 22.45; H, 1.06; I, 67.91. Found: C, 22.49; H, 1.09; I, 67.94.

## 3,5-DIIODO-4-METHOXY-BENZALDEHYDE

IR ($cm^{-1}$): 2845 (C-H *stretch of* CHO ), 1652 (C=O), 1548 (C=C stretch). 1H NMR ($CDCl_3$): δ 3.27 (s, 3H, $OCH_3$), 7.85 (s, 2H, Ar-H), 9.97 (s, $^1H$, CHO). $^{13}C$ NMR (75 MHz, $DMSO_{d6}$, δ, ppm): 190.78 (C=O), 137.60, 88.23 (*C of Aromatic ring*), 60.92 (OCH3). MS m/z: 388 ($M^+$). Anal. calcd. for: $C_8H_6O_2I_2$: C, 24.74; H, 1.54; I, 65.46. Found: C, 24.68; H, 1.51; I, 65.42.

## Dihydroxy-3,5-diiodoacetophenone

IR ($cm^{-1}$): 3428 (OH), 1660 (C=O), 1555 (C=C stretch). 1H NMR (CDCl3): δ 2.37 (s, 3H, $CH_3$), 7.90 (s, 1H, Ar-H), 12.62 (s, $^1H$, OH), 8.92 (s, 1H, OH). 13C NMR (75 MHz, $DMSO_{d6}$, δ,ppm): 24.17 (C *of methyl group*) 76.14 (C *of two* Ar-I), 141.29 (C of Ar-H), 178.24 (C *of two* Ar-OH), 197.68 (C *of carbonyl group*). MS m/z: 404 ($M^+$). Anal.calcd. for: $C_8H_6O_3I_2$: 23.76; H, 1.48; I, 62.87. Found: C, 23.80; H, 1.51; I, 62.93.

## (1-Hydroxy-4-iodo-naphthalen-2-yl)-ethanone

IR ($cm^{-1}$): 3438 (OH), 1664 (C=O), 1560 (C=C stretch). $^1H$ NMR (CDCl3): δ 2.35 (s, 3H, $CH_3$), 6.25-7.91 (m, 5H, Ar-H), 14.12 (s, $^1H$, OH). 13C NMR (75 MHz, $DMSO_{d6}$, δ, ppm): 200.37 (C=O), 164.95, 135.29, 133.20, 130.73, 127.28, 124.62, 122.47, 91.82(*C of Aromatic ring*), 26.62 ($CH_3$). MS m/z: 312 ($M^+$). Anal.calcd for: $C_{12}H_9O_2I$: C, 46.15; H, 2.88; I, 40.70. Found: C, 46.10; H, 2.86; I, 40.67.

## Diiodo-4-nitroaniline

IR ($cm^{-1}$): 3348 (NH), 1553 (C=C), 1368 ($NO_2$). $^1H$ NMR ($CDCl_3$): δ 4.08 (s, 2H, $NH_2$), 7.89 (s, 2H, Ar-H). 13C NMR (75 MHz, $DMSO_{d6}$, δ, ppm): 156.20, 145.89, 135.33, 86.90 (*C of Aromatic ring*). MS m/z: 390 (M+). Anal.calcd. for: $C_6H_4N_2O_2I_2$: C, 18.46; H, 1.02; I, 65.12. Found: C, 18.52; H, 1.05; I, 65.17.

## Diiodo-4-nitrophenol

IR ($cm^{-1}$): 3442 (OH), 1557 (C=C), 1352 ($NO_2$). $^1H$ NMR ($CDCl_3$): δ 7.82 (s, 2H, Ar-H). 13C NMR (75 MHz, $DMSO_{d6}$, δ, ppm): 161.86, 140.18, 134.70, 90.85. MS m/z: 391 ($M^+$). Anal.calcd. for: $C_6H_3O_3I_2$: C, 18.41; H, 0.76; I, 64.96. Found: C, 18.38; H, 0.75; I, 64.92.

## Results and Discussion

2-methoxyethanol is non-halogenated, inexpensive and water soluble which facilitate its removal from reaction product. In view of this observation it was thought worthwhile to carry out iodination of reactive aromatics using iodine and iodic acid as iodinating agent in 2-methoxyethanol as an efficient and alternative reaction solvent (Scheme-1).

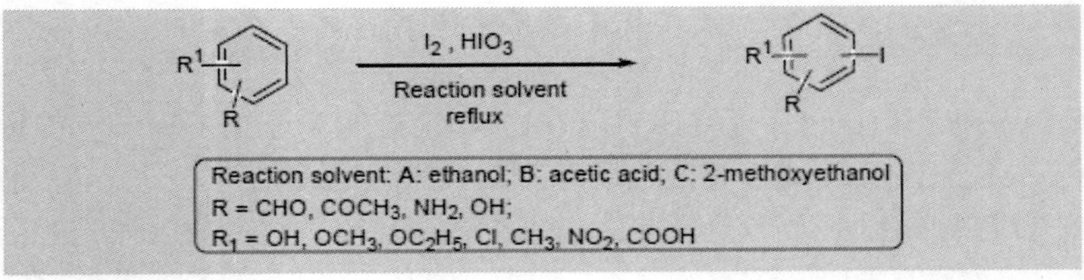

**Scheme 1.** Iodination of reactive aromatics.

In order to optimize the reaction conditions in terms of solubility, clean reaction conditions, isolation of product, time required for completion of reaction, yield and purity of product, we carried out the above reaction in ethanol and acetic acid and our results summarized in Table:1-4. We found that 2-methoxyethanol is remarkably effective reaction solvent consuming shorter reaction time besides increasing yield of product (Table.1-4). Encouraged by the results, we turned our attention towards variety substituted reactive aromatics. In all cases, the reaction proceeded smoothly in high yields at 120oC using 2-methoxyethanol as an attractive reaction solvent for iodination reaction.

## CONCLUSION

We reported remarkably efficient reaction medium for modified practical procedure for iodination of aromatics using iodine-ionic acid in 2-methoxyethanol. Present method, offer additional advantages such as comparatively least requirement of amount of solvent, simple reaction conditions, no need of catalyst, economical process with easier setup and workup procedure giving high yields of desired product.

### Acknowledgements

The authors gratefully acknowledge UGC-New Delhi for sanctioning major research grant (No. 38-267/2009). The authors are also thankful to principal, Yeshwant Mahavidyalaya, Landed, for providing laboratory facilities and Director IICT, Hyderabad, for providing necessary instrumental facilities.

## REFERENCES AND NOTES

1. Diederich, F. J.; Stang, P. J. Weinheim, Germany: Wiely-VCH, 1998.
2. Wakefield, B. J. London: Academic press, 1988.
3. Olivera, R.; Sanmartin, R.; Dominguez, E. Tetrahedron Lett. 2000, 41, 4357. [CrossRef] b) Qiang, L.; Juan, N.; Fan, Y.; Rui, Z.; Gang, Z.; Jie, T. Chin. J. Chem. 2004, 22, 419.
4. Radner, F. J. Org. Chem. 1988, 53, 3548. [CrossRef]

5. Joseph, R.; Pallan, P.; Sudalai, A.; Ravindranathan T. Tetrahedron Lett. 1995, 36, 609. [CrossRef]
6. Mattern, D. L. J. Org. Chem. 1983, 48, 4772. [CrossRef]
7. Hatanaka, Y.; Hashimoto, M.; Kurihara, H.; Nakayama, H.; Kanaoka, Y. J. Org. Chem. 1994, 59, 383. [CrossRef]
8. Edgar, K. J.; Falling, S. N. J. Org .Chem. 1990, 55, 5287. [CrossRef]
9. Suresh Kumar R. K.; Narender, N.; Rohitha, C. N.; Kulkarni, S. J. Synth.Commun. 2008, 38, 3894. [CrossRef]
10. Sathiyapriya, R.; Joel Karunakaran, R. Synth. Commun. 2006, 36, 1915. [CrossRef]
11. Bhilare, S. V.; Deorukhkar, A. R.; Darvatkar, N. B.; Salunkhe, M. M. Synth. Commun. 2008, 38, 2881. [CrossRef]
12. Chen, J. J.; Spear, S. K.; Huddieston, J. G.; Robin, D. R. Green Chemistry 2005, 7, 64. [CrossRef] b) Zang, Z. H.; Yin, L.; Wang, Y. M.; Liu, J. Y.; Li, Y. Green Chemistry 2004, 6, 563. [CrossRef] c) Kumar, R.; Choudhary, P; Nimesh, S; Chandra, R. Green Chemistry 2006, 8, 356. [CrossRef]
13. Helderant, D. J.; Jessop, P. G. J. Am. Chem. Soc. 2003, 125, 5600. [CrossRef]
14. Chandrasekhar, S.; Narsihmulu, Ch.; Sultana, S. S.; Ramakrishna Reddy, N. Org. Lett. 2002, 4, 4399. [CrossRef]
15. Jiang, R.; Kuang, Y. Q.; Sun, X. L.; Zhang, S. Tetrahedron Asymmetry, 2004, 15, 743. [CrossRef]
16. Namboodiri, V. V.; Varma, R. S. Green Chemistry 2001, 3, 146. [CrossRef]
17. Shinde, A. T.; Zangade, S. B.; Chavan, S. B.; Vibhute, A. Y.; Nalwar, Y. S.; Vibhute, Y. B. Synth. Commun. 2010, 40, 3506. [CrossRef] b) Patil, B. R.; Bhusare, S. R.; Pawar, R. P.; Vibhute, Y. B. Tetrahedron Lett. 2005, 46, 7179. [CrossRef] c) Patil, B. R.; Bhusare, S. R.; Pawar, R. P.; Vibhute, Y. B. Arkivoc 2006, i, 104. d) Zangade, S.; Mokle, S.; Shinde, A.; Vibhute, Y. Eur. J. Chem. 2012, 3, 314. [CrossRef]
18. Khansole, S. V.; Mokle, S. S.; Sayyed, M. A.; Vibhute, Y. B. J. Chin. Chem. Soc. 2008, 55, 871.
19. Aldrich Chemical Company. Handbook of Fine Chemicals; Aldrich Chemical Company: Bangalore, India, 2007-2008.

20. Maciej, S.; Lech, S.; Katarzyna, W. Molecules 2004, 9, 617. [CrossRef]
21. Gallo, R. D. C.; Gebara, K. S.; Muzzi, R. M.; Raminelli, C. J. Braz. Chem. Soc. 2010, 21, 770. [CrossRef]
22. Agnieszka, Z.; Lech, S. Molecules 2005, 10, 1307. [CrossRef]

# Chapter 6

# SYNTHESIS AND AROMATIZATION OF HANTZSCH 1,4-DIHYDROPYRIDINES UNDER MICROWAVE IRRADIATION. AN OVERVIEW

Jean Jacques Vanden Eynde[1,2] and Annie Mayence[1]

[1]Xavier University of Louisiana, College of Pharmacy, Division of Basic Pharmaceutical Sciences, 1 Drexel Drive, New Orleans, LA 70125, USA.

[2]Department of Organic Chemistry, University of Mons-Hainaut, 20 Place du Parc, B-7000 Mons, Belgium.

## ABSTRACT

Domestic microwave ovens as well as laboratory reactors have been successfully employed to prepare dialkyl 1, 4-dihydropyridine-3,5-dicarboxylates and to induce the synthesis of the corresponding aromatic derivatives. In that latter particular case, unexpected results have been reported.

## INTRODUCTION

Described more than one century ago by Hantzsch [1], dialkyl 1,4-dihydro-2,6-dimethylpyridine-3,5-dicarboxylates (1; 1,4-DHP; Figure 1) have now been recognized as vital drugs in the treatment of angina and hypertension. Some of them (Amlodipine 2, Felodipine 3, Isradipine 4, Lacidipine 5, Nicardipine 6, Nifedipine 7, Nimodipine

8, Nitrendipine 9) have been commercialized and it has been proven that their therapeutic success is related to their efficacy to bind to calcium channels and consequently to decrease the passage of the transmembrane calcium current, associated in smooth muscle with a long lasting relaxation and in cardiac muscle with a reduction of contractility throughout the heart [2-4].

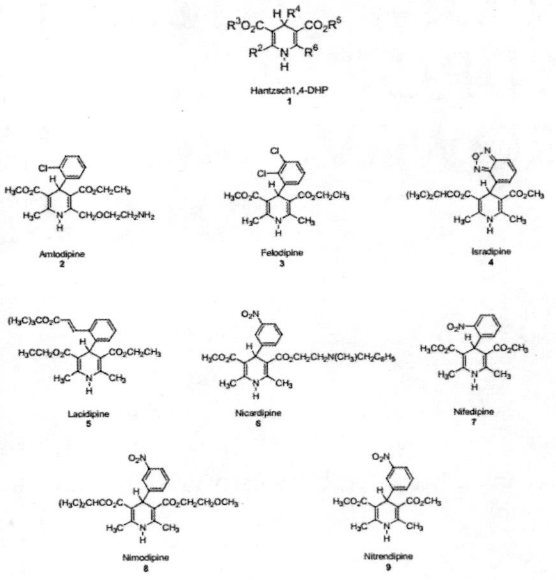

Figure.1

The usefulness of those calcium antagonists has led to the development of novel synthetic strategies to improve classical methods of preparation [5-7] and microwave activation stands among the alternative routes proposed the past decade.

Aromatization of 1,4-DHP has also attracted considerable attention in recent years as Böcker [8] has demonstrated that metabolism of those drugs involves a cytochrome P-450 catalyzed oxidation in the liver. The so-obtained pyridines are devoid of the pharmacological activity of the parent heterocycles and are further transformed by additional chemical modifications. Due to the biological importance of the oxidation step of 1,4-DHP, that reaction has been the subject of a large number of studies and a plethora of reagents has been utilized to mimic the in vivo transformation. In that field, surprising results have been collected when the reactions are performed under microwave irradiation.

## Hantzsch 1, 4-DHP Synthesis

In 1882, Hantzsch [1] reported the first synthesis of dialkyl 1,4-dihydro-2,6-dimethylpyridine-3,5-dicarboxylates from a refluxing mixture of an aldehyde, a β-ketoester, and aqueous ammonium hydroxide in ethanol (Scheme 1). That multicomponent reaction often affords the target compounds in good yields and this experimental method remains the most widely used protocol to access to 1,4-DHP differently substituted in position 4. Some modified procedures have later been proposed and they involve the use of preformed Knoevenagel adducts between the aldehyde and the ketoester or the use of preformed enaminoesters (represented in the E form to clarify the scheme).

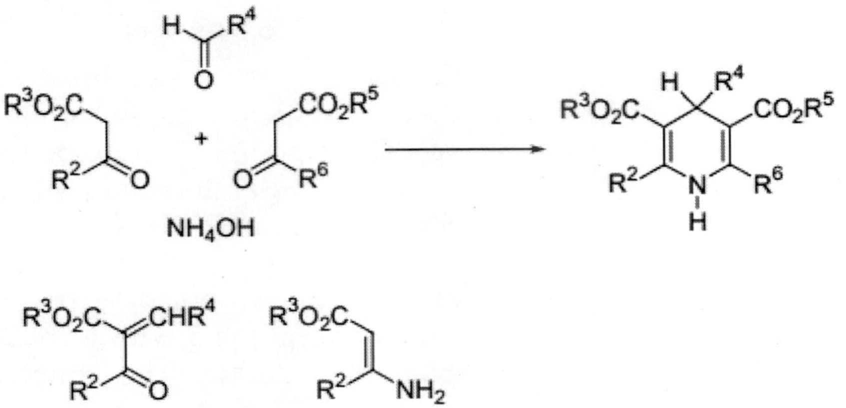

Scheme 1

The pioneering report on the use of microwave activation to obtain Hantzsch 1,4-DHP was published by Alajarin et al. in 1992 [9]. This group prepared a series of 4-aryl derivatives in a domestic oven by the classical multicomponent method (aldehyde: 15 mmol; alkyl acetoacetate: 43 mmol; ammonia: 30 mmol; ethanol: 3 mL). Yields ranged from 15 to 52 % for a reaction time of 4 minutes.

The authors claim that classical protocols for the formation of the same compounds require a reflux period of 12 hours but they did not notice any yield improvement when microwave irradiation was applied. Three years later the same group extended its work [10] to the preparation of 3,5-unsymmetrically substituted 1,4-DHP starting from arylmethyleneacetoacetate (8 mmol) and 3-aminocrotonate (4 mmol) in ethanol (4.5 mL). This report again emphasizes that the rapidity of the microwave-assisted syntheses does not affect the isolated yields. The same year Zhang [11] obtained four 4-aryl 1,4-DHP from 3-aminocrotonate (20 mmol), methyl acetoacetate (20 mmol) and arylaldehydes (20 mmol) in a domestic oven. For the first time, the preparations were conducted in the absence of solvent. Yields ranging from 59 to 77 % are reported and optimized heating periods do not exceed 10 minutes. To avoid any loss of volatile material, the reaction flasks were fitted with a condenser containing xylene. Also in 1995, Khadilkar [12] used the same building blocks as Zhang but in the presence of a solvent (ethanol, volume not reported; 3-aminocrotonate: 10 mmol; methyl acetoacetate: 14 mmol; arylaldehyde: 10 mmol). The heterocycles were prepared in a domestic oven within 3 to 5 minutes in 32 to 80 % yield.

Interestingly, Khadilkar [13] also described the formation, in a domestic oven, of 1,4-DHP in an aqueous hydrotrope solution (50% butylmonoglycolsulphate : 5 ml). The experiments were performed with 3-aminocrotonate (10 mmol), methyl acetoacetate (14 mmol) and aliphatic or aromatic aldehydes (10 mmol). The final products were obtained within 3 to 6 minutes in 35 to 97 % yield. All reactions described by Khadilkar [12, 13] were carried out by exposing the reactants to microwaves in containers equipped with a condenser charged with precooled carbon tetrachloride. The coupling of microwave heating (in a domestic oven) with the use of a mineral solid support (alumina: 2 g) has later been exploited by Suarez [14] to synthesize, within 6 minutes and with a yield higher than 85 %, an unsymmetrical 1,4-DHP from methyl 3-aminocrotonate (3 mmol), ethyl acetoacetate (3 mmol) and benzaldehyde (3 mmol). A catalytic amount of DMF (0.5 mL), as an energy transfer medium to attain higher temperatures, was added to the reaction mixture.

In 2001, a single-mode microwave reactor (SmithSyntheziser from Personal Chemistry, Uppsala, Sweden) was used for the first time to accelerate the preparation of series of 1,4-DHP from various alkyl acetoacetates (12.5 mmol), aldehydes (2.5 mmol) and 25 % aqueous ammonium hydroxide (10.0 mmol) [15]. In comparison with experiments performed in domestic ovens, use of a laboratory synthesizer does not appear to provide improved results. Indeed, for irradiation times varying from 10 to 15 minutes, reported yields fluctuate from "not determined" to 92 % whereas purity, evaluated by LC/MS, ranges from values as modest as 53% to 95%. Among all those reports, let us point out that, independently of the experimental conditions, Nifedipine (7) has only been obtained in moderate yields (32 % [12], 34 % [9], 35 % [13], 58 % [10]). On the other hand Nimodepine (8) was prepared nearly quantitatively (94 % [10]) whereas Nitrendipine (9) was synthesized in 72 % [13] and 98 % [10] yields. Recently, 1,4-DHP anchored to a soluble polymer - poly(styrene-co-allyl alcohol) - have been described [16, 17]. Such polymers, as well as the parent 4-unsubstituted 1,4-DHP, can be synthesized (Scheme 2) under solvent-free conditions and microwave irradiation in a monomode oven (Synthewave 402 from Prolabo). The experimental procedure only requires 100 seconds. The original strategy involves ethyl acetoacetate (20 mmol) and hexamethylenetetramine (8 mmol) as the source of the ammonia-formaldehyde mixture. Ammonium acetate (10 mmol) was added to the reaction medium in order to obtain the stoichiometric balance between ammonia and methanol.

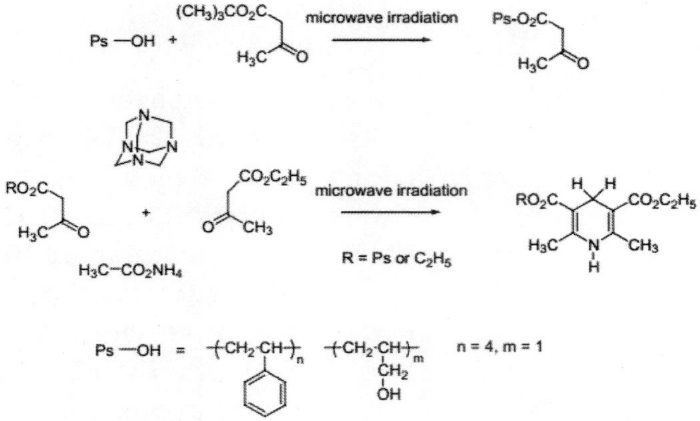

Scheme 2

## Aromatization of Hantzsch 1,4-DHP

From the numerous results published on the aromatization of Hantzsch 1,4-DHP (excluding the papers dealing with microwave-mediated experiments) the following rules can be established (Scheme 3):

- Heterocycles bearing an aryl (or heteroaryl) group in position 4 always undergo a dehydrogenation process [18];
- Heterocycles bearing a linear alkyl group in position 4 undergo a dehydrogenation process, except in vivo where dealkylation occurs and is accompanied by inhibition of cytochrome P-450 [8, 18-20];
- Heterocycles bearing a benzylic or secondary alkyl group in position 4 undergo a dealkylation process except when the oxidizing species is sulfur [21] or 2,3-dichloro-5,6-dicyano-1,4-benzoquinone [22].

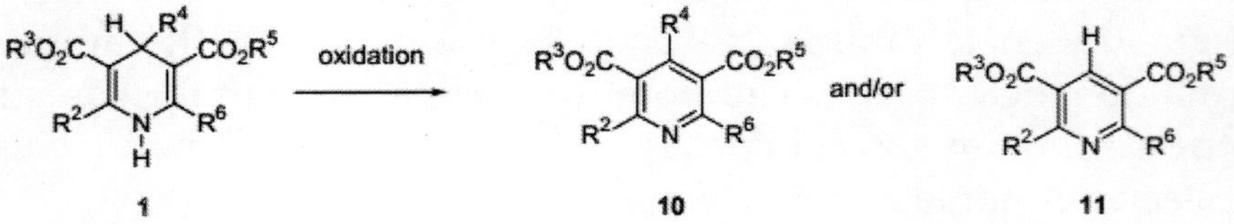

Scheme 3

Oxidation of 1,4-DHP under microwave irradiation was reported for the first time in 1991 by Alvarez et al. [23, 24]. They oxidized a series of 1,4-DHP (0.5 g) in a domestic oven by treatment on a mixture of manganese dioxide and Mexican bentonite clay (5.0 g, prepared from 1:2 or 1:4 mixtures of potassium permanganate and clay) in the absence of solvent. The procedure is characterized by short reaction times (10 minutes) and fair to quantitative yields (47-100 %). The most noticeable results were observed when starting from 1,4-DHP bearing a methyl, ethyl, or propyl group in position 4.

Indeed those reactions afforded, unexpectedly, mixtures of 4-alkylpyridines (10) and 4-unsubstituted pyridine (11). In contrast, the same group related [25], two years later, that those 4-alkyl 1,4-

DHP (0.25 g) do not undergo the dealkylation process when they are treated for 1 minute in a domestic microwave oven in the presence of a HNO3/Mexican bentonite clay system (2.5 g; prepared from a 1:1 mixture of the components). Aromatization of 1,4-DHP has also been studied by Varma [26]. He observed that solid state oxidation of 1, 4-DHP (1mmol) using elemental sulfur (1.3 mmol) and microwave activation in a domestic oven affords the dehydro derivatives, whichever the 4-substituent is.

Our experience in the field of 1,4-DHP chemistry [7, 22, 27] led us to initiate a systematic study of the solvent-free aromatization of Hantzsch 1,4-DHP upon microwave and classical heating in the absence of any external oxidant, except air [28]. In order to compare our results, we determined the temperature increase profile of a bath of alumina under given power settings in a domestic microwave oven and we managed to maintain another bath of alumina at the same final temperature (200 °C) on a hot plate. Then, taking care to use fresh alumina in each experiment, we put a round-bottom flask filled with 10 mmol of the 1,4-DHP in the bath and submitted it either to microwave irradiation or to classical heating for the same given period of time (10 min). The 4-phenyl DHP derivative appeared to be quite stable under microwave and classical conditions. The 4-isopropyl DHP derivative afforded (around 50 % conversion) the 4-unsubstituted pyridine (11) under both kinds of heating. Interestingly the 4-propyl DHP derivative yielded a mixture of the (expected) 4-alkylated pyridine (10) and the (less expected) 4-unsubstituted pyridine (11) under both experimental conditions. We observed that slight modifications of the temperature of the bath were accompanied by dramatic variations of the overall yields: yields increased from 25 % to 80 % when the temperature rose from 180 to 200 °C. However the ratio of concentrations 10:11 remains fairly constant and in each case in favor of the dealkylated product (10:11 = 1 : 4). We also noticed that the 4-propylpyridine (10) is stable upon heating and therefore formation of the 4-unsubstituted heterocycle (11) cannot be explained by a dealkylation of 10. Those preliminary results suggest that specific microwave effects can hardly be claimed to account for the aromatization of 1,4-DHP. On the other hand our observations reveal a particular behavior of Hantzsch 1,4-DHP upon heating and a dramatic influence of the temperature on their susceptibility to be oxidized by air (oxygen) above their melting point. Therefore when considering the aromatization of 1,4-DHP in dry media, the rate of the oxidation process induced by air should not be neglected. That rate being highly temperature-dependent, the presence of an inorganic solid in the reaction mixture emerge as an additional factor that can modify the yields, due to the thermal conductivity properties of that inorganic substance especially under microwave irradiation.

Finally, mention should be made that Barbry [17] utilized the parent 1,4-DHP (6 mmol) in the reduction (Scheme 4) of activated carbon-carbon double bonds (4 mmol) on silica gel (4 g) under microwave irradiation and without solvent. The reactions were performed with comparable success in a domestic multimode oven (8 min) and in the laboratory Synthewave monomode oven (4-6 min). The efficiency of the process appeared to be dependent on steric effects in the DHP as the bulky tert-butyl ester derivatives failed to react.

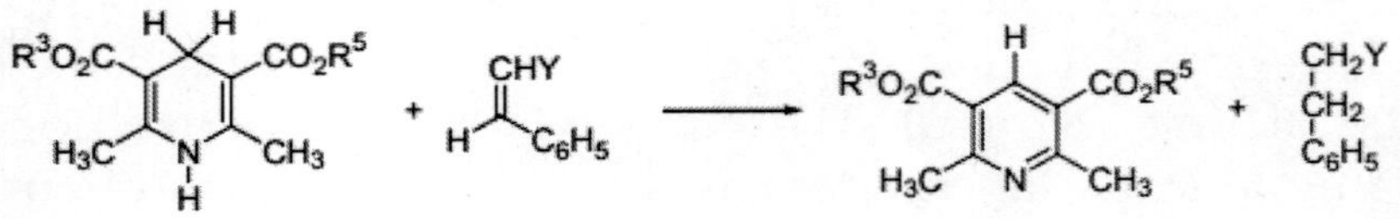

Scheme 4

## Domino Synthesis of Hantzsch Pyridines

To the best of our knowledge, only two papers [29, 30] describe the domino synthesis of Hantzsch pyridines (Scheme 5). Both groups start from a mixture of a bentonite clay, a β-ketoester, and an aldehyde. They used ammonium nitrate as the source of ammonia and oxidizing species. The experiments were carried out in domestic ovens.

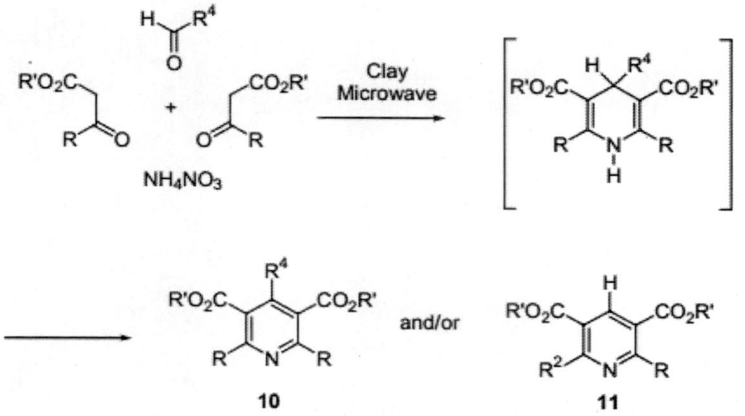

Scheme 5

The published results are intriguing as they are somehow contradictory. In the absence of solvent, Penieres [29] isolated from isobutyraldehyde (20 mmol), 40 mmol of the ketoester and 20 mmol of $NH_4NO_3$ on 5 g of clay the alkylated pyridine (10) as the major product. On the other hand he obtained the 4-unsubstituted pyridine (11), in substantial yield, when starting from n-butyraldehyde or benzaldehyde. That latter observation also contrasts with the claims of Cotterill [30] who carried out the syntheses in DMF (35 μL) from diverse arylaldehydes (0.1 mmol; 2 eq. of ketoester; 100 mg of a 5:1 w/w mixture of bentonite and $NH_4NO_3$) in a combinatorial approach and does not mention the presence of the parent pyridine 11 in the final mixtures.

In addition to those two papers, let us also describe the domino preparation of pyridinones structurally related to Hantzsch pyridines (Scheme 6) under microwave irradiation in a domestic oven [31]. The authors found that the pyridinones could readily be obtained from 4-(ethoxymethylene)-2-phenyloxazol-5(4H)-one and enaminocarbonyl derivatives in a dry medium, whereas the classical procedure required to heat the reactants in boiling dichlorobenzene for several hours.

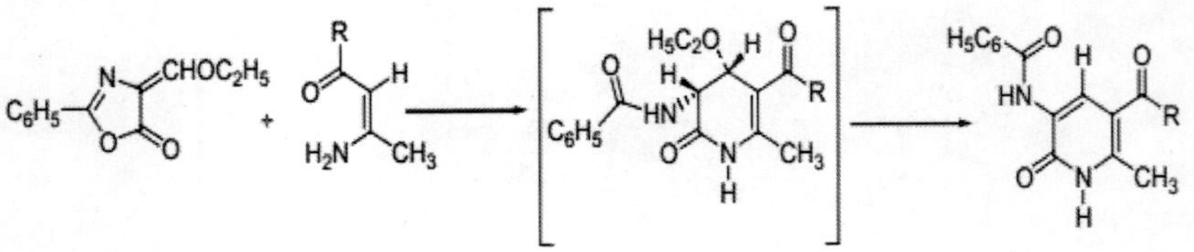

$R = OC_2H_5, NHC_6H_5, CH_3$

Scheme 6

## High-Throughput Synthesis

Due to shorter reaction times associated with experiments performed with the help of microwave irradiation, it is tempting to evoke combinatorial chemistry and to claim preparation of libraries of compounds. Therefore, let us mention in that sense that Cotterill [30] has reported the domino synthesis of Hantzsch pyridines in the 96-well microtiter plate format in a domestic microwave oven and that Ohberg [15] has utilized an automated laboratory reactor for this purpose.

## CONCLUSIONS

In recent years microwave irradiation has been successfully used to promote reactions yielding Hantzsch 1,4-DHP and the corresponding pyridines. A survey of the literature indicates that construction of 1,4-DHP in a microwave oven has been performed in solution as well as in dry media. Protocols yielding the pyridine derivatives, however, involve solvent-free conditions only and have led to unexpected and somehow contradictory results.

## REFERENCES AND NOTES

1. Hantzsch, A. Condensationprodukte aus Aldehydammoniak und Ketoniartigen Verbindungen. Ber. 1881, 14, 1637-1638.
2. Love, B.; Goodman, M.; Snader, K.; Tedeschi, R.; Macko, E. "Hantzsch-type" dihydropyridine hypotensive agents. 3. J. Med. Chem. 1974, 17, 956-965.

3. Bossert, F.; Meyer, H.; Wehinger, E. 4-Aryldihydropyridines, a new class of highly active calcium antagonists. Angew. Chem. Int. Ed. Engl. 1981, 20, 762-769.

4. Katzung, B.G. in Basic & Clinical Pharmacology 1998, Appleton & Lange, Stamford, CT (USA).

5. Gordeev, M.F.; Patel, D.V.; Gordon, E.M. Approaches to combinatorial synthesis of heterocycles: a solid-phase synthesis of 1,4-dihdropyridines. J. Org. Chem. 1996, 61, 924-928.

6. Breitenbucher, J.G.; Figliozzi, G. Solid-phase synthesis of 4-aryl-1,4-dihydropyridines via the Hantzsch three component condensation. Tetrahedron Lett. 2000, 41, 4311-4315.

7. Vanden Eynde, J.J.; Mayence, A. New methodologies for the preparation of drugs and their metabolites. Application to 1,4-dihydropyridines structurally related to calcium channel modulators of the Nifedipine-type. Intl. J. Med. Biol. Environ. 2000, 28, 25-31.

8. Böcker, R.H.; Guengerich, F.P. Oxidation of 4-aryl- and 4-alkyl-substituted 2,6-dimethyl-3,5-bis(alkoxycarbonyl)-1,4-dihydropyridines by human liver microsomes and immunochemical evidence for the involvement of a form of cytochrome P-450. J. Med. Chem. 1986, 29, 1596-1603.

9. Alajarin, R.; Vaquero, J.J.; Garcia Navio, J.L.; Alvarez-Builla, J. Synthesis of 1,4-dihydropyridines under microwave irradiation. Synlett 1992, 297-298.

10. Alajarin, R.; Jordan, P.; Vaquero, J.J.; Alvarez-Builla, J. Synthesis of unsymmetrically substituted 1,4-dihydropyridines and analogous calcium antagonists by microwave heating. Synthesis 1995, 389-391.

11. Zhang, Y.-W.; Shan, Z.-X.; Pan, B.; Lu, X.-H.; Chen, M.-H. Research on the synthesis of 1,4-dihydropyridines under microwave. Synth. Commun. 1995, 25, 857-862.

12. Khadilkar, B.M.; Chitnavis, A.A. Rate enhancement in the synthesis of some 4-aryl-1,4-dihydropyridines using methyl 3-aminocrotonate under microwave irradiation. Ind. J. Chem. 1995, 34B, 652-653.

13. Khadilkar, B.M.; Gaikar, V.G.; Chitnavis, A.A. Aqueous hydrotrope solution as a safer medium for microwave enhanced Hantzsch dihydropyridine ester synthesis. Tetrahedron Lett. 1995, 36, 8083-8086.

14. Suarez, M.; Loupy, A.; Perez, E. ; Moran, L. ; Gerona, G. ; Morales, A. ; Autié, M. An efficient procedure to obtain hexahydroquinoleines and unsymmetrical 1,4-dihydropyridines using solid inorganic supports and microwave activation. Heterocycl. Commun. 1966, 2, 275-280.

15. Ohberg, L.; Westman, J. An efficient and fast procedure for the Hantzsch dihydropyridine synthesis under microwave conditions. Synlett 2001, 1296-1298

16. Vanden Eynde, J.J.; Rutot, D. Microwave-mediated derivatization of poly(styrene-co-allyl alcohol), a key step for the soluble polymer-assisted synthesis of heterocycles. Tetrahedron 1999, 55, 2687-2694.

17. Torchy, S.; Cordonnier, G.; Barbry, D.; Vanden Eynde, J.J. Hydrogen transfer

from Hantzsch 1,4-dihydropyridines to carbon-carbon double bonds under microwave irradiation. Molecules 2002, 7, 528-533.

18. Love, B.; Snader, K. The Hantzsch reaction. I. Oxidation dealkylation of certain dihydropyridines. J. Org. Chem. 1965, 30, 1914-1916.

19. de Matteis, F.; Hollands, C.; Gibbs, A.H.; de Sa, N.; Rizzardini, M. Inactivation of cytochrome P-450 and production of N-alkylated porphyrins caused in isolated hepatocytes by substituted dihydropyridines. Structural requirements for loss of haem and alkylation of the pyrrole nitrogen atom. FEBS Lett. 1982, 145, 87-92.

20. Augusto, O.; Beilan, H.S.; Ortiz de Montellano, P.R. The catalytic mechanism of cytochrome P-450. Spin-trapping evidence for one-electron substrate oxidation. J. Biol. Chem. 1982, 257, 11288-11295.

21. Ayling, E.E. The influences of alkyl groups in carbonyl compounds. J. Chem. Soc. 1938, 1014-1023.

22. Vanden Eynde, J.J.; Delfosse, F.; Mayence, A.; Van Haverbeke, Y. Old reagents, new results: aromatization of Hantzsch 1,4-dihydropyridines with manganese dioxide and 2,3-dichloro-5,6-dicyano-1,4-benzoquinone. Tetrahedron 1995, 51, 6511-6516.

23. Alvarez, C.; Delgado, F.; Garcia, O.; Medina, S.; Marquez, C. MnO2/ bentonite: a new reactive for the oxidation of Hantzsch's dihydropyridines using microwave irradiation in the absence of solvent (I). Synth. Commun. 1991, 21, 619-624.

24. Delgado, F.; Alvarez, C.; Garcia, O.; Penieres, G.; Marquez, C. Unusual oxidative dealkylation of certain 4-alkyl-1,4-dihydropyridines with MnO2/ bentonite using microwave irradiation in the absence of solvent (II). Synth. Commun. 1991, 21, 2137-2141.

25. Garcia, O.; Delgado, F.; Cano, A.C.; Alvarez, C. Oxydation d'esters de Hantzsch, par le nouveau système HNO3/bentonite, et irradiation aux micro-ondes. Tetrahedron Lett. 1993, 34, 623-625.

26. Varma, R.S.; Kumar, D. Solid state oxidation of 1,4-dihydropyridines to pyridines using phenyliodine (III) bis(trifluoroacetate) or elemental sulfur. J. Chem. Soc., Perkin Trans. 1 1999, 1755-1757.

27. Vanden Eynde, J.J.; Mayence, A.; Maquestiau, A. A novel application of the oxidizing properties of pyridinium chlorochromate: aromatization of Hantzsch 1,4-dihydropyridines. Tetrahedron 1992, 48, 463-468.

28. Vanden Eynde, J.J. Unpublished results.

29. Penieres, G.; Garcia, O.; Franco, K.; Hernandez, O.; Alvarez, C. A modification to the Hantzsch method to obtain pyridines in a one-pot reaction: use of a bentonitic clay in a dry medium. Heterocycl. Commun. 1996, 2, 359-360.

30. Cotterill, I.C.; Usyatinsky, A.Y.; Arnold, J.M.; Clark, D.S.; Dordick, J.S.; Michels, P.C.; Khmelnitsky, Y.L. Microwave assisted combinatorial chemistry, synthesis of substituted pyridines. Tetrahedron Lett. 1998, 39, 1117-1120.

31. Vanden Eynde, J.J.; Labuche, N.; Van Haverbeke, Y. Microwave-mediated domino reaction in dry medium. Preparation of dihydropyridinones and pyridinones structurally related to Hantzsch esters. Synth. Commun. 1997, 27, 3683-3690.

# Chapter 7

# AN INNOVATIVE APPROACH TO FUNCTIONALITY TESTING OF ANALYSERS IN THE CLINICAL LABORATORY

Wolfgang Stockmann, Werner Engeldinger, Albert Kunst, and Margaret McGovern
Roche Diagnostics GmbH, Professional Diagnostics, Clinical Trials, Mannheim 68305, Germany

## ABSTRACT

The established protocols for evaluating new analytical systems produce indispensable information with regard to quality characteristics, but in general they fail to ana¬lyse the system per¬formance under routine-like conditions. We describe a model which allows the testing of a new analytical system under conditions close to the routine in a controlled and systematic manner by using an appropriate software tool. Performing routine simulation experiments, either reflecting imprecision or method comparison characteristics, gives the user essential information on the overall system performance under real intended-use conditions.

## INTRODUCTION

Conventionally, the evaluation of new analytical systems is conducted on the basis of established protocols related to analytical performance like those published by CLSI (NCCLS) [1], ECCLS [2], or other national organizations. Sometimes additional exploratory testing is performed in the hope of gaining some insight into the routine behaviour of the system. While the standard protocols produce indispensable information with regard to the quality characteristics, they fail to analyse the system performance under routine conditions. Similarly, random testing only gives a chance opportunity to detect system malfunctions.

Obviously, there is no easy way to experimentally test the course of events that lead to an erroneous assay result and to verify its incorrectness under nonstandardized, that is, routine conditions. This situation is caused by the increasingly complex interactions of hardware, software, and chemistry which are found on modern analysis systems. The manual generation of experiments that describe a sample sequence with variable specimens and request patterns is feasible, but it is cumbersome to produce and provides no information on the correctness of the measurements. A better approach can be obtained by developing a software tool which generates appropriate experimental request lists, allowing the testing of a new system under conditions close to the routine in a controlled and systematic manner, and which provides sufficient data reduction for the analysis of the results.

## METHODS

We have integrated this functionality in our evaluation software tool Windows-based computer-aided evaluation (WinCAEv) [3, 4] in such a way, that the generation of simulation experiments, the transfer of requests to the instrument, the on-line data capture, and the result evaluation, can be easily achieved with the available programme functions [5]. The routine simulation (hereafter referred to as RS) module allows for the definition and generation of typical test request patterns.

A request list that reflects a routine laboratory workload can be simulated by WinCAEv using appropriately defined parameters. The required input data embraces typical test distributions, sample materials, and sample request profiles. As an alternative to this programme supported simulation, laboratory specific request lists captured electronically from the laboratory information system or directly from the routine analysers are automatically converted by WinCAEv to a corresponding worklist for the system under evaluation.

Three main types of RS experiments were designed to allow systematic testing of an analytical system. In this way, different types of routine situations can be modelled and the respective performance situation evaluated.

The RS-Precision experiment type is used to test for systematic and/or random errors using the imprecision characteristics of the system. The goal is to compare the analyte recovery and precision generated during randomized processing with that produced during batch analysis. Pooled quality control and pooled human materials are used as samples. A typical request list is shown in Table 1.

Table 1: Basic structure of a routine simulation precision experiment.

| | | | Assays | | | | | | | | | |
|---|---|---|---|---|---|---|---|---|---|---|---|---|
| | Sample no. | Material | A | B | C | D | E | F | G | H | I | J |
| | 1 | Pool A | x | x | | | x | x | x | x | x | |
| | 2 | " | x | x | | | x | x | x | x | x | |
| | ... | " | x | x | | | x | x | x | x | x | |
| | ... | " | ... | ... | | | ... | ... | ... | ... | ... | |
| | ... | " | x | x | | | x | x | x | x | x | |
| | 11 | " | x | x | | | x | x | x | x | x | |
| | 12 | Pool B | | x | x | x | x | | x | | | |
| | 13 | " | | x | x | x | x | | x | | | |
| | ... | " | | x | x | x | x | | x | | | |
| Batch part | ... | " | | ... | ... | ... | ... | | ... | | | |
| | ... | " | | x | x | x | x | | x | | | |
| | 22 | " | | x | x | x | x | | x | | | |
| | 23 | Pool C | x | | x | | x | | | | | x |
| | 24 | " | x | | x | | x | | | | | x |
| | ... | " | x | | x | | x | | | | | x |
| | ... | " | ... | | ... | | ... | | | | | ... |
| | ... | " | x | | x | | x | | | | | x |
| | 33 | " | x | | x | | x | | | | | x |
| | ... | ... | ... | ... | ... | ... | ... | ... | ... | ... | ... | ... |
| | n | ... | ... | ... | ... | ... | ... | ... | ... | ... | ... | ... |
| | n+1 | Pool B | | x | | x | | | x | | | |
| | n+2 | Pool A | x | | | | x | x | x | x | | |
| | n+3 | Pool C | x | | x | | x | | | | | x |
| | n+4 | Pool B | | x | | | x | | | | | |
| | n+5 | Pool A | x | | | | | x | | | | |
| | n+6 | Pool A | x | | | | | | | x | x | |
| | n+7 | Pool B | | x | | x | x | | | | | |
| | ... | ... | | x | x | x | x | | x | | | |
| | ... | ... | x | x | x | | x | x | x | | | |
| | ... | ... | | | | | x | x | | | | |
| | ... | ... | x | x | | | | x | x | | x | |
| Random part | ... | ... | | | x | | x | | | | | x |
| | ... | ... | | x | | x | | | x | | | |
| | ... | ... | | | | | | | | | x | |
| | ... | ... | | | | x | | | x | | | |
| | ... | ... | | x | | | x | | x | | x | |
| | ... | ... | | | x | | | | | | | x |
| | ... | ... | x | | | | x | | x | | x | |
| | ... | ... | | | x | | | | | | | |
| | ... | ... | | | | x | | | x | | | |
| | ... | ... | | x | x | | x | | | | | |
| | ... | ... | x | x | | | | x | x | | x | |
| | ... | ... | | x | | | x | x | | x | | |
| | n+x | ... | ... | ... | ... | ... | ... | ... | ... | ... | ... | ... |

In repetitions of the same experiment, routine provocations are introduced during the randomized processing to further challenge the system's performance under various conditions. The type and number of provocations depend on the system under evaluation, but generally include items regularly encountered during operation in a routine laboratory, like calibration and quality control measurements, reagent switchover or exchange, sample short, STAT analysis, provocation of various data flags, sample reruns, and so forth.

Errors related to instrument malfunctions or chemistry problems can be deducted from the experimental data by comparing the batch and random results. The mean, median, CV, relative 68%-median distance (md68% describes a robust measure of variation) [6], and minimum and maximum of the random part are compared with those from the batch part for every analyte measured. Random and/or systematic errors will result in significant deviations like elevated CVs and differences of the means. One can expect that the imprecision in a simulated routine run will result in somewhat higher CVs due to more interactions of the analytical system than during a standard batch run. Based on experience from various system evaluations, we use the following expected CV in the random part: $CV_{exp,rand} = CV_{exp,ref} + CV$, where we set $CV = 1/2(CV_{exp,ref})$; therefore $CV_{exp,rand} = 1.5\,CV_{exp,ref}$; $CV_{exp,ref}$ is the expected CV in the reference part.

Usually the routine simulation experiment is performed with many different methods, and the high number of results produced has to be assessed for relevant deviant results. This can easily be done by comparing the CV and the relative 68%-median distance.

The system handling of the routine provocations is assessed for correctness, and the analytical results produced during and after provocations are checked for marked deviations which may represent systematic and/or random errors.

Recently we extended this experiment in order to run the routine simulation precision experiment via a host download procedure (see below) so that the real routine request pattern is reflected and a simulation by the WinCAEv software is not necessary.

RS-Series 1/2 is used for the comparison of randomized test processing in two runs. Fresh human specimens are used as sample materials with request patterns reflecting the evaluation sites typical routine workloads. The sequence of sample processing is identical in both runs and the same samples are used, not placing fresh samples for the second run.

Random errors can be deducted from the experimental data by comparing the deviation of the second from the first run results. The results are grouped in 7 five percent categories between ±15% deviation. Each sample pair is categorized; a summary shows the number of samples per analyte in each category as well as a total statistic per category for the complete experiment (see Table 2 and Figure 1). Random errors will result in marked deviations between both run results for one or more samples.

**Table 2**: Routine simulation series 1/2, cobas 6000.

| Method | N | Categories | | | | | | |
|---|---|---|---|---|---|---|---|---|
| | | --- | -- | - | | + | ++ | +++ |
| | | <-15% | <-10% | <-5% | ±5% | >5% | >10% | >15% |
| Na | 52 | | | | 52 | | | |
| K | 60 | | | | 60 | | | |
| Cl | 11 | | | | 11 | | | |
| ALP_2 | 26 | | | | 26 | | | |
| ALT | 36 | | | 4 | 31 | | 1 | |
| AMYL_2 | 10 | | | | 10 | | | |
| AST | 25 | | | | 25 | | | |
| CK | 21 | | | | 21 | | | |
| LDH_2 | 34 | | | | 34 | | | |
| LIP | 8 | | | | 7 | 1 | | |
| BIL-T | 36 | 5 | | 6 | 20 | 3 | 1 | 1 |
| CREA_2 | 35 | 1 | | | 32 | | 2 | |
| GLUC_3 | 44 | | | | 44 | | | |
| TP_2 | 64 | | | | 64 | | | |
| UREA | 52 | | | | 50 | 2 | | |
| LACT_2 | 10 | | | | 10 | | | |
| Ca | 19 | | | | 19 | | | |
| Mg | 16 | | | | 15 | 1 | | |
| PHOS_2 | 15 | | | | 15 | | | |
| CRP | 38 | | | | 37 | 1 | | |
| DIG | 9 | | | 1 | 3 | 1 | 1 | 3 |
| Li | 10 | | | | 8 | 2 | | |
| AMYL_2 (urine) | 10 | | | | 10 | | | |
| FT4 | 10 | | | | 10 | | | |
| TSH | 8 | | | | 6 | 2 | | |
| Pro-BNP | 9 | | | | 8 | 1 | | |
| TNT | 9 | | | 1 | 8 | | | |
| MYO_2 | 10 | 1 | | | 9 | | | |
| ßHCG | 10 | | | | 7 | 3 | | |
| TOTAL abs | 697 | 7 | 0 | 12 | 652 | 17 | 5 | 4 |
| TOTAL rel | | 1% | 0% | 2% | 94% | 2% | 1% | 1% |

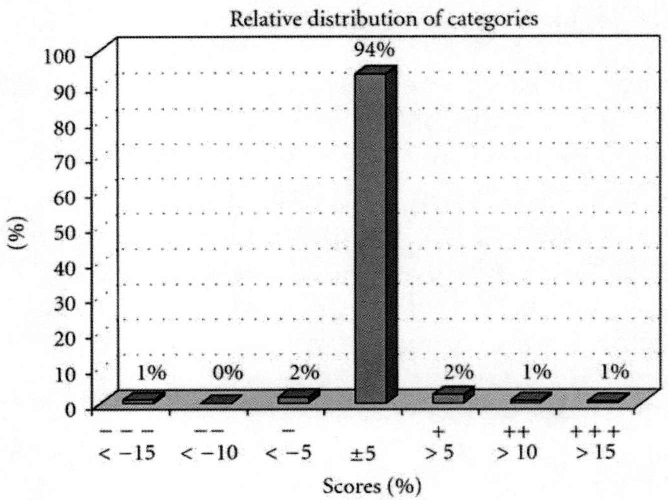


**Figure** 1: Routine simulation series 1/2, cobas 6000.

RS-Method Comparison Download allows direct comparison of the routine analyser methods (reference data) with those of the instrument under evaluation processed in a randomized routine-like fashion. The test results and sampling patterns from the routine laboratory analyser(s) are electronically captured by WinCAEv via file import (host-download) or simply with a batch upload. Using the host-download option, sample identification numbers, requests, and results are exported from the laboratory host in a text format file (e.g., comma separated values (CSV)) and then imported in WinCAEv. No patient demographics are transmitted to WinCAEv. Method comparison statistics and graphs are generated per analyte and comparison instrument.

## APPLICATIONS

Over the last decade, routine simulation experiments have become an integral part of inhouse and multicentre system evaluations at Roche Diagnostics. Here, we outline some typical areas of use based on practical experience, as well as examples of errors difficult to produce using conventional procedures yet observed using these experiments.

RS-Precision is an extremely effective means of testing the interaction of software with all other system components under stressed conditions. During a multicentre study of Roche/Hitachi 917 in the early nineties for example, these experiments yielded CVs of up to 4% for test applications using low-sample volume (2 $\mu$L). An example is shown in Figure 2 for cholesterol. Of the 48 runs performed during this experiment, 83% (=40 series) were found with a CV higher than the expected 2%; 24 series had a relative 68-median distance of more than 2%. The difference between CV and md68% indicated that several series had clear deviant results. Further inhouse investigations revealed that a software malfunction in the sample pipetting process under certain conditions was the root cause for these conspicuous results. After correction of the software and repetition of this experiment, the CV of the cholesterol assay was in all cases below 2%.

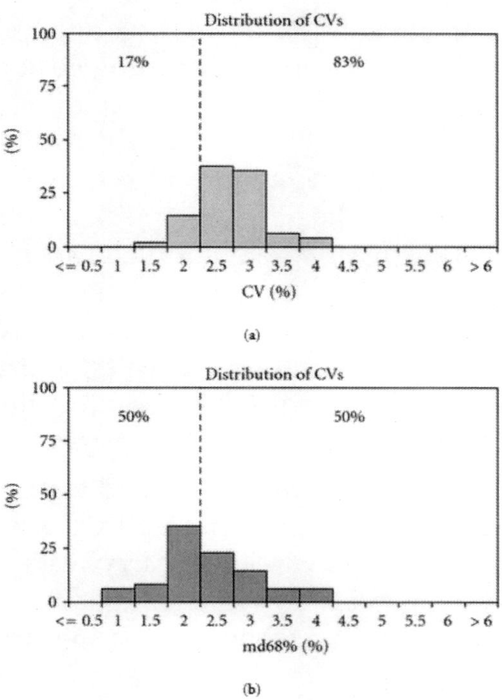


**Figure** 2: Routine simulation precision Roche/Hitachi 917, distribution of CV and md68% for cholesterol (n = 48 series).

On Roche/Hitachi 912, we found that introduction of STAT samples during operation led to intermittent incorrect data flags on STAT sample results when a sample material other than serum was selected. Investigation showed that if a STAT sample was requested by sample disk position only on the analyzer, and the sample type downloaded from the laboratory host is one other than serum or plasma, the sample was correctly measured for the specified biological material but the data flagging was done as if the sample was serum. In this case, an incorrect rerun of the sample was indicated by the generated data flag.

With the new generation of Roche systems MODULAR ANALYTICS from the late nineties, combining multiple analyser modules, this experiment became indispensable and gained many new areas of use. An Intelligent Process Manager distributes the sample carriers to the various analyser modules in a way that ensures most efficient operation, and background maintenance features on these systems allow the operator to perform maintenance on one or more modules while continuing routine operation on the other modules. The RS-Precision experiment allowed us to check these, among other complex functions, in a systematic manner under numerous simulated routine-like conditions. A typical provocation on such systems is the deactivation and reactivation of a module during routine operation. The goal is to check that the samples with requests for tests on the deactivated module(s) are handled correctly, and that the reactivated module performs as expected after return to operation. During the MODULAR ANALYTICS SWA (serum work area) evaluation, these provocations revealed sporadic errors after module reactivation like wrong reagent inventory and bottle change-over to standby reagents although current reagents were not empty.

Also, RS-Precision is an effective tool to test the interaction of reagents on a selective access analyser. During the recent multicentre evaluation of cobas 6000, a new reagent carryover—not observed on other Roche analysers—caused by the reagent probe was found for the creatinine enzymatic assay, when the creatine kinase reagent (CK) was pipetted just before the creatinine assay. Creatine phosphate of the CK reagent (vial 2) may partly be hydrolysed to creatine which can influence the creatine concentration of the enzymatic creatinine assay, when contaminated by the reagent probe. As shown in Table 3, the CV changes from 0.9% in the batch part to 2.1% in the random part. Looking to the single results, the creatinine concentration of sample number 68 is increased and classified as an outlier (255 compared to the median of 225 mmol/L, which is increased by more than the fivefold md68 [20 mmol/L]). By monitoring the "cuvette history" data base of the analyser, one could confirm for that sample that the CK reagent was pipetted just before the creatinine assay. Consequentially, an extra probe wash cycle was installed on cobas 6000 for this reagent combination in order to avoid carryover. Table 3 shows also that the analyser works correctly after a provocation with an expired reagent pack (example ß-HCG).

Table 3: Routine simulation precision cobas 6000.

| Test | [Study unit] | Material | N | Mean | CV [%] | CV diff. [%] | Remarks |
|---|---|---|---|---|---|---|---|
| Na | [mmol/L] | CSP | 144 | 138.9 | 0.7 | 0.5 | |
| K | [mmol/L] | CSP | 144 | 5.0 | 0.8 | 0.5 | |
| ALAT | [U/L] | CSP | 53 | 88.6 | 1.0 | 0.3 | |
| ASAT | [U/L] | CSP | 18 | 105.8 | 0.9 | 0.1 | |
| CK | [U/L] | CSP | 11 | 308.4 | 1.1 | 0.3 | |
| GGT_2 | [U/L] | CSP | 12 | 124.1 | 0.9 | −0.3 | |
| CHOL_2 | [mmol/L] | CSP | 70 | 3.6 | 0.9 | 0.3 | |
| CREA_2 enz | [mol/L] | CSP | 80 | 225.0 | 2.1 | 1.2 | Outlier |
| GLUC_3 | [mmol/L] | CSP | 61 | 9.6 | 1.0 | 0.3 | |
| TP_2 | [g/L] | CSP | 55 | 58.0 | 0.9 | −0.1 | |
| UREA | [mmol/L] | CSP | 23 | 15.1 | 0.5 | −0.3 | |
| UA_2 | [mol/L] | CSP | 19 | 449.3 | 1.0 | 0.2 | |
| Ca | [mmol/L] | CSP | 39 | 2.9 | 1.1 | 0.6 | |
| Fe_2 | [mol/L] | CSP | 75 | 26.6 | 1.6 | 0.9 | |
| CRP | [mg/L] | HSP | 70 | 14.5 | 1.2 | 0.3 | |
| IGG_2 | [g/L] | CSP | 11 | 6.8 | 1.7 | 0.8 | |
| IGA_2 | [g/L] | CSP | 11 | 1.6 | 1.4 | 0.2 | |
| IGM_2 | [g/L] | CSP | 12 | 0.7 | 3.3 | 1.0 | |
| RF_II | [kU/L] | HSP | 10 | 19.2 | 2.4 | 1.0 | |
| TRSF_2 | [mol/L] | CSP | 12 | 24.4 | 1.3 | 0.2 | |
| DIG | [mmol/L] | CSP | 10 | 4.6 | 2.2 | 1.0 | |
| Li | [mmol/L] | CSP | 10 | 1.6 | 1.5 | 0.4 | |
| CREA_2 enz | [mol/L] | HUP | 19 | 7474.3 | 2.5 | 1.7 | |
| TP_U-CSF_3 | [mg/L] | HUP | 20 | 99.4 | 2.7 | 1.4 | |
| FT4 | [pmol/L] | CSP | 26 | 16.9 | 1.6 | 1.0 | |
| TSH | [mU/L] | CSP | 31 | 1.2 | 1.6 | 0.6 | |
| Pro-BNP | [pmol/L] | CSP | 10 | 277.9 | 1.7 | 1.0 | |
| TNT | [g/L] | HSP | 10 | 0.3 | 2.0 | 0.9 | |
| CKMBm | [g/L] | CSP | 10 | 29.2 | 1.6 | 0.8 | |
| PSA | [g/L] | CSP | 10 | 20.9 | 1.3 | 0.9 | |
| FPSA | [g/L] | CSP | 10 | 5.1 | 1.3 | 0.4 | |
| ßHCG | [U/L] | CSP | 0 | | | | Flag"FL" |

Conventionally used to test for reproducibility of results in two runs with the goal to check for random errors, the RS-Series 1/2 experiment can also be easily adopted to address numerous system specific functionalities. For modular systems, this experiment is used to compare consistency between modules for example [7]. On the other hand, during the cobas Integra 800 evaluation [8], it was used to check the analyser's clot and bubble detection function. Samples with clots and bubbles were included in the first run and prior to the second run; bubbles were removed and those with clots were centrifuged. The results were checked for correct flagging of problematic samples in the first run and analyte recovery compared with the second run.

Having a practically automated procedure to immediately repeat the workload from one or more routine instruments on the system under evaluation and to directly compare the results, the RS-Method Comparison Download is an invaluable tool. It allows the investigator to assess the new system from an analytical performance as well as from an overall system performance perspective, under real life conditions.

A total of 187 method comparisons were processed for 50 analytes in nine laboratories under site-specific routine conditions during the MODULAR ANALYTICS evaluation [9] for example. Analysis of approximately 27000 measurements for fresh human specimens gave the final proof that the system was ready for market launch.

## CONCLUSION

Conducting such routine simulation experiments gives the manufacturer essential information on the overall system performance under real intended-use conditions. This innovative, realistic, and thorough approach to system testing has won the approval of principle investigators during international multicentre evaluations over more than ten years.

With the experience gained, we strongly recommend to focus more on routine simulation type experiments during the performance evaluation of a new analytical system, and above all to derive the various quality characteristics like imprecision and method comparison from these results.

# REFERENCES


1. CLSI evaluation protocols, http://www.nccls.org/.
2. R. Haeckel, E. W. Busch, R. D. Jennings, and A. Trucheaud, Eds., Guidelines for the Evaluation of Analysers in Clinical Chemistry, vol. 3 of ECCLS Document, Beuth, Berlin, Germany, 1986.
3. A. Kunst, A. Busse Grawitz, W. Engeldinger, et al., "WinCAEv– a new program supporting evaluations of reagents and analysers," Clinica Chimica Acta, vol. 355S, (Abstr-No WP6.04), p. 361, 2005.
4. W. Bablok, R. Barembruch, W. Stockmann, et al., "CAEv– a program for computer aided evaluation," Journal of Automatic Chemistry, vol. 13, no. 5, pp. 167–179, 1991. View at Publisher· View at Google Scholar
5. W. Bablok and W. Stockmann, "An alternative approach to a system evaluation in the field,"Quimica Clinica, vol. 14, p. 239, 1995.
6. W. Bablok, R. Haeckel, W. Meyers, and W. Wosniok, "Biometrical methods," in Evaluation Methods in Laboratory Medicine, R. Haeckel, Ed., pp. 203–241, VCH, Weinheim, Germany, 1993.
7. C. Bieglmayer, D. W. Chan, L. Sokoll, et al., "Multicentre evaluation of the E 170 Module forMODULAR ANALYTICS," Clinical Chemistry and Laboratory Medicine, vol. 42, no. 10, pp. 1186–1202, 2004. View at Publisher · View at Google Scholar
8. F. L. Redondo, P. Bermudez, C. Cocco, et al., "Evaluation of Cobas Integra ® 800 under simulated routine conditions in six laboratories," Clinical Chemistry and Laboratory Medicine, vol. 41, no. 3, pp. 365–381, 2003. View at Publisher · View at Google Scholar
9. G. L. Horowitz, Z. Zaman, N. J. C. Blanckaert, et al., "MODULAR ANALYTICS: a new approach to automation in the clinical laboratory," Journal of Automated Methods and Management in Chemistry, vol. 2005, no. 1, pp. 8–25, 2005.

# Chapter 8

# SYNTHESIS OF BRANCH FLUORINATED CATIONIC SURFACTANT AND SURFACE PROPERTIES

Hongke Wu, Jiaqi Zhong, Haimin Shen, and Hongxin Shi

State Key Laboratory Breeding Base of Green Chemistry Synthesis-Technology, Zhejiang University of Technology, Hangzhou 310032, China

## ABSTRACT

A novel fluorinated quaternary ammonium salt cationic surfactant N,N,N-trimethyl-2-[[4-[[3,4,4,4-tetrafluoro-2-[1,2,2,2-tetrafluoro-1-(trifluoromethyl)ethyl]-1,3-bis(tri-fluoromethyl)-1-buten-1-yl]oxy]-benzoyl]amino]-iodide (FQAS) was synthesized successfully, and its structure was characterized by FTIR,$^1$H-NMR, $^{19}$F-NMR, and MS. The surface activities of FQAS and the effect of temperature, electrolyte, and combination with hydrocarbon surfactant were investigated. The results showed that FQAS exhibited excellent surface activity and combination with hydrocarbon surfactant.

## INTRODUCTION

A surfactant is an amphiphilic molecule bearing both hydrophobic and hydrophilic parts. The fluorinated surfactants have generated much interest for high surface activity, chemical resistance, thermal stabilization, and low critical micelle concentration. Cationic fluorinated surfactants were generally used in coatings [1–3], paints

[4], inks [5, 6], waxes, additives for etching, leather, firefighting foams [7, 8], and other applications [9–16].

The most known are perfluorooctanoate (PFOA) and perfluorooctane sulfonate (PFOS) (Scheme 1). However they are persistent in the chemosphere, toxic, bioaccumulable pollutants and can be transferred into the environment such as soil and water as well as in the animal body through the food chain [17]. The US Environmental Protection Agency launched the PFOA Stewardship Program to decrease the production of PFOA and PFOS. However, decomposition of perfluorooctanoic acid photocatalyzed by $TiO_2$ was reported recently [18]. Hexafluoropropylene oligomers are an important class of compounds which can be used to prepare useful PFOA and PFOS replacements fluorinated anionic, cationic, nonionic, and amphoteric surfactants [19, 20].

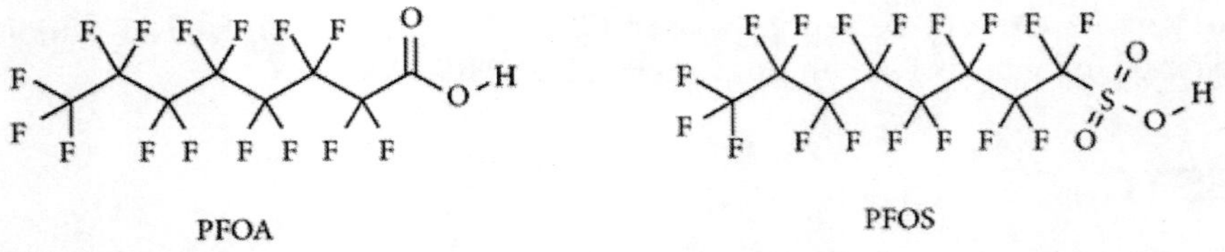

PFOA PFOS

Scheme 1: Structure of PFOA and PFOS.

In this paper, we describe a new fluorinated quaternary ammonium salt cationic surfactant N,N,N-trimethyl-2-[[4-[[3,4,4,4-tetrafluoro-2-[1,2,2,2-tetrafluoro-1-(trifluoromethyl)ethyl]-1,3-bis(tri-fluoromethyl)-1-buten-1-yl]oxy]benzoyl]amino]-iodide(4, FQAS) using the 4-[[3,4,4,4-Tetra fluoro-2-[1,2,2,2-tetrafluoro-1-(trifluoromethyl)ethyl]-1,3-bis(trifluoromethyl)-1-buten-1-yl]oxy]-benzoyl chloride (1) (Scheme 2). The surface activity of FQAS, such as surface tension and critical micelle concentrations (cmc), the effect of temperature and electrolyte on the surface activity of FQAS, and the surface activity of FQAS mixed with nonionic hydrocarbon surfactant polyethylene glycol 400 (PEG400) were investigated in detail.

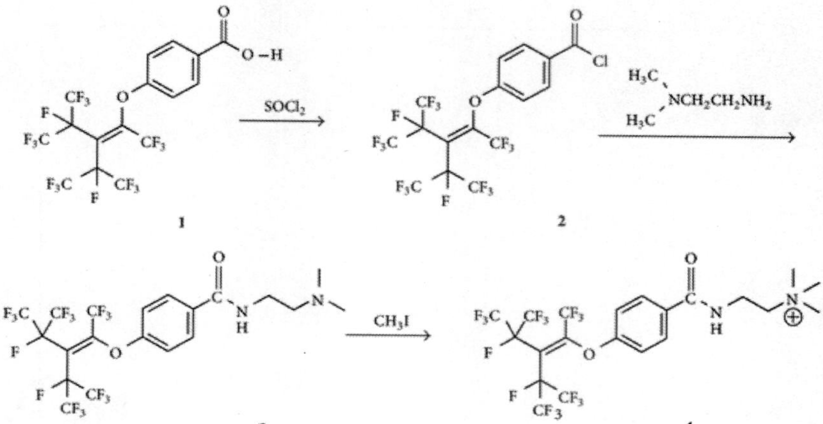

Scheme 2: Synthetic route of FQAS.

# RESULTS AND DISCUSSION

## Surface Tension

Unlike many fluorinated surfactants, the branch fluorinated cationic surfactant (4) (FQAS) exhibits excellent solubility in water at room temperature; thus, it could be employed to reduce the surface tension of water. The curve of surface tension was obtained by Wihemlmy (Figure 1). It can be seen in Figure 1 that the cmc value of FQAS in water is $6.0 \times 10^{-5}$ mol·L$^{-1}$ at 293 K and the surface tension of the aqueous solution is 21.1 mNm$^{-1}$ at the cmc. However, the value of cmc for hexadecyl trimethyl ammonium chloride (HTAC) is $1.6 \times 10^{-3}$ molL$^{-1}$, and the surface tension is 42.3 mN/m at the same temperature (in the literature Applied Surfactants, John Wiley & Sons Inc.). The cmc value of branch fluorinated cationic surfactant FQAS is remarkably lower than that of the hydrocarbon surfactants HTAC for the C-F group introduced [21, 22]. We infer that the decrease results from the formation of a closely packed film of fluorinated cationic surfactant caused by further adsorption of the surfactant to the air/aqueous solution interface.

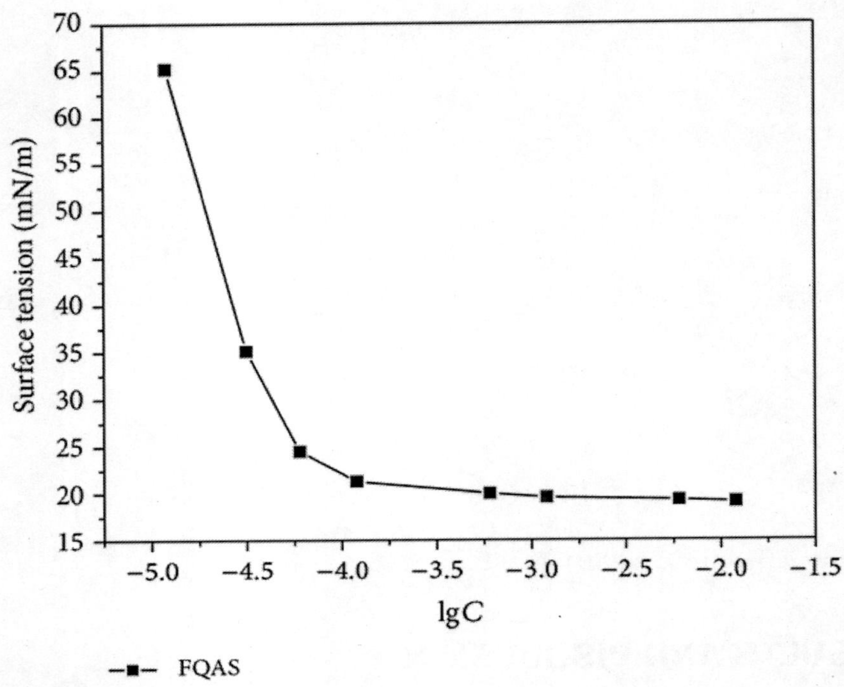


**Figure** 1: Surface tension of FQAS versus logarithm concentration/ $molL^{-1}$ (lg C).

## Effect of Temperature on Fqas Surface Tension

Temperature is a significant effect factor for the surface tension of ionic fluorinated surfactant.

At $1.2 \times 10^{-5}$ $molL^{-1}$, the surface tension of FQAS decreased from 67.0 $mNm^{-1}$ to 54.8 $mNm^{-1}$, when the temperature was increased from 293 K to 313 K (Figure 2). The present data shows that the effect of temperature was small when the concentration of FQAS is high ($\geqq 1.3 \times 10^{-4}$ $molL^{-1}$). This phenomenon can be explained by the following facts. The fluorinated surfactant tends to be adsorbed at the interface, driving the system to a thermodynamically more stable state. The effect of adsorption form both sides at the vapor/liquid interface was considered simultaneously. The result is a reduction in Gibbs energy, thereby a decrease in surface tension [23–25].

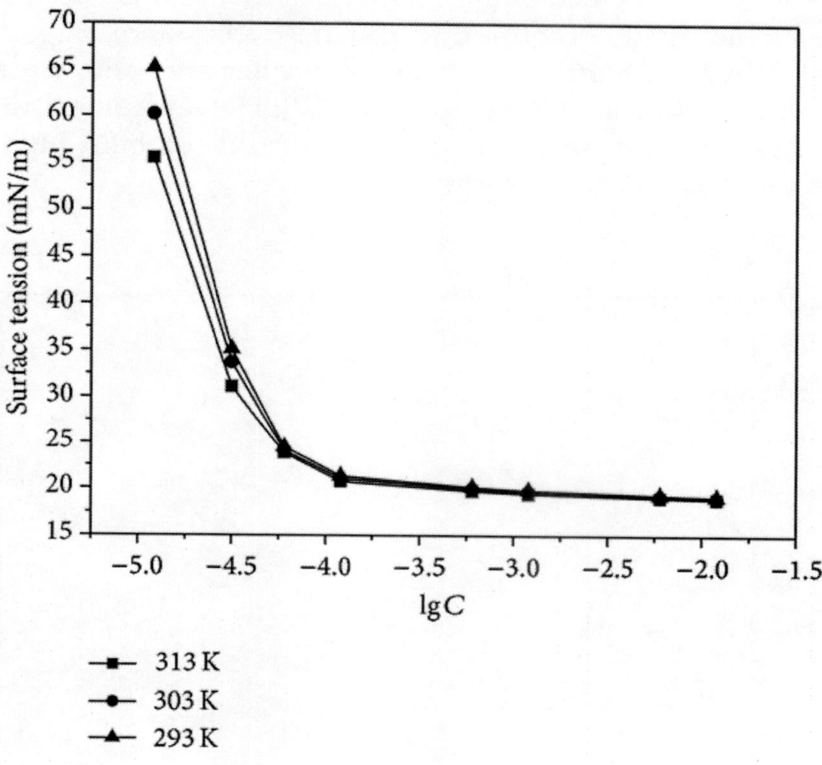


**Figure** 2: Effect of temperature on surface tension of FQAS solution.

## Effect of Electrolyte on the Surface Tension of FQAS

The FQAS exhibited lower surface tension at different electrolyte concentrations of NaCl (1.0 $gL^{-1}$ and 5.0 $gL^{-1}$) aqueous solution (Figure 3). It can be seen in Figure 3 that the effect of electrolyte on the surface tension of FQAS is small. That shows FQAS has good salt resistance. The electrolyte remains in many aspects enigmatic. Although the phase at which the solution is at equilibrium does neither attract nor repel electrolytes, the solutes absorb either positively or negatively depending on their nature. It has strong attraction with just one water molecule under ions free. In double-layer terminology, they should be classified as "indifferent." Rather, their hydration with more water molecules keeps them out of the surface region. The observed ionic specificities are relatively minor and do not exhibit a clear ionic trend. According to general expectation, bigger ions would dehydrate more easily than the smaller ones and hence are more inclined to enrich the surface [26, 27]. However, the $Ca^{2+}$ and $Mg^{2+}$ may vary dramatically with changes in the solubility of FQAS, which can make the surface activity poor.

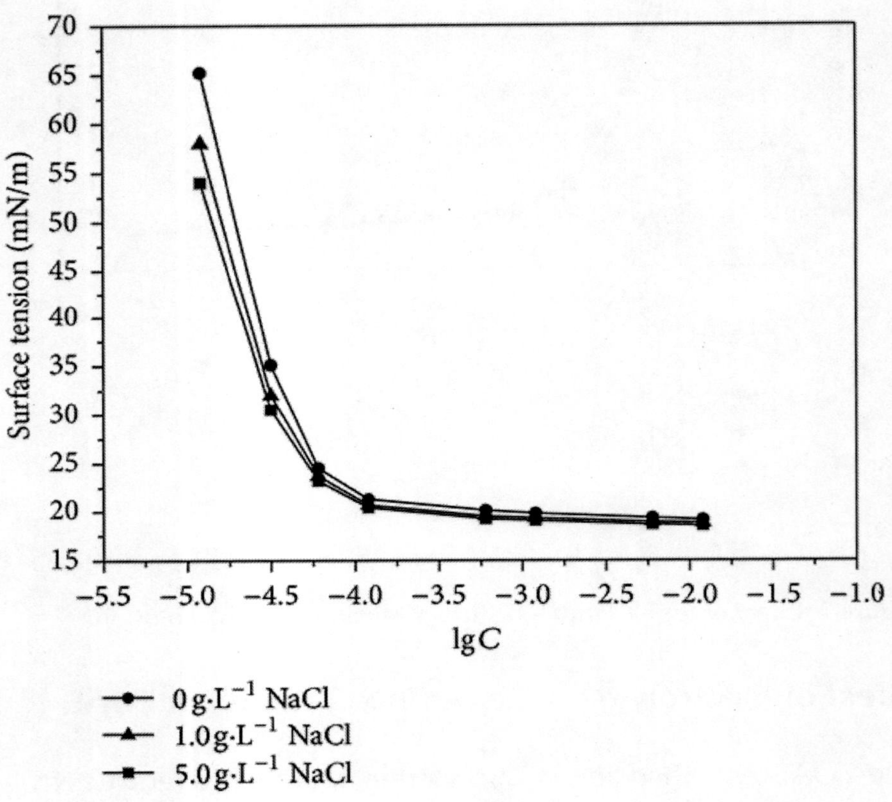


Figure 3: Effect of electrolyte on surface tension of FQAS solution.

## Combination with Hydrocarbon Surfactant

The different mass ratio nonionic surfactant PEG600 was mixed with FQAS. Surface tension was measured for mixed surfactant solution of different concentrations at 293 K (Figure 4). The surface tension was decreased with the increase in the total concentration and the mass ratio of FQAS of combined surfactants. This suggests that when micelles begin to from at the critical micelle concentration, the impurity becomes soluble in the micelles because of its hydrophobic character. The concentration in bulk and surface concentration have been reduced. Therefore, the interaction between the combined surfactants is strengthened [28–32].

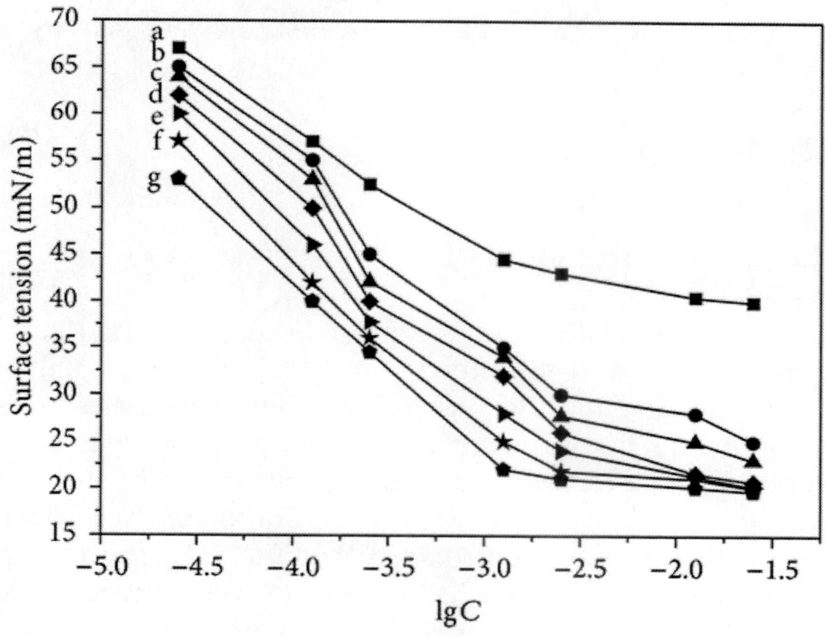


a: m(PEG600) : m(FQAS) = 5 : 0
b: m(PEG600) : m(FQAS) = 4 : 1
c: m(PEG600) : m(FQAS) = 3 : 2
d: m(PEG600) : m(FQAS) = 1 : 1
e: m(PEG600) : m(FQAS) = 2 : 3
f: m(PEG600) : m(FQAS) = 1 : 4
g: m(PEG600) : m(FQAS) = 0 : 5

**Figure** 4: Effect of combination with hydrocarbon surfactant on surface tension of FQAS solution.

## CONCLUSION

The novel branch fluorinated cationic surfactant N,N,N-trimethyl-2-[[4-[[3,4,4,4-tetrafluoro-2-[1,2,2,2-tetrafluoro-1-(trifluoromethyl) ethyl]-1,3-bis(tri-fluoromethyl)-1-buten-1-yl]oxy]-benzoyl]amino]-iodide (FQAS) was synthesized successfully. The structure was characterized by FTIR, $^1$H-NMR, $^{19}$F-NMR, and MS. The surface activities of FQAS, the effect of temperature, electrolyte, and combination with hydrocarbon surfactant were investigated. The results showed that FQAS exhibited excellent surface activity and combined properties.

## EXPERIMENTAL

### Materials and Instruments

4-Perfluoro-(1, 3-dimethyl-2-isopropyl)-but-1-enyloxy]-benzoic acid was purchased from Juhua Group Corporation, Quzhou, China. Sulphoxide chloride, N,N-dimethylethylenediamine, and iodomethane, which were analytical grade, were purchased from Shuanglin Chemical Reagent Company, Hangzhou, China. Melting points were determined using an X-4 apparatus and uncorrected. FTIR were recorded of Nicolet Avata 370 Fourier transform infrared spectrometer from KBr pellets. NMR spectra were measured on a Bruker AV500 instrument using TMS as an internal standard and $CDCl_3$ as solvent. Mass spectra were performed on a Thermo Finnigan LCQ Advantage LC/mass detector instrument. The characterization of the surface tension was carried out on DCA-315 equipped with a Wilhelmy plate made of platinum.

### Synthesis of FQAS

4-[[3,4,4,4-Tetrafluoro-2-[1,2,2,2-tetrafluoro-1-(trifluoromethyl) ethyl]-1,3-bis(trifluoromethyl)-1-buten-1-yl]oxy]-benzoyl chloride (2)

To a solution of 4-[[3,4,4,4-tetrafluoro-2-[1,2,2,2-tetrafluoro-1-(trifluoromethyl)ethyl]-1,3-bis(trifluoro-methyl)-1-buten-1-yl]oxy]-

benzoic acid (1) (28.4 g, 0.05 mol) in $CCl_4$ (50 mL) and sulphoxide chloride (11.90 g, 0.10 mol) was added stepwise slowly at room temperature [33–35]. After refluxing for 3 h, the solvent and excess of sulphoxide chloride were evaporated to give the viscous liquid.

4.2.2. N-(2-Dimethylamino-ethyl)-4-[3,4,4,4-tetrafluoro-2-(1,2,2,2-tetrafluoro-1-trifluoromethyl-ethyl)-1,3-bis-trifluoromethyl-buten-1-enyloxy]-benzamide (3)

To a solution of compound (2) (2.93 g, 0.005 mol) and $K_2CO_3$ (0.69 g, 0.005 mol) in $CH_3CN$ (40 mL), N,N-dimethylethylenediamine (0.58 g, 0.006 mol) was added dropwise at room temperature [36–38]. After refluxing for 5 h, it was quenched with water (50 mL). Then the resultant solution was extracted by ethyl acetate (3 × 50 mL). The combined organic phase was dried by $Na_2SO_4$. The solvent was evaporated to give a white residue, which was recrystallized in ethyl acetate and petroleum ether to afford a light yellow solid (3) (2.71 g, 85%). IR ($cm^{-1}$): (N–H) 3255, (C=O) 1655, (C=C) 1601,1502, (C–F) 1185,1134; $^1$H-NMR ($CD_3Cl$) : 7 .81 (d, .80 Hz, 2H, ArH), 6.95 (d, 8.75, 2H, ArH), 3.56–3.58 (m, 2H, $CH_2$), 2.50-2.51 (m, 2H, $CH_2$), 2.30 (s, 6H, $CH_3$); and MS (EI) m/z 638.09 (M-1).

4.2.3. N,N,N-Trimethyl-2-[[4-[[3,4,4,4-tetrafluoro-2-[1,2,2,2-tetrafluoro-1-(trifluoromethyl)ethyl]-1,3-bis(trifluoromethyl)-1-buten-1-yl]oxy]benzoyl]amino]-iodide (4)

Compound (3) (3.19 g, 0.005 mol) and iodomethane (1.42 g, 0.01 mol) were refluxed in MeCN (50 mL) for 5 h under the atmosphere of $N_2$, and then the resultant mixture was cooled to room temperature. After the solution was evaporated under reduced pressure, the residue was recrystallized in acetic ether and petroleum ether. The resulting white solid (4) (3.52 g, 90%) was dried under reduced pressure. IR ($cm^{-1}$):$\upsilon$(N–H) 3446, (C=C) 1626,1590,1491, (C–O) 1242, (S=O) 1014, (C–F) 1240,979,742. $^1$H-NMR ($CDCl_3$) (d, Hz, 2H, ArH), 6.95 (d, Hz, 2H, ArH), 3.87–3.85 (m, 2H, $CH_2$), 3.63–3.65 (m, 2H, $CH_2$), 2.67 (s, 9H, $CH_3$); $^{19}$F-NMR ($CDCl_3$): −188.5 ppm (1F, F), −186.5 (1F, F), −91.3 (6F, $CF_3$), −90.2 (6F, $CF_3$), −75.0 (3F, $CF_3$); MS (EI) m/z 654.1 (M-I).

## Measurements

The static surface tension was measured using Thermo Cahn surface tension requirements (DCA-315) with platinum plate. The measurement were performed under 293 K after equilibration for 8 h. The solution preparation of FQAS used distilled water.

## REFERENCES

1. G. Para, A. Hamerska-Dudra, K. A. Wilk, and P. Warszyński, "Mechanism of cationic surfactant adsorption—effect of molecular structure and multiple charge," Colloids and Surfaces A, vol. 383, no. 1–3, pp. 67–72, 2011. View at Publisher · View at Google Scholar · View at Scopus
2. L. J. Chen, H. X. Shi, H. K. Wu, and J. Xiang, "Preparation and characterization of a novel fluorinated acrylate resin," Journal of Fluorine Chemistry, vol. 131, no. 6, pp. 731–737, 2010. View at Publisher · View at Google Scholar · View at Scopus
3. K. Ghosh, H.-J. Lehmler, S. E. Rankin, and B. L. Knutson, "Supercritical carbon dioxide swelling of fluorinated and hydrocarbon surfactant templates in mesoporous silica thin films," Journal of Colloid and Interface Science, vol. 367, no. 1, pp. 183–192, 2012. View at Publisher · View at Google Scholar · View at Scopus
4. A. J. Kessman, E. E. Defusco, A. W. Hoover, K. A. Sierros, and D. R. Cairns, "Structural, mechanical, and tribological properties of fluorinated mesoporous silica films: effect of functional moiety and surfactant template concentrations," Thin Solid Films, vol. 520, no. 11, pp. 3896–3903, 2012. View at Publisher · View at Google Scholar · View at Scopus
5. V. C. Malshe, S. Elango, S. S. Bhagwat, and S. S. Maghrabi, "Fluorinated acrylic copolymers: part II: polymeric surfactants," Progress in Organic Coatings, vol. 53, no. 3, pp. 212–216, 2005. View at Publisher · View at Google Scholar · View at Scopus
6. L. Chen, H. Shi, H. Wu, and J. Xiang, "Influence of the reactive emulsifier-HPMA on the properties of the acrylate emulsion," Journal of Dispersion Science and Technology, vol. 32, no. 2, pp. 235–240, 2011. View at Publisher · View at Google Scholar · View at Scopus
7. P. M. Karlsson, A. Baeza, A. E. C. Palmqvist, and K. Holmberg, "Surfactant inhibition of aluminium pigments for waterborne printing inks," Corrosion Science, vol. 50, no. 8, pp. 2282–2287, 2008. View at Publisher · View at Google Scholar · View at Scopus
8. M. Pabon and J. M. Corpart, "Fluorinated surfactants: synthesis, properties, effluent treatment,"Journal of Fluorine Chemistry, vol. 114, no. 2, pp. 149–156, 2002. View at Publisher · View at Google Scholar · View at Scopus
9. W. Shen, L.-M. Wang, and H. Tian, "Quaternary ammonium salt gemini

surfactants containing perfluoroalkyl tails catalyzed one-pot Mannich reactions in aqueous media," Journal of Fluorine Chemistry, vol. 129, no. 4, pp. 267–273, 2008. View at Publisher · View at Google Scholar · View at Scopus

10. D. J. Holt, R. J. Payne, and C. Abell, "Synthesis of novel fluorous surfactants for microdroplet stabilisation in fluorous oil streams," Journal of Fluorine Chemistry, vol. 131, no. 3, pp. 398–407, 2010. View at Publisher · View at Google Scholar · View at Scopus
11. D. Rana and T. Matsuura, "Surface modifications for antifouling membranes," Chemical Reviews, vol. 110, no. 4, pp. 2448–2471, 2010. View at Publisher · View at Google Scholar · View at Scopus
12. J. Tan, D. Ma, and S. Feng, "Effect of headgroups on the aggregation behavior of cationic silicone surfactants in aqueous solution," Journal of Colloid and Interface Science, vol. 417, pp. 146–153, 2013.
13. K. Sakai, M. Kaji, Y. Takamatsu et al., "Fluorocarbon-hydrocarbon gemini surfactant mixtures in aqueous solution," Colloids and Surfaces A, vol. 333, no. 1–3, pp. 26–31, 2009. View at Publisher ·View at Google Scholar · View at Scopus
14. S. C. Sharma, D. P. Acharya, Y. Itami, M. García-Roman, and H. Kunieda, "Phase behavior and surface tensions of amphiphilic fluorinated random copolymer aqueous solutions," Colloids and Surfaces A, vol. 280, no. 1–3, pp. 140–145, 2006. View at Publisher · View at Google Scholar · View at Scopus
15. P. Thebault, E. T. D. Givenchy, S. Géribaldi, R. Levy, Y. Vandenberghe, and F. Guittard, "Surface and antimicrobial properties of semi-fluorinated quaternary ammonium thiol surfactants potentially usable for Self-Assembled Monolayers," Journal of Fluorine Chemistry, vol. 131, no. 5, pp. 592–596, 2010. View at Publisher · View at Google Scholar · View at Scopus
16. H. Wu, J. Zhong, and H. Shi, "Synthesis of a novel branch fluorinated cationic surfactant and its surface activity," Journal of Fluorine Chemistry, vol. 156, pp. 5–8, 2013. View at Publisher · View at Google Scholar
17. P. M. Murphy, C. S. Baldwin, and R. C. Buck, "Syntheses utilizing n-perfluoroalkyl iodides [RFI, $CnF_{2n+1}$-I] 2000–2010," Journal of Fluorine Chemistry, vol. 138, pp. 3–23, 2012. View at Publisher ·View at Google Scholar · View at Scopus
18. M. Sansotera, F. Persico, C. Pirola, W. Navarrini, A. di Michele, and C. L. Bianchi, "Decomposition of perfluorooctanoic acid: chemical modification of the catalyst surface induced by fluoride ions,"Applied Catalysis B, vol. 148-149, pp. 29–35, 2014.
19. L. Chen, H. Shi, H. Wu, and J. Xiang, "Synthesis and combined properties of novel fluorinated anionic surfactant," Colloids and Surfaces A, vol. 384, no. 1–3, pp. 331–336, 2011. View at Publisher ·View at Google Scholar · View at Scopus
20. L. Chen, H. Shi, H. Wu, and J. Xiang, "Study on the double fluorinated modification of the acrylate latex," Colloids and Surfaces A, vol. 368, no.

1–3, pp. 148–153, 2010. View at Publisher · View at Google Scholar · View at Scopus

21. T. Yoshimura, A. Ohno, and K. Esumi, "Equilibrium and dynamic surface tension properties of partially fluorinated quaternary ammonium salt gemini surfactants," Langmuir, vol. 22, no. 10, pp. 4643–4648, 2006. View at Publisher · View at Google Scholar · View at Scopus
22. N. Bongartz, S. Patil, and D. Blunka, "A new fluorinated inositol-based surfactant," Colloids and Surfaces A, vol. 414, pp. 320–326, 2012.
23. K. Lunkenheimer, A. Lind, and M. Jost, "Surface tension of surfactant solutions," Journal of Physical Chemistry B, vol. 107, no. 31, pp. 7527–7531, 2003. View at Scopus
24. S. Azizian and N. Bashavard, "Surface tension of dilute solutions of alkanes in cyclohexanol at different temperatures," Journal of Chemical and Engineering Data, vol. 53, no. 10, pp. 2422–2425, 2008. View at Publisher · View at Google Scholar · View at Scopus
25. A. Firooz and P. Chen, "Effect of temperature on the surface tension of 1-hexanol aqueous solutions," Langmuir, vol. 27, no. 6, pp. 2446–2455, 2011. View at Publisher · View at Google Scholar · View at Scopus
26. Y.-G. Li, "Reply to 'comments on 'Surface Tension Model for Concentrated Electrolyte Solutions by the Pitzer Equation''," Industrial and Engineering Chemistry Research, vol. 38, no. 10, pp. 4137–4138, 1999. View at Scopus
27. J. Drzymala and J. Lyklema, "Surface Tension od Aqueous electrolyte solutions. Thermodynamics," Journal of Physical Chemistry A, vol. 116, pp. 6465–6472, 2012.
28. M. Pisarcik, M. Polakovicova, and M. Lukac, "Molecular structure-surface property relationships for heterocyclic and dipropylamino derivatives of hexadecylphosphocholine and cetyltrimethylammonium bromide in NaBr and salt-free aqueous solutions," Colloids and Surfaces A, vol. 407, pp. 169–176, 2012.
29. K. Sakai, M. Kaji, Y. Takamatsu et al., "Fluorocarbon-hydrocarbon gemini surfactant mixtures in aqueous solution," Colloids and Surfaces A, vol. 333, no. 1–3, pp. 26–31, 2009. View at Publisher ·View at Google Scholar · View at Scopus
30. E. Blanco, C. Rodriguez-Abreu, P. Schulz, and J. M. Ruso, "Effect of alkyl chain asymmetry on catanionic mixtures of hydrogenated and fluorinated surfactants," Journal of Colloid and Interface Science, vol. 341, no. 2, pp. 261–266, 2010. View at Publisher · View at Google Scholar · View at Scopus
31. D. Tikariha, K. K. Ghosh, N. Barbero, P. Quagliotto, and S. Ghosh, "Micellization properties of mixed cationic gemini and cationic monomeric surfactants in aqueous-ethylene glycol mixture,"Colloids and Surfaces A, vol. 381, no. 1–3, pp. 61–69, 2011. View at Publisher · View at Google Scholar · View at Scopus
32. S. C. Sharma, D. P. Acharya, M. García-Roman, Y. Itami, and H. Kunieda, "Phase behavior and surface tensions of amphiphilic fluorinated random

copolymer aqueous solutions," Colloids and Surfaces A, vol. 280, no. 1–3, pp. 140–145, 2006. View at Publisher · View at Google Scholar · View at Scopus

33. H.-J. Liu, J.-Q. Weng, C.-X. Tan, and X.-H. Liu, "1-Cyano-N-(2,4,5-trichlorophenyl)cyclo-propane-1-carboxamide," Acta Crystallographica Section E, vol. 67, no. 8, article o1940, 2011. View at Publisher · View at Google Scholar · View at Scopus

34. X.-H. Liu, C.-X. Tan, J.-Q. Weng, and H.-J. Liu, "(E)-(4-Bromobenzylidene) amino cyclopropanecarboxylate," Acta Crystallographica Section E, vol. 68, no. 2, article o493, 2012. View at Publisher · View at Google Scholar · View at Scopus

35. Y. L. Xue, Y. G. Zhang, and X. H. Liu, "Synthesis, crystal structure and biological activity of 1-cyano-N-(2, 4-dichlorophenyl)cyclopropanecarboxamide," Asian Journal of Chemistry, vol. 24, pp. 5087–5089, 2012.

36. R. Wu, C. Zhu, X. J. Du et al., "Synthesis, crystal structure and larvicidal activity of novel diamide derivatives against Culex pipiens," Chemistry Central Journal, vol. 6, no. 99, 2012.

37. G. X. Sun, Z. H. Sun, M. Y. Yang, X. H. Liu, Y. Ma, and Y. Y. Wei, "Design, synthesis, biological activities and 3D-QSAR of new N, N'-diacylhydrazines containing 2, 4-dichlorophenoxy moieties,"Molecules, vol. 18, pp. 14876–14891, 2013.

38. C. Cui, Z. P. Wang, X. J. Du et al., "Synthesis and antiviral activity of hydrogenated ferulic acid derivatives," Journal of Chemistry, vol. 2013, Article ID 269434, 5 pages, 2013. View at Publisher ·View at Google Scholar

# Chapter 9

# ANALYSIS OF LABORATORY INTERCOMPARISON DATA. A MATTER OF INDEPENDENCE

Mauro F. Rebelo[I]; José M. Monserrat[II]; Wanderley G. Bastos[III]

[I]Instituto de Biofísica Carlos Chagas Filho, Centro de Ciências da Saúde, Universidade Federal do Rio de Janeiro, 21949-900 Rio de Janeiro - RJ

[II]Departamento de Ciências Fisiológicas, Universidade do Rio Grande, CP 474, 96201-900 Rio Grande - RS

[III]Departamento de Medicina, Universidade Federal de Rondônia, Br364 Km 9,5, 78900-970 Porto Velho - RO

## ABSTRACT

When laboratory intercomparison exercises are conducted, there is no *a priori* dependence of the concentration of a certain compound determined in one laboratory to that determined by another(s). The same applies when comparing different methodologies. A existing data set of total mercury readings in fish muscle samples involved in a Brazilian intercomparison exercise was used to show that correlation analysis is the most effective statistical tool in this kind of experiments. Problems associated with alternative analytical tools such as mean or paired ‹t›-test comparison and regression analysis is discussed.

## INTRODUCTION

There is a clear need for intercomparison exercises when studies involving more than one laboratory are conducted[1,2]. When several species and compounds are analyzed, such as mercury[3-5] and cholinesterase activity[6], this provides confidence results as verifies the accuracy of the measurements.

However, it must be recognize that there is no *a priori* dependence of the concentration of a certain compound determined in one laboratory to that determined by others. A mercury concentration datum, for example, is dependent only on the mercury content of a tissue or blood or whatever the sample matrix is. On the other hand, the readings data supplied by one or more laboratories could be related or not, depending on the degree of intercalibration.

Regression analysis, correlation tests and the t-test, are commonly used to evaluate the degree of likelihood between two data sets, as in intercomparison exercises.

## REGRESSION ANALYSIS

Whenever there are two data sets, in which one is dependent (Y variable) on the other (X variable), regression analysis is used to estimate the percentage variability of Y explained by its relationship with X. This quantity is expressed by the $r^2$ value (the closer to 1, the better the estimation). If there are two data sets from two laboratories to be intercalibrated, the regression model is not suitable, since it is impossible to say *a priori* which set of data is the response variable and which independent one. In this way, none of the equations can represent better the relation between the readings (Figure 1).

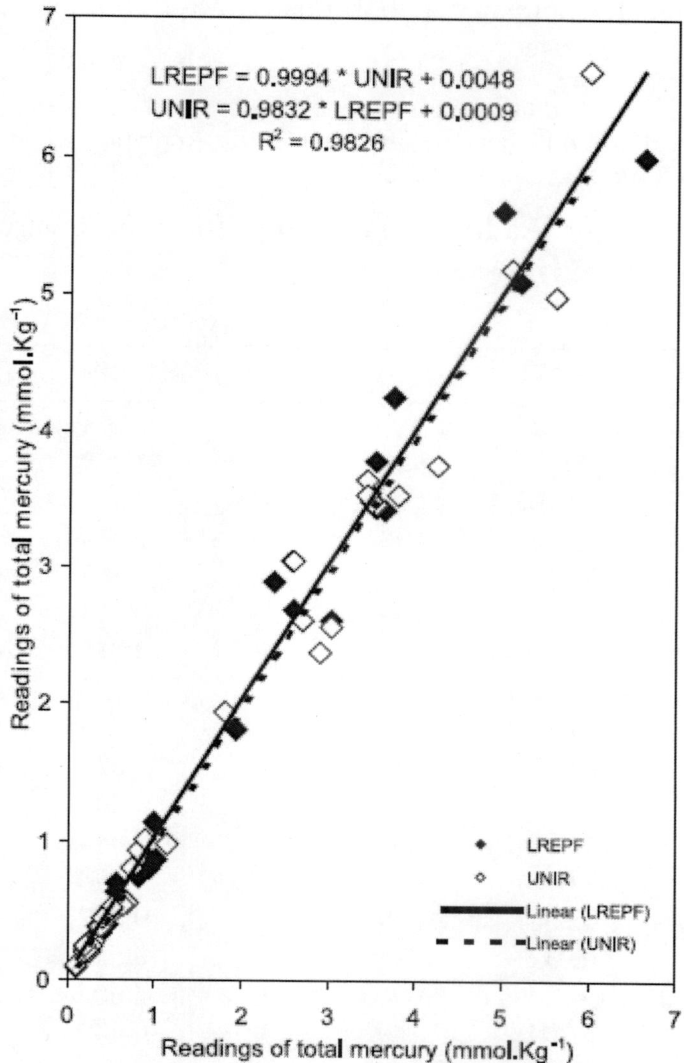


**Figure 1.** Estimated regression functions using UNIR or LREPF as response variables. Since is not possible to establishes which is the independent and the response variable, either equation could be chosen to represent the relationship. But there is no right criteria to do so, showing that this model is not suitable

As an example, an existing data set of total mercury readings in fish muscle samples (Table 1) analyzed by the Laboratory of Radioisotopes from Rio de Janeiro Federal University (LREPF) and the Analytical Chemistry Laboratory from Rondonia University (UNIR), has been used.

**Table 1.** Readings of total mercury (mmol.Kg-1) on fish muscle samples used for the intercalibration laboratory exercise. UNIR: Analytical Chemistry Laboratory from Rondonia University. LREPF:

Laboratory of Radioisotopes from Rio de Janeiro Federal University

| Reading | UNIR | LREPF | Reading | UNIR | LREPF |
|---|---|---|---|---|---|
| 1 | 3.744 | 4.252 | 20 | 0.204 | 0.199 |
| 2 | 3.470 | 3.540 | 21 | 0.259 | 0.249 |
| 3 | 0.279 | 0.299 | 22 | 0.429 | 0.399 |
| 4 | 0.110 | 0.100 | 23 | 1.939 | 1.810 |
| 5 | 3.535 | 3.789 | 24 | 0.563 | 0.698 |
| 6 | 0.224 | 0.199 | 25 | 0.244 | 0.274 |
| 7 | 0.813 | 0.748 | 26 | 0.219 | 0.229 |
| 8 | 0.484 | 0.499 | 27 | 0.548 | 0.648 |
| 9 | 6.620 | 5.992 | 28 | 0.518 | 0.499 |
| 10 | 2.562 | 3.041 | 29 | 1.012 | 0.882 |
| 11 | 0.508 | 0.499 | 30 | 0.249 | 0.214 |
| 12 | 4.990 | 5.603 | 31 | 3.644 | 3.425 |
| 13 | 3.041 | 2.602 | 32 | 3.480 | 3.485 |
| 14 | 0.384 | 0.349 | 33 | 0.389 | 0.379 |
| 15 | 0.937 | 0.818 | 34 | 0.523 | 0.513 |
| 16 | 5.195 | 5.090 | 35 | 2.602 | 2.692 |
| 17 | 0.204 | 0.249 | 36 | 0.449 | 0.399 |
| 18 | 0.254 | 0.249 | 37 | 2.373 | 2.891 |
| 19 | 0.992 | 1.147 | 38 | 3.530 | 3.440 |

## MEAN COMPARISONS TEST

The method most commonly used to evaluate the differences in the means of two groups is the t-test. However, it is clear that two means can be statistically similar despite there being no correspondence between the two data sets. Therefore this approach does not apply.

Alternatively, if each laboratory analyzed *n* samples, there will be *n* pairs coming from the same sample. In this case, it is common practice to analyze these data by the means of paired t-test[7,8]. Still, this is not the appropriate tool. It must be pointed out that, if this analysis is conducted over the original data set (Table 1), no significant difference is detected between the readings that came from UNIR and those from the LREPF ($p= 0.567$). However, this result must be considered as proof of absence of a consistent higher (or lower) reading from one laboratory, with respect to the other. It is not possible to imply that there is a linear relationship between the readings of the two laboratories, which would be the expected result if the two laboratories were intercalibrated.

## CORRELATION COEFFICIENT

When the objective is to show a linear relationship between two data sets, correlation coefficients are the correct tool. Note that this kind of relationship is what we would expect if the two laboratories were equally calibrated. The Pearson's correlation coefficient, r, (or a non-parametric equivalent) will quantify how similar the laboratories readings are. The parametric correlation coefficient assumes that the two data sets belong to a bivariate normal population and, if this is true, the distribution of each variable must be normal, although the converse does not apply[7].

As can be seen in Figure 2, the distribution for both UNIR and LREPF variables are not normal. Formal statistical test like Shapiro-Wilk's W test[5] rejected the null hypothesis of normality for both variables ($p< 0.0001$). So, the non-parametric Kendall Tau coefficient was calculated[9] and estimated as 0.931.

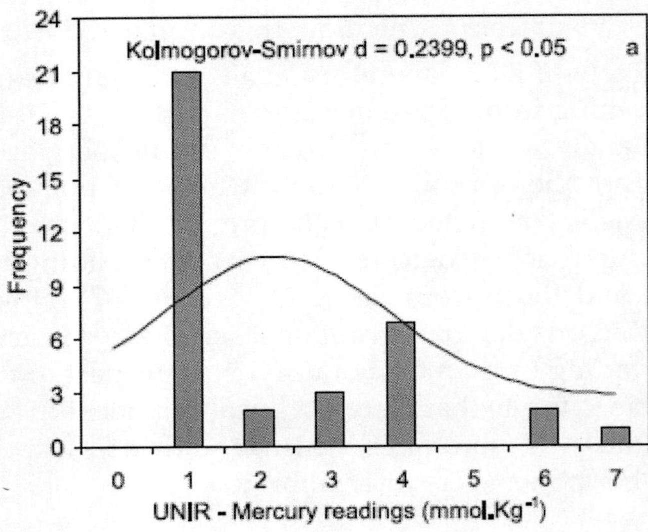


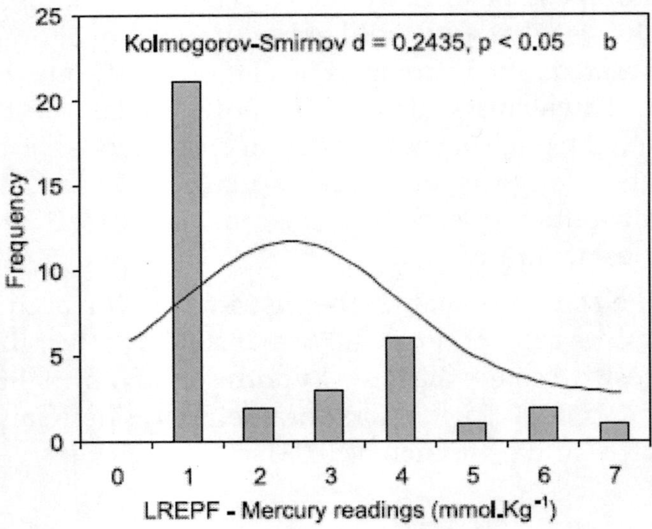


**Figure 2.** Distribution of (a) UNIR and (b) LREPF mercury readings. The expected normal distribution is also shown

The null hypothesis of absence of correlation between the two variables was tested using the normal approximation ($Z= 8.22$; $p< 0.0001$) and rejected. As shown in Figure 3, the conclusion is that the mercury readings from the two laboratories are linearly related.

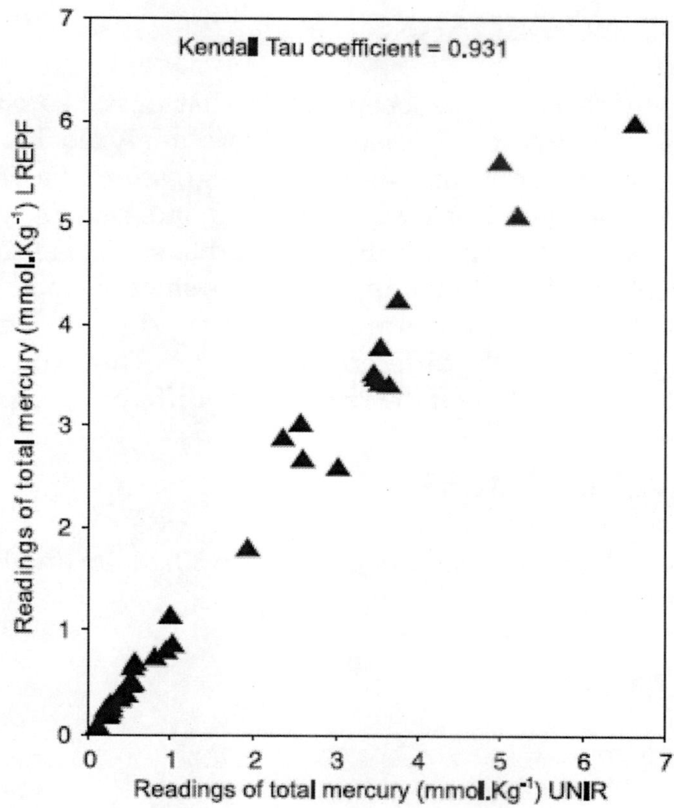


**Figure 3.** Relationships between UNIR and LREPF mercury readings, using the original data set showed in Table 1. The Kendall Tau coefficient was estimated as 0.931

Although our data exhibits linearity, this is not the only criterion need to demonstrate that the laboratories are intercalibrated. Is also necessary to prove that the slope (b) of the line is not significantly different from 1 and that the intercept (a) is not significantly different from 0. Note that if these conditions are fulfilled, the same regression line will be obtained, whatever which variable (UNIR or LREPF) was used as response variable. The hypothesis mentioned above can be tested using a t-test[7,8]. The two hypotheses (b = 1 and a = 0) were not rejected when using the data shown in Table 1, suggesting that the two laboratories are, in fact, intercalibrated.

## CONCLUDING REMARKS

Using a real data set, it has been shown that mistaken conclusions can be drawn if intercomparison data are analyzed by means of paired test comparisons, and also when regression analysis is used to estimate functions. Correlation analysis and tests of hypotheses about slope and intercept are more desirable statistical tools in the analysis of laboratory intercomparison data, since it is possible to test if the laboratories involved are intercarlibrated (ie: if a correlation coefficient is statistically different from zero). This would imply a linear relationship between the readings of different samples.

## ACKNOWLEDGEMENT

The authors would like thanking J. Valentin, O. Malm and B. Howard for the careful reading of this manuscript.

## REFERENCES

1. Cofino, W. P.; Baarsma, J. P.; Peerboom, R. A. L.; *Eur. Water Pollut. Control* **1995**, *5*, 8.
2. Arri, E.; Cabiati, F.; D'Emilio, S.; Gonella, L.; *Meas.* **1995**, *16*, 51.
3. Bloom, N. S.; Horvat, M.; Watras, C.J.; *Water, Air, Soil Pollut.* **1995**, *80*, 1314.
4. Kehrig, H. A.; Malm, O.; Akagi, H.; *Water, Air, Soil Pollut.* **1997**, *97*, 29.
5. Nott, B. R.; *Water, Air, Soil Pollut.* **1995**, *80*, 1314.
6. Marden, B. T.; Fairborther, A.; Bennet, J. K.; *Environ. Toxicol. Chem.* **1994**, *13*, 1768.
7. Zar, J, H.; *Biostatistical analysis*, Prentice Hall: New Jersey, 1984.
8. Snedecor, G. W.; Cochran, W. G.; *Statistical methods*, Iowa State University Press: Iowa. 1980.
9. Conover, W. J.; *Practical Nonparametric Statistics*, John Wiley & Sons: New York, 1980.

# Chapter 10

# DESIGN AND SYNTHESIS OF TWO OXAZINE DERIVATIVES USING SEVERAL STRATEGIES

Figueroa-Valverde Lauro,[1] Díaz-Cedillo Francisco,[2] Rosas-Nexticapa Marcela,[3] García-Cervera Elodia,[1] Pool-Gómez Eduardo,[1] López-Ramos María,[1] Hau-Heredia Lenin,[1] and Sarabia-Alcocer Betty[4]

[1]Laboratory of Pharmaco-Chemistry, Faculty of Chemical Biological Sciences, University Autonomous of Campeche, Avenida Agustín Melgar s/n, Col Buenavista, 24039 San Francisco de Campeche, CAM, Mexico

[2]Escuela Nacional de Ciencias Biológicas del Instituto Politécnico Nacional, Prol, Carpio y Plan de Ayala s/n, Col Santo Tomas, 11340 Mexico, DF, Mexico

[3]Facultad de Nutrición, Universidad Veracruzana, Médicos y Odontólogos s/n, 91010 Xalapa, VER, Mexico

[4]Faculty of Medicine, University Autonomous of Campeche, Avenida Patricio Trueba de Regil s/n, Col Lindavista, 24090 San Francisco de Campeche, CAM, Mexico

## ABSTRACT

Several oxazine derivatives have been synthesized; nevertheless, expensive reagents and special conditions are required. Therefore, in this study, two oxazine derivatives (2-chloro-3-{{2-[-(3-chloro-2-oxo-cyclobutyl)-(2,3-dimethoxy-9,10-dihydrostrychnid-10-yl)-amino]-ethyl}-[1,5-dimethyl-4-(1H-naphtho[1,2-e][1,3-oxazin-2-yl)-2-phenyl-2,3 dihydro-1H-pyrazol-3-yl]-amino}-cyclobutanone and 2-chloro-3-{{2-[(3-chloro-2-oxo-cyclobutyl)-(1,7,7-trimethyl-bicyclo[2.2.1]hept-2-yl)-amino]-ethyl}-[1,5-dimethyl-4-(1Hnaphtho[1,2-e][1,3]oxazin-2-yl)-2-phenyl-2,3-dihydro-1H-pyrazol-3-yl]-amino} cyclobutanone) were synthesized using several strategies. The structure of compounds obtained was confirmed by elemental analysis, spectroscopy, and spectrometry data. In conclusion, the methods used offer some advantages such as good yields, simple procedure, low cost, and ease of workup.

## INTRODUCTION

Oxazine derivatives are very important heterocyclic compounds with several biological activities [1, 2]. Several years ago, some oxazine derivatives have been synthesized; for example, the synthesis of 1,2-dihydro-1-arylnaphtho[1,2-e][1, 3]oxazine-3-oneusing the three-component system($\beta$-naphthol, benzaldehyde, and urea) in presence of $HClO_4$-$SiO_2$ [3]. Other report [4] showed the preparation of dirithromycin (9-N-11-O-oxazine derivative) by condensation of 9(S)- erythromycylamine with 2-(2-methoxy-ethoxy)acetaldehyde.

Other studies showed the condensation of 2-naphthol with heteroarylaldehydes or substituted benzaldehydes in the presence of ammonia [5]. In addition, a study [6] showed the synthesis of 3,4-dihydro-3-substituted-2H-naphtho[2,1-e][1,3] oxazine derivatives catalyzed by zirconium (IV) chloride. Other data [7] showed the reaction of substituted aryl and heteroarylaldehydes to give a series of 8-bromo-1,3-diaryl- 2,3-dihydro-1H-naphth[1,2e][1,3]oxazines. Also another oxazine derivative (1,3-diphenyl-1H-naphtho[1,2-e][1,3]oxazine) was developed by the reaction of N-arylidene-1-($\alpha$- aminoarylbenzyl)-2-naphthol with iodobenzene diacetate [8]. All these experimental results show several procedures which are available for synthesis of several oxazine derivatives; nevertheless, expensive reagents and special conditions are required. Therefore, in this study two oxazine derivatives were synthesized using some strategies.

## EXPERIMENTAL

The $N^1$-(2,3-dimethoxystrychnidin-10-yliden)-ethane-1,2-diamine (7) was synthesized according to a previously reported method [9]. The other compounds evaluated in this study were purchased from Sigma-Aldrich Co. Ltd. The melting points for the different compounds were determined on an electrothermal (900 model). Infrared spectra (IR) were recorded using KBr pellets on a Perkin Elmer Lambda 40 spectrometer. $^{1}H$ and $^{13}C$ NMR spectra were recorded on a Varian VXR-300/5 FT NMR spectrometer at 300 and 75.4MHz in $CDCl_3$ using TMS as internal standard. EIMS spectra were obtained with a Finnigan Trace GCPolaris Q. spectrometer. Elemental analysis data were acquired from a Perkin Elmer Ser. II CHNS/0 2400 elemental analyzer.

## 1,5-Dimethyl-4-(1H-naphtho[1,2,e][1,3]oxazin-2-(3H)-yl)-2-phenyl-1,2-dihydro-3H-pyrazol-3-one (3)

A solution of naphthol (100mg, 0.69mmol) and 4-aminoantipyrine (141 mg, 0.69mmol) formaldehyde (1 mL) in 10mL of methanol was stirred for 72 h at room temperature. The reaction mixture was evaporated to a smaller volume. After the mixture was diluted with water and extracted with chloroform. The organic phase was evaporated to dryness under reduced pressure; the residue was purified by crystallization from methanol: water (3: 1) yielding 80% of product, m.p. 176–178°C; IR ($V_{max}$, $cm^{-1}$): 1718, 1228, 1140; $^{1}H$ NMR (300MHz, $CDCl_3$) $\delta_H$: 2.10 (s, 3H), 2.80 (s, 3H), 5.20–5.70 (m, 4H), 7.01 (m, 1H), 7.10 (m, 1H), 7.30 (m, 1H), 7.40 (m, 2H), 7.44 (m, 2H), 7.58 (m, 2H), 7.60–7.66 (m, 3H) ppm. $^{13}C$ NMR (75.4MHz, $CDCl_3$) $\delta_C$: 15.02, 34.20, 47.42, 81.88, 110.12, 110.20, 118.22, 120.90, 123.26, 126.30, 126.42, 127.90, 128.11, 128.25, 128.40, 133.84, 133.90, 134.18, 134.38, 151.80, 166.30 ppm. EI-MS m/z: 371.10 ($M^{+}$10). Anal. Calcd, for $C_{23}H_{21}N_3O_2$: C, 74.39; H, 5.70; N, 11.31; O, 8.61. Found: C, 74.36; H, 5.68.

## *N-1-[1,5-Dimethyl-4-(1H-naphtho[1,2-e][1,3]oxazin-2-yl)-2-phenyl-1,2-dihydro pyrazol-3-ylidene]-ethane-1,2-diamine (4)*

A solution of **3** (100 mg, 0.27mmol), ethylenediamine (100 $\mu$L, 1.50mmol), and boric acid (90mg, 1.45mmol) in 10mL of methanol was stirred for 72 h at room temperature. The reaction mixture was evaporated to a smaller volume, after the mixture was diluted with water and extracted with chloroform. The organic phase was evaporated to dryness under reduced pressure; the residue was purified by crystallization from methanol :water (2 : 1) yielding 62% of product, m.p. 102–104∘C; IR ($V_{max}$, cm−1): 3382, 1228, 1140; 1H NMR (300MHz, $CDCl_3$) $\delta_H$: 1.10 (s, 3H), 2.02 (s, 3H), 3.06 (s, 3H), 3.18 (t, 2H, $J$ = 6.44Hz), 3.70 (t, 2H, $J$ = 6.44 Hz), 4.34 (broad, 2H), 5.20–5.76 (m, 4H), 6.98–7.04 (m, 3H), 7.30–7.48 (m, 3H) 7.52 (m, 2H), 7.60–8.06 (m, 3H) ppm. $^{13}C$ NMR (75.4 MHz, $CDCl_3$) $\delta_C$: 15.50, 34.70, 40.24, 47.24, 54.92, 81.38, 105.28, 109.66, 119.30, 121.68, 121.80, 123.30, 123.90, 126.30, 126.80, 127.91, 128.40, 131.90, 133.80, 134.18, 139.50, 142.60, 151.48 ppm. EI-MS *m/z:* 413.20 (M+10). Anal. Calcd. for $C_{25}H_{27}N_5O$: C, 72.61; H, 6.58; N, 16.94; O, 3.87. Found: C, 72.60; H, 6.56.

## N-[1,5-Dimethyl-4-(1H-naphtho[1,2-e][1,3]oxazin-2-yl)-2-phenyl-1,2-dihydro-pyrazol-3-ylidene]-2,3-dimethoxystrychnidin-10-ylidene-ethane-1,2-diamine (6)

Method A. A solution of 4 (100mg, 0.24mmol), brucine (95 mg, 0.24mmol), and boric acid (30mg, 0.48mmol) in 10mL of methanol was stirred for 72 h at room temperature. The reaction mixture was evaporated to a smaller volume, after the mixture was diluted with water and extracted with chloroform. The organic phase was evaporated to dryness under reduced pressure; the residue was purified by crystallization from methanol :water (3 : 2) yielding 80% of product, m.p. 158–160°C; IR ($V_{max}$, $cm^{-1}$): 3322, 1228, 1140; 1H NMR (300MHz, $CDCl_3$) $\delta_H$: 1.40–1.84 (m, 3H), 2.02 (s, 3H), 2.22– 3.10 (m, 7H), 3.12 (s, 3H), 3.26–3.69 (m, 6H), 3.80 (s, 3H), 3.90 (t, 2H, $J$ = 6.54Hz), 3.91 (s, 3H), 3.92 (t, 2H, $J$ = 6.54Hz), 4.70 (m, 1H), 5.20–5.78 (m, 4H), 5.83–5.88 (m, 2H), 7.00–7.04 (m, 3H), 7.32–7.44 (m, 3H), 7.52 (m, 2H), 7.55 (m, 1H), 7.60–8.04 (m, 3H) ppm. $^{13}C$ NMR (75.4MHz, $CDCl_3$) $\delta_C$: 15.48, 27.37, 29.20, 34.52, 34.75, 40.90, 45.82, 47.26, 50.74, 53.02, 53.09, 55.95, 56.20, 56.60, 63.20, 64.62, 65.02, 79.26, 81.70, 99.23, 105.30, 109.70, 119.34, 120.89, 121.69, 121.76, 123.26, 123.30, 123.90, 126.22, 126.82, 127.87, 127.90, 128.38, 130.60, 131.98, 133.82, 134.12, 139.59, 140.30, 142.65, 145.48, 145.78, 147.70, 151.50 ppm. EI-MS m/z: 789.38 (M+10). Anal. Calcd., for $C_{48}H_{51}N_7O_4$: C, 72.98; H, 6.51;N, 12.41;O, 8.10. Found: C, 72.96; H, 6.50.

### *Method B*

A solution of 3 (100 mg, 0.24mmol), brucine derivative (105mg, 0.24mmol), and boric acid (30mg, 0.48mmol) in 10mL of methanol was stirred for 48h at room temperature. The organic phase was evaporated to dryness under reduced pressure; the residue was purified by crystallization from methanol :water yielding 50% of product, m.p. 158–160°C; 1H and $^{13}C$ NMR data were similar to method A product.

## 2-Chloro-3-{{2-[-(3-chloro-2-oxo-cyclobutyl)-(2,3-dimethoxy-9,10-dihydrostrychnid-10-yl)-amino]-ethyl}-[1,5-dimethyl-4-(1H-naphtho[1,2-e][1,3oxazin-2-yl)-2-phenyl-2,3-dihydro-1H-Pyrazol-3-yl]-amino}-cyclobutanone (9)

A solution of 6 (100mg, 0.10mmol), chloroacetyl chloride (35 $\mu$L, 0.44mmol), and triethylamine (60 $\mu$L, 0.43mmol) in 10mL of methanol was stirred for 72 h at room temperature. The reaction mixture was evaporated to dryness under reduced pressure, after the mixture was diluted with water and extracted with chloroform. The organic phasewas evaporated to dryness under reduced pressure; the residue was purified by crystallization from methanol : water (3 : 2) yielding 65% of product, m.p. 74–76°C; IR ($V_{max}$, $cm^{-1}$): 3410, 1716, 1228, 1150, 1140; 1H NMR (300MHz, CDCl3) $\delta_H$: 1.12–1.46 (m, 2H), 1.54 (m, 2H), 1.85 (m, 1H), 1.86–1.87 (m, 2H), 1.96 (s, 3H), 2.04– 2.70 (m, 5H), 2.70 (s, 3H), 2.82–2.86 (m, 4H), 2.87 (m, 1H), 2.90–3.42 (m, 3H), 3.54 (m, 1H), 3.62–3.78 (m, 3H), 3.80 (s, 3H), 3.90 (m, 1H), 3.93 (s, 3H), 3.98 (m, 1H), 4.02 (m, 1H), 4.38 (m, 1H), 4.62 (m, 2H), 5.16–5.70 (m, 4H), 5.88–5.97 (m, 2H), 6.90 (m, 1H), 6.95 (m, 1H), 7.00m(m, 2H), 7.04 (m, 1H), 7.20 (m, 1H), 7.30 (m, 2H), 7.38–7.80 (m, 5H) ppm. $^{13}C$ NMR (75.4MHz, $CDCl_3$) $\delta_C$: 15.02, 27.34, 31.88, 32.40, 32.66, 34.28, 34.32, 40.86, 46.50, 47.72, 50.30, 50.38, 50.72, 50.74, 55.95, 56.20, 56.29, 62.90, 63.20, 64.16, 66.32, 70.72, 71.00, 73.10, 74.29, 78.30, 82.20, 96.38, 100.40, 108.30, 109.20, 113.70, 114.66, 118.60, 119.30, 119.90, 121.06, 121.60, 123.28, 126.30, 127.80, 127.90, 128.11, 128.40, 132.70, 132.90, 134.89, 140.30, 143.40, 146.30, 146.78, 150.54, 203.44, 204.20 ppm. EI-MS m/z: 997.40 (M+10).Anal.Calcd. for$C_{56}H_{61}Cl_2N_7O_6$: C, 67.33;H, 6.15;Cl, 7.10, N, 9.81;O, 9.61. Found: C, 67.30; H, 6.12.

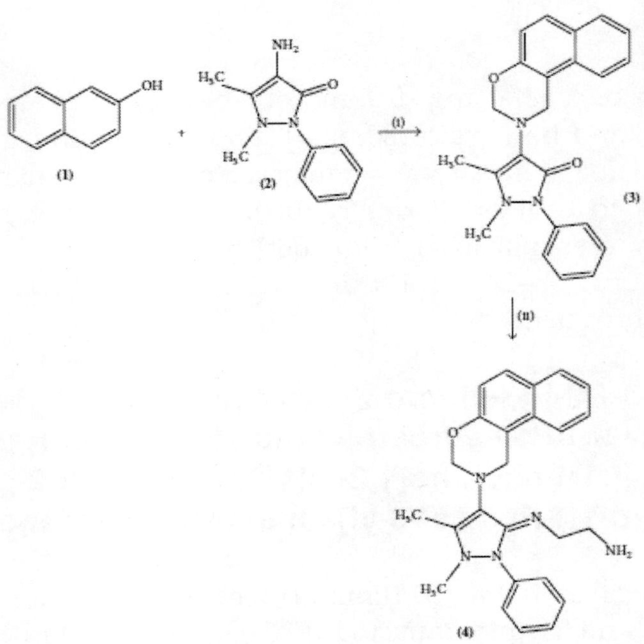

**Figure 1.** Synthesis of N-1-[1,5-dimethyl-4-(1H-naphtho[1,2-e][1,3]oxazin-2-yl)-2-phenyl-1,2-dihydro-pyrazol-3-ylidene]-ethane-1,2-diamine (4). The first stage involves the reaction of naphthol (1) with 4-aminoantipyrine (2) to form the compound 1,5-dimethyl-4-(1Hnaphtho[1,2,e][1,3]oxazin-2-(3H)-yl)-2-phenyl-1,2-dihydro-3H-pyrazol-3-one (3) in presence of formaldehyde (i). The second stage was achieved by the reaction of 3 with ethylenediamine (ii) to form 4.

## N-[1,5-Dimethyl-4-(1H-naphtho[1,2-e][1,3]oxazin-2-yl)-2-phenyl-1,2-dihydro-pyrazol-3-ylidene]-N[1]-(1,7,7-trimethylbicyclo[2.2.1]hept-2-ylidene)ethane-1,2-diamine (11)

A solution of 4 (100mg, 0.24mmol), 1,7,7-trimethylbicyclo[2.2.1] heptan-2-one (40mg, 0.26mmol), and boric acid (30mg, 0.48mmol) in 10mL of methanol was stirred for 48 h at room temperature. The reaction mixture was evaporated to a smaller volume, after the mixture was diluted with water and extracted with chloroform. The organic phasewas evaporated to dryness under reduced pressure;

the residue was purified by crystallization from methanol : water (4 : 1) yielding 75% of product, m.p. 90–92°C; IR ($V_{max}$, cm−1): 3320, 1228, 1138; 1H NMR (300MHz, $CDCl_3$) $\delta_H$: 0.80 (s, 3H), 0.84 (s, 6H), 1.12–1.74 (m, 4H), 2.04 (s, 3H), 2.08–2.70 (m, 3), 3.12 (s, 3H), 3.72 (t, 2H, $J$ = 6.54Hz), 3.84 (t, 2H, $J$ = 6.54Hz), 5.20–5.74 (m, 4H), 7.01–7.06 (m, 3H), 7.32–7.46 (m, 3H), 7.52 (m, 2H), 7.60–8.04 (m, 3H) ppm. 13C NMR (75.4 MHz, CDCl3) $\delta$C: 11.20, 15.48, 19.04, 19.40, 27.40, 31.18, 34.70, 38.66, 43.70, 47.24, 47.40, 49.90, 52.50, 53.35, 81.72, 105.31, 109.60, 119.38, 121.66, 121.78, 123.30, 123.88, 126.22, 126.80, 127.90, 128.40, 131.98, 133.86, 134.10, 139.57, 142.62, 151.50, 172.02 ppm. EI-MS m/z: 547.30 (M+10). Anal. Calcd. for $C_{35}H_{41}N_5O$: C, 76.75; H, 7.54; N, 12.79;O, 2.92. Found: C, 76.74; H, 7.52.

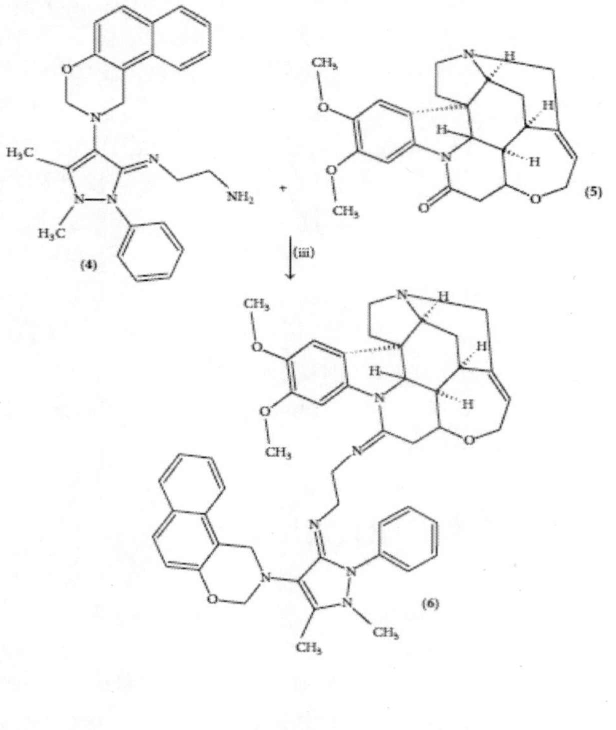

**Figure 2.** Synthesis of N-[1,5-dimethyl-4-(1H-naphtho[1,2-e][1,3]oxazin-2-yl)-2-phenyl-1,2-dihydro-pyrazol-3-ylidene]-2,3-dimethoxy-strychnidin-10-ylidene-ethane-1,2-diamine (6). Reaction of N-1-[1,5-dimethyl-4-(1H-naphtho[1,2-e][1,3]oxazin-2-yl)-2-phenyl-1,2-dihydropyrazol-3-ylidene]-ethane-1,2-diamine (4) with brucine (5) to form the compound 6. iii: boric acid/rt.

## 2-Chloro-3-{{2-[(3-chloro-2-oxo-cyclobutyl)-(1,7,7-trimethyl-bicyclo[2.2.1]hept-2-yl)-amino]-ethyl}-[1,5-dimethyl-4-(1H-naphtho[1,2-e][1,3]oxazin-2-yl)-2-phenyl-2,3-dihydro-1H-pyrazol-3-yl]-amino}-cyclobutanone (12)

A solution of 11 (100mg, 0.18mmol), chloroacetyl chloride (35 $\mu$L, 0.44 mmol), and triethylamine (60 $\mu$L, 0.43mmol) in 10mL of methanol was stirred for 48 h at room temperature. The reaction mixture was evaporated to a smaller volume, after the mixture was diluted with water and extracted with chloroform. The organic phase was evaporated to dryness under reduced pressure; the residue was purified by crystallization from methanol :water (2 : 1) yielding 75% of product, m.p. 98–100°C; IR ($V_{max}$, cm−1): 1716, 1228, 1196, 1138; 1H NMR (300MHz, $CDCl_3$) $\delta_H$: 0.68 (s, 3), 0.72 (s, 3H), 0.81 (s, 3H), 1.12–1.28 (m, 4H), 1.58 (m, 1H), 1.68–1.70 (m, 2H), 1.71 (m, 1H), 1.75 (m, 1H), 1.86 (m, 2H), 1.98 (s, 3H), 1.99 (m, 1H), 2.60 (m, 1H), 2.62–2.68 (t, 2H, $J$ = 6.51Hz), 2.74 (s, 3H), 2.83–2.84 (t, 2H, $J$ = 6.51Hz), 3.64–4.04 (m, 2H), 4.38 (m, 1H), 4.62 (m, 2H), 5.16–5.70 (m, 4H), 6.90–7.00 (m, 3H), 7.09 (m, 1H), 7.30 (m, 2H), 7.38–7.78 (m, 5H) ppm. $^{13}$C NMR (75.4 Hz, CDCl3) $\delta$C: 13.36, 15.06, 19.86, 20.20, 26.40, 32.66, 32.70, 34.30, 36.50, 36.72, 44.40, 47.12, 47.78, 50.66, 51.40, 53.88, 61.70, 62.92, 71.00, 72.22, 73.10, 82.24, 100.40, 109.20, 113.70, 114.66, 118.60, 119.90, 121.09, 123.30, 126.28, 127.90, 128.12, 128.40, 132.76, 134.80, 140.86, 150.50, 202.02, 203.40 ppm. EI-MS m/z: 755.30 (M+10). Anal. Calcd. for $C_{43}H_{51}Cl_2N_5O_3$: C, 68.24; H, 6.79; Cl, 9.37; N, 9.25; O, 6.34. Found: C, 68.20; H, 6.78.

## RESULTS AND DISCUSSION

In this study were synthesized two oxazine derivatives using some strategies; it is important to mention that there are reports which indicate that condensation of naphthol with formaldehyde and amino groups result in naphthoxazines [10, 11]. Therefore, in the first stage was synthesized the compound 1,5-dimethyl-4-(1H-naphtho[1,2,e][1,3]oxazin-2-(3H)- yl)-2-phenyl-1,2-dihydro-3H-pyrazol-3-one (3) by the reaction of naphthol, 4-aminoantipyrine, and formaldehyde (Figure 1). The1 H NMR spectrum of 3 show signals at 2.10 and 2.80 ppmformethyl groups; at 5.00–5.70 ppmfor oxazine ring; at 7.01, 7.30, 7.44, and 7.60–7.66 ppm for naphthalene group bound to oxazine ring; at 7.10, 7.40, and 7.58 ppm for phenyl group. The 13CNMR spectrumof 3 contains peaks at 15.02 and 34.20 ppm for methyl groups; at 110.12 and 134.38 ppm for carbons involved in the cyclopentene ring; at 47.42, 81.88, 110.20, and 151.80 ppm for oxazine ring; at 118.52– 126.30, 127.90, 128.40, and 133.90–134.18 ppmfor naphthalene group; at 126.42, 128.11–128.25, and 133.84 ppm for phenyl group; at 166.30 ppm for ketone group. Finally, the presence of compound 3 was further confirmed from mass spectrum which showed a molecular ion at m/z 371.10.

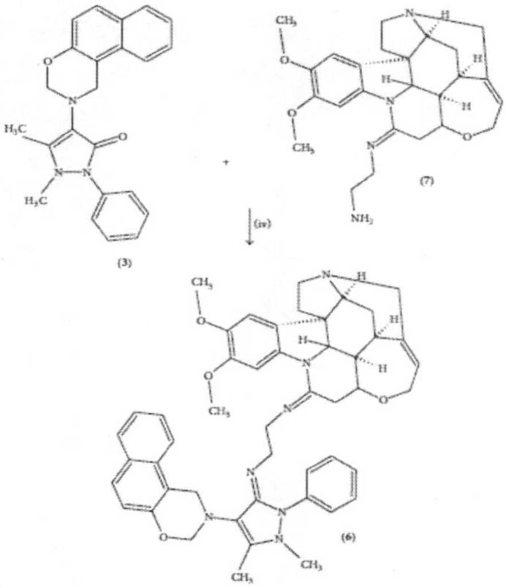

**Figure 3.** Synthesis ofN-[1,5-dimethyl-4-(1H-naphtho[1,2-e][1,3]oxazin-2-yl)-2-phenyl-1,2-dihydro-pyrazol-3-ylidene]-2,3-dimethoxystrychnidin-10-ylidene-ethane-1,2-diamine (6). Reaction of 1,5-dimethyl-4-(1H-naphtho[1,2,e][1,3]oxazin-2-(3H)-yl)-2-phenyl-1,2-dihydro-3H-pyrazol-3-one (3) with N1-(2,3-dimethoxystrychnidin-10-yliden)-ethane-1,2-diamine (7) to form the compound 6. iv: boric acid/rt.

The second stage was achieved by reaction of the compound 3 with ethylenediamine (Figure 1) resulting in imino bond formation involved in the compound 4 (N-1-[1,5- dimethyl-4-(1H-naphtho[1,2-e][1,3]oxazin-2-yl)-2-phenyl-1,2-dihydro-pyrazol-3-ylidene]-ethane-1,2-diamine). Many procedures for the synthesis of imino groups are described in the literature [12–14]; nevertheless, in this study boric acid was used as a catalyst, because it is not an expensive reagent and no special conditions for its use are required [15]. The results of $^{1}$H NMR spectrum of 4 show signals at 2.02 and 3.06 ppm for methyl groups; at 3.18 and 3.70 ppm for methylene groups involved in the arm bound to cyclopentene ring; at 4.34ppm for amino group; at 5.20–5.76ppm for oxazine ring; at 6.98–7.04 and 7.52 for phenyl group; at 7.30–7.48 and 7.60–8.06 ppm for naphthalene group. The $^{13}$C NMR spectrum of 4 contains peaks at 15.50 and 34.70 ppm for methyl groups; at 40.24 and 54.92ppm form ethylene groups involved in the arm bound to cyclopentene ring; at 47.24, 81.38–81.38, 109.66, and 151.48 ppmfor oxazine ring; at 105.28 and 121.68 ppm for methylene groups of cyclopentene ring; at 119.30, 121.80–123.30, 126.30, 127.91–128.40, and 133.80– 134.18 ppm for naphthalene group; at 123.90, 126.80, 131.90, and 142.60 ppm for phenyl group; at 139.50 ppm for imino group. Finally, the presence of compound 4 was further confirmed from mass spectrum which showed a molecular ion at m/z 413.20.

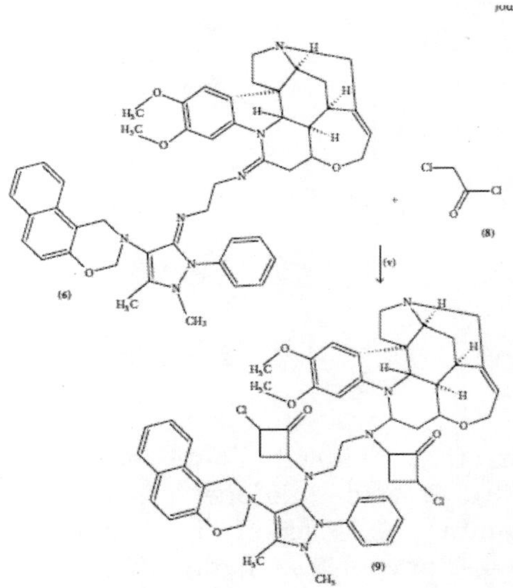


**Figure 4.** Synthesis of 2-chloro-3-{{2-[-(3-chloro-2-oxo-cyclobutyl)-(2,3-dimethoxy-9,10-dihydrostrychnid-10-yl)-amino]-ethyl}-[1,5-dimethyl-4-(1H-naphtho[1,2-e][1,3oxazin-2-yl)-2-phenyl-2,3-dihydro-1H-pyrazol-3-yl]-amino}-cyclobutanone (9). Reaction of N-[1,5-dimethyl-4-(1H-naphtho[1,2-e][1,3]oxazin-2-yl)-2-phenyl-1,2-dihydro-pyrazol-3-ylidene]-2,3-dimethoxystrychnidin-10-ylidene-ethane-1,2-diamine (6) with chloroacetyl chloride (8) to form the compound 9. v: triethylamine/rt.

In the third stage two different methods for synthesis of the oxazine-brucine derivative (6) were employed; the first step was achieved by reaction of the compound 4 with brucine to form the oxazine-brucine derivative (6) using boric acid as catalyst (method A, Figure 2). The results of $^{1}$H NMR spectrum of 6 show signals at 2.02 and 3.12 ppm for methyl groups bound to cyclopentene ring; at 1.40–1.84, 2.22–3.10, 3.26–3.69, 4.70, 5.83–5.88, and 7.55 ppmfor brucine nucleus; at 3.90 and 3.92 ppm for methylene groups involved in arm bound to cyclopentene ring; at 3.80–3.91 for methyl groups of brucine; at 5.20–5.78 for protons of oxazine ring; at 7.00–7.04 and 7.52 ppm for phenyl group; at 7.32–7.44 and 7.60–8.04 ppm for naphthalene group.

The 13C NMR spectrum of 6 contains peaks at 15.48 and 34.75 ppm for methyl groups bound to cyclopentene ring; at 27.37-34.52, 40.90- 45.82, 56.20, 63.20–79.26, 99.23, 123.26, 127.87, 130.60, 140.30, and 145.48–147.70 ppm for brucine nucleus; at 47.26, 81.70, 109.70–109.74, and 151.50 ppmfor oxazine ring; at 55.02–55.09 for arm bound both imino groups; at 105.30, 121.69, and 139.59 ppm for cyclopentene ring; at 119.34, 121.76, 123.30–126.22, 127.90–128.38, and 133.82–134.12 ppmfor naphthalene group; at 123.90, 126.82, 131.98, and 142.65 ppm for phenyl group; at 55.95 and 56.60 for methyl groups of brucine.

Finally, the presence of compound 6 was further confirmed from mass spectrum which showed a molecular ion at m/z 789.38. The second step was achieved by reaction of the compound 3 with 7 to form the compound 6 using boric acid as catalyst (Figure 3). Similar $^{1}$H and $^{13}$C NMR data were obtained compared with method A product. However, it is noteworthy that with this method the yield was low as compared to method A; this phenomenon possibly is due to time of reaction required with this methodology.

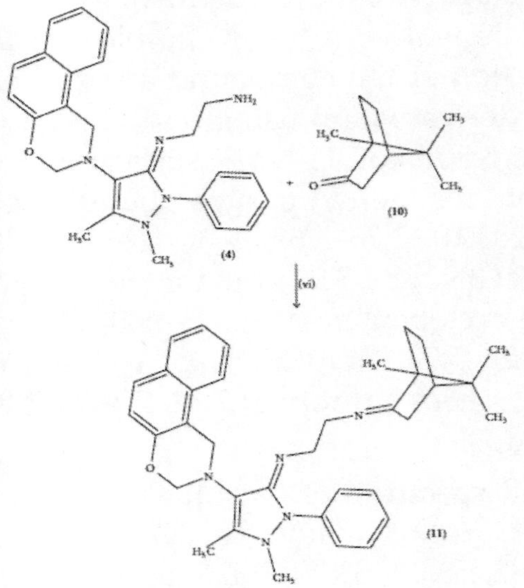

**Figure 5.** Synthesis of N-[1,5-dimethyl-4-(1H-naphtho[1,2-e][1,3]oxazin-2-yl)-2-phenyl-1,2-dihydro-pyrazol-3-ylidene]-N□-(1,7,7-trimethylbicyclo[ 2.2.1] hept-2-ylidene)ethane-1,2-diamine (11). Reaction of N-1-[1,5-dimethyl-4-(1H-naphtho[1,2-e][1,3]oxazin-2-yl)-2-phenyl-1,2- dihydro-pyrazol-3-ylidene]-ethane-1,2-diamine (4) with 1,7,7-trimethyl bicyclo[2.2.1]heptan-2-one (10) to form the compound 11. vi: boric acid/rt.

On the other hand, the compound 9 was synthesized (Figure 4); this compound has two chlorocyclobutane groups bound to both amino groups involved in their chemical structure. It is important to mention that there are several reports for preparation of chlorocyclobutenones using several techniques [16–19]; nevertheless, expensive reagents and special conditions are required. Therefore, in this study a new chlorocyclobutenone was formed in the chemical structure of 9 using chloroacetyl chloride in presence of triethylamine.

The results of $^{1}$H NMR spectrum of 9 show signals at 1.12– 1.46, 1.85, 2.04–2.70, 2.87–3.42, 3.62–3.78, 3.90, 3.98, 5.88–5.97, and 7.20 ppm for brucine fragment; at 1.54, 1.86–1.87, 3.54, 4.02, and 4.62 ppm for cyclobutanone groups; at 1.96 and 2.74 ppm for methyl groups bound to cyclopentene ring; at 2.82–2.86 for methylene bound to both amino groups; at 3.80 and 3.93 ppm for methyl groups of brucine fragment; at 4.38 ppm for cyclopentene ring; at 5.16–5.70 ppm for oxazine group; at 6.90–7.00 and 7.30 ppm for phenyl group; at 7.04

and 7.38–7.80 ppm for naphthalene group. The $^{13}C$ NMR spectrum of 9 contains peaks at 15.02 and 34.28 ppm for methyl groups bound to cyclopentene ring; at 27.34–31.88, 34.32–46.50, 50.72–50.74, 56.20, 63.20–66.32, 74.29–78.30, 96.38, 108.30, 121.06, 132.90, and 140.30–146.30 ppmfor brucine nucleus; at 32.40–32.66, 62.90, 70.72–71.00, 203.44, and 204.20 ppm for cyclobutanone groups; at 47.72, 82.20, and 109.20 ppm for oxazine group; at 50.30 and 50.38 for methylene groups bound to both amino groups; at 55.95 and 56.29 ppm for methyl groups of brucine fragment; at 73.10, 100.40, and 114.66 ppmfor cyclopentene ring; at 113.70, 119.30, 128.11, and 146.78 ppm for phenyl group; at 118.60, 119.90, 121.60–126.30, 127.90, 128.40–132.70, 134.89, and 150.54 ppm for naphthalene group. Finally, the presence of compound 9 was further confirmed from mass spectrum which showed a molecular ion at m/z 997.40.

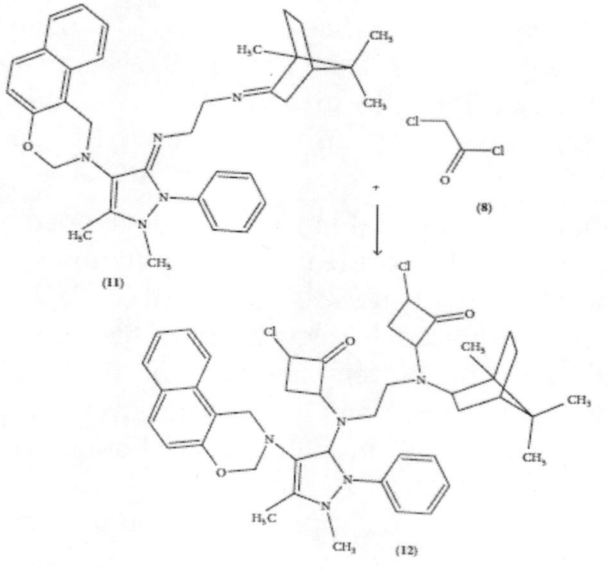

**Figure 6.** Synthesis of 2-chloro-3-{{2-[(3-chloro-2-oxo-cyclobutyl)-(1,7,7-trimethyl-bicyclo[2.2.1]hept-2-yl)-amino]-ethyl}-[1,5-dimethyl-4-(1H-naphtho[1,2-e][1,3]oxazin-2-yl)-2-phenyl-2,3-dihydro-1H-pyrazol-3-yl]-amino}-cyclobutanone (12). Reaction of N-[1,5-dimethyl-4-(1H-naphtho[1,2-e][1,3]oxazin-2-yl)-2-phenyl-1,2-dihydro-pyrazol-3-yli dene]-$N^1$-(1,7,7-trimethyl-bicyclo[2.2.1]hept-2-ylidene)ethane-1,2-diamine (11) with chloroacetyl chloride (8) to form the compound 12. v: triethylamine/rt.

The fourth stage was achieved by the reaction of 4 with 1,7,7-trimethylbicyclo[2.2.1]heptan-2-one to form the compound 11 using boric acid as catalyst (Figure 5). The $^{1}H$ NMR spectrum of 11 shows signals at 0.80 and 0.84 ppm for methyl groups bound to bicyclic ring; at 1.12–1.74, 2.08–2.70 ppm for protons of bicyclic ring; at 2.04 and 3.12 ppm for methyl groups bound to cyclopentene ring; at 3.72 and 3.84 ppm for methylene groups bound to both imino groups; at 5.20–5.74 ppm for oxazine ring; at 7.01–7.06 and 7.52 ppm for phenyl group; at 7.32–7.46 and 7.60–8.04 ppm for naphthalene group. The 13CNMRspectrumof 11 contains peaks at 11.20 and 19.04–19.40 ppm for methyl groups bound to bicyclic ring; at 15.48 and 34.70ppm for methyl groups bound to cyclopentene ring; at 27.40, 31.18, 38.66–43.70, 47.40, 49.90, and 172.02 ppm for bicyclic ring; at 47.24, 81.72, and 109.60 ppm for carbons involved in oxazine ring; at 105.31, 121.66, and 139.57 ppmfor cyclopentene ring; at 52.50 and 53.35 ppm for arm bound to both imino groups; at 119.38, 121.78–123.30, 126.22, 127.90–128.40, 133.86–134.10, and 151.54 ppm for naphthalene group; at 123.88, 126.80, 131.98, and 142.62 ppm for phenyl group. In addition, the presence of compound 11 was further confirmed from mass spectrum which showed a molecular ion at m/z 547.30.

Finally, the compound 12 (Figure 6) was developed by the reaction of 11 with chloroacetyl chloride using triethylamine as catalyst. The 1H NMR spectrum of 12 shows signals at 0.68, 0.72, and 0.81 ppm for methyl groups bound to bicyclic ring; at 1.12–1.28, 1.68–1.70, 1.75, and 2.60 ppm for bicyclic ring; at 1.58, 1.71, 1.86, 1.99, 3.64–4.04, and 4.62 ppm for protons involved in cyclobutanone groups; at 1.98 and 2.74 ppm for methyl groups bound to cyclopentene ring; at 2.62–2.68 and 2.83–2.84 ppm for methylene groups bound to both amino groups; at 4.38 ppm for proton of cyclopentene ring; at 5.16–5.70 ppm for oxazine ring; at 7.09 and 7.38–7.78 ppm for naphthalene group; at 6.90–7.00 and 7.30 ppm for phenyl group.The $^{13}C$NMR spectrumof 12 contains peaks at 13.36, 19.86, and 20.20ppm for methyl groups bound to bicyclic ring; at 15.06 and 34.30methyl groups bound to cyclopentene ring; at 26.40, 36.50, 47.12 and 53.88–61.70 ppmfor carbons of bicyclic ring; at 73.10, 100.40, and 114.66 ppmfor cyclopentene ring; at 47.78, 82.24, 109.20, and 150.50ppmfor oxazine ring; at 50.66 and 51.40 ppm for methylene groups bound to both amino groups; at 32.66, 32.70, 62.92, and 71.00–72.22 ppm for cyclobutanone groups; at 118.60, 121.09–127.90, and 128.40–134.80 ppm for naphthalene group; at 113.70, 119.90, and 128.12–146.86 ppm for phenyl group; at 202.02 and 203.40 ppm for ketone groups. Finally, the presence of compound 12 was further confirmed frommass spectrum which showed a molecular ion at m/z 755.30.

## CONCLUSIONS

In in this study two oxazine derivatives were synthesized using some strategies; the methods used offer some advantages such as goodyields, simpleprocedure, lowcost, and ease of workup.

## CONFLICT OF INTERESTS

The authors declare that they do not have any financial relations with any of the commercial entities mentioned in the paper that could lead to a conflict of interests.

## REFERENCES

1. M. E. Kuehne, E. A. Konpka, and B. F. Lambert, "Steroidal dihydro-1, 3-oxazines as antitumor agents," Journal of Medicinal and Pharmaceutical Chemistry, vol. 5, no. 2, pp. 281–296, 1962.
2. A. Chaskar, V. Vyavhare, V. Padalkar, K. Phatangare, and H. Deokar, "An environmentallybenignone-potsynthesisof1,2-dihydro-1-arylnaphtho[1,2-e][1,3]oxazine-3-one derivatives catalysed by phosphomolybdic acid," Journal of the Serbian Chemical Society, vol. 76, no. 1, pp. 21–26, 2011.
3. H. A. Ahangar, G. H.Mahdavinia, K.Marjani, and A. Hafezian, "A one-pot synthesis of 1,2-Dihydro-1-arylnaphtho[1,2-e][1,3] oxazine-3-one derivatives catalyzed by perchloric acid supported on silica (HClO4/SiO2) in the absence of solvent," Journal of the Iranian Chemical Society, vol. 7, no. 3, pp. 770–774, 2010.
4. F. T. Counter, P. W. Ensminger, D. A. Preston et al., "Synthesis and antimicrobial evaluation of dirithromycin (AS-E 136; LY237216), a new macrolide antibiotic derived from erythromycin," Antimicrobial Agents and Chemotherapy, vol. 35, no. 6, pp. 1116–1126, 1991.
5. Z. Turgut, E. Pelit, and A. K¨oyc¨u, "Synthesis of new 1,3-disubstituted-2,3-dihydro-1H-naphth-[1,2e][1,3] oxazines,"Molecules, vol. 12, no. 3, pp. 345–352, 2007.
6. A. H. Kategaonkar, S. S. Sonar, R. U. Pokalwar, A. H. Kategaonkar, B. B. Shingate, and M. S. Shingare, "An efficient synthesis of 3, 4-dihydro-

3-substituted-2H-naphtho[2, 1-e][1,3]oxazine derivatives catalyzed by zirconyl(IV) chloride and evaluation of its biological activities," Bulletin of the Korean Chemical Society, vol. 31, no. 6, pp. 1657–1660, 2010.

7. A.Mayekar, H. Yathirajan, B. Narȧyana, B. Sarojini, N. Kumari, and W. T. A. Harrison, "Synthesis and antimicrobial study of new 8-bromo-1, 3-diaryl-2, 3-dihydro-1H-naphtho[1, 2e][1,3]oxazines," International Journal of Chemistry, vol. 3, no. 1, pp. 74–86, 2011.

8. V. Verma, K. Singh, D. Kumar et al., "Synthesis, antimicrobial and cytotoxicity study of 1, 3-disubstituted-1H-naphtho[1, 2-e][1, 3]oxazines," European Journal of Medicinal Chemistry, vol. 56, pp. 195–202, 2012.

9. L. Figueroa-Valverde, F. D´ıaz-Cedillo, E. Garc´ıa-Cervera, E. Pool-G´omez, and A. Camacho-Luis, "Design and synthesis of two brucine derivatives," Asian Journal of Chemistry, vol. 25, no. 12, pp. 6783–6786, 2013.

10. W. J. Burke, R. P. Smith, and C.Weatherbee, "N,N-Bis-(hydroxybenzyl)-amines: synthesis from phenols, formaldehyde and primary amines," Journal of the American Chemical Society, vol. 74, no. 3, pp. 602–605, 1952.

11. W. J. Burke, M. J. Kolbezen, and C.W. Stephens, "Condensation of naphthols with formaldehyde and primary amines," Journal of the American Chemical Society, vol. 74, no. 14, pp. 3601–3605, 1952.

12. A. K. Shirayev, I. K. Moiseev, and S. S. Karpeev, "Synthesis and cis/trans isomerism of N-alkyl-1,3-oxathiolane-2-imines," Arkivoc, vol. 2005, no. 4, pp. 199–207, 2005.

13. D. J.-N. Uppiah, M. G. Bhowon, and S. J. Laulloo, "Solventless synthesis of imines derived from diphenyldisulphide diamine or p-Vanillin," E-Journal of Chemistry, vol. 6, no. 1, pp. S195–S200, 2009.

14. M. M. Hania, "Synthesis of some imines and investigation of their biological activity," E-Journal of Chemistry, vol. 6, no. 3, pp. 629–632, 2009.

15. L. Figueroa-Valverde, F. D´ıaz-Cedillo, E. Garc´ıa-Cervera, E. Pool-G´omez, and M. L´opez-Ramos, "Design and Synthesis of N-[2- (2, 3-dimethoxy-strychnidin-10-ylidenamino)-ethyl]- succinamic acid 4-allyl-2-methoxy-phenyl ester," Bulgarian Chemical Communication, vol. 45, no. 1, pp. 71–76, 2013.

16. G.K.Kole, G. K. Tan, andJ. J.Vittal, "Anion-controlled stereoselective synthesis of cyclobutane derivatives by solid-state [2 + 2] cycloaddition reaction of the salts of trans-3-(4-Pyridyl) acrylic acid," Organic Letters, vol. 12, no. 1, pp. 128–131, 2010.

17. J. Panda and S. Ghosh, "Intramolecular [2 + 2] photocycloaddition for the direct stereoselective synthesis of cyclobutane fused $\gamma$-lactols," Tetrahedron Letters, vol. 40, no. 36, pp. 6693–6694, 1999.

18. G.K.Kole, G. K. Tan, and J. J.Vittal, "Role of anions in the synthesis of cyclobutane derivatives via [2 + 2] cycloaddition reaction in the solid state and their isomerization in solution," Journal of Organic Chemistry, vol. 76, no. 19, pp. 7860–7865, 2011.

19. Y. Okada, T. Minami, S. Yahiro, and K. Akinaga, "A new synthesis of cyclobutane annelated compounds by the use of a (1-cyclobutenyl) triphenylphosphonium salt," Journal of Organic Chemistry, vol. 54, no. 4, pp. 974–977, 1989.

# Chapter 11

# AN EFFICIENT SOLVENT-FREE SYNTHESIS OF MESO-SUBSTITUTED DIPYRROMETHANES USING $SNCL_2 \cdot 2H_2O$ CATALYSIS

Kabeer A. Shaikh,[a*] Vishal A. Patila and B. P. Bandgar [b*]

[a]Organic synthesis laboratory, P. G. Department of Chemistry, Sir Sayyed College of Arts, Commerce and Science. Aurangabad 431001, M.S., India.

[b]Medicinal Chemistry Research Laboratory, Solapur University, Solapur-413255, M.S., India.

## ABSTRACT

Highly rapid and simple methodology has been developed for the quantitative synthesis of meso-substituted dipyrromethanes from lowest pyrrole/aldehyde ratio. The method was carried out by using $SnCl_2 \cdot 2H_2O$ as a catalyst under solvent free condition. The method is environmentally friendly, easy to workup, and gives excellent yield of the products

## INTRODUCTION

Dipyrromethanes are important building blocks for the synthesis of porphyrins [1], Calixpyrrols [2], and Corroles [3]. Dipyrromethanes are compounds known for more than a century [4]. In the past decades, a variety of conditions have been established for the synthesis of dipyrromethanes in the presence of various catalysts such as p-toluenesulfonic acid [5], $TiCl_4$ [6], $CF_3COOH$ [7] and pyrrolidinium tetrafluoroborate [8]. In the synthesis of dipyrromethanes most of the conditions are based on the acid catalyzed condensation of pyrrole with aldehyde. Recently, several methods have been developed, for the synthesis of dipyrromethanes in various catalyst such as ionic liquid [Hmim] BF4 [9], HCl/water [10], cation exchange resin [11], metal triflate catalysis [12], HCl [13], iodine/$CH_2Cl_2$ [14] and $InCl_3$ [15]. However, all of the synthetic protocols

reported so far suffer from disadvantages such as, use of metal [12], expensive reagent [11], prolonged reaction time[13], use of organic solvent [14], harsh reaction condition [13], use of excess pyrrole [12] and low yield [9]. Because of that, the researcher still continuous to have a better methodology for the synthesis of dipyrromethanes in terms of simplicity, eco-friendly, economic viability, high yielding at lowest pyrrole/aldehyde ratio which is achieved by using stannous chloride dehydrate.

In recent years, $SnCl_2 \cdot 2H_2O$ is frequently used in organic synthesis [16] as a catalyst due to its properties such as nontoxic nature, easy availability, inexpensiveness and easiness for work up. It played a great role for the synthesis of biologically active heterocycles such as benzimidazoles [17], quinoxalines [18] and functionalization of 4,5-diaminopyrazoles [19].

## MATERIAL AND METHODS

Purity of the compounds was checked by thin layer chromatography (TLC) on Merck silica gel 60 F254 pre-coated sheets. Melting points of the synthesized compounds were determined in open-glass capillaries on a stuart-SMP10 melting point apparatus. IR absorption spectra were recorded on a Perkin Elmer 1650 FTIR using KBr pellets in the range of 4,000-450 $cm^{-1}$. $^1$H-NMRs were recorded on a Bruker spectrometer operating at 400 MHz. The $^1$H-NMR chemical shifts are reported as parts per million (ppm) downfield from TMS ($Me_4Si$) used as an internal standard. Mass spectra were recorded on LCQ ion trap mass spectrometer. All compounds were known, and obtained physical and spectroscopic data were compared with literatures data.

## General procedure

A mixture of pyrrole (2 mmol), aldehyde (1 mmol) and $SnCl_2 \cdot 2H_2O$ (0.2 mmol) was crushed in a mortar with a pestle at room temperature. Progress of reaction was monitored by TLC. After completion of reaction (< 1 min) the crude product was washed with water, dried and purified by column chromatography using silica gel with petroleum ether/chloroform as the eluent. Pure products were obtained as solids.

*5-(4-nitrophenyl)dipyrromethane*: Yellow powder; mp: 159–160°C, IR (KBr) 3394, 3360, 3100, 1597, 1516, 1348, 1120, 1025, 807, 737, 660, 565 $cm^{-1}$; $^1H$ NMR(400 MHz, $CDCl_3$): d 5.58 (s, 1H, mesoH), 5.85 (br s, 2H, 2C3–H), 6.16 (dd, J = 2.8, 5.7, 2H, 2C4–H), 6.74 (dd, J = 2.8, 4.2, 2H, 2C5–H), 7.36 (d, J = 8.6, 2H, H-Ar), 8.0 (br s, 2H, N–H), 8.15 (d, J = 8.8, 2H, Ar-H).; MS (ES) Found [Calcd.]: m/z 267.30 [267.28] ($MH^+$).

*5-(2-Nitrophenyl) dipyrromethane* **(1)**: Brown oily liquid; IR (KBr) 3402, 1602, 1534, 1340, 1134, 1029, 815, 730, 661, 567 $cm^{-1}$; $^1H$ NMR (400 MHz, $CDCl_3$): δ 5.87 (s, 1H, mesoH), 6.11–6.24 (m, 4H), 6.65–6.70 (m, 2H), 7.24–7.55 (m, 3H), 7.87–7.90 (m, 1H), 8.18 (brs, 2H, NH); MS (ES) Found [Calcd.]: m/z 267.26 [267.29] ($MH^+$).

*5-(4-Fluorophenyl)dipyrromethane* **(2)**: brown crystals; mp: 80–81°C, IR (KBr) 3416, 2930, 1608, 1496, 1450, 1287, 1174, 1100, 964, 764, 558 $cm^{-1}$; $^1H$ NMR (400 MHz, $CDCl_3$): d 5.47 (s, 1H, mesoH), 5.88 (br s, 2H, 2C3-H), 6.17 (dd, J 2.7, 5.8, 2H, 2C4-H), 6.67 (br s, 2H, 2C5-H), 7.02–7.07 (m, 2H, Ar-H), 7.18–7.24 (m, 2H, Ar-H), 7.80 (br s, 2H, 2N-H); MS (ES) Found [Calcd.]: m/z 240.15 [240.27] ($MH^+$).

*5-Phenyldipyrromethane* **(3)**: Pale yellow crystals; mp: 100°C, IR (KBr) 3448, 2950, 1634, 1512, 1412, 1291, 1227, 1048, 761, 704, 607 $cm^{-1}$; $^1H$ NMR (400 MHz, $CDCl_3$): d 5.47 (s, 1H, mesoH), 5.94 (br s, 2H, 2C3-H), 6.14 (dd, J 2.8, 5.8, 2H, 2C4-H), 6.67 (dd, J 2.6, 4.2, 2H, 2C5-H), 7.20–7.37 (m, 5H, Ar-H), 7.87 (br s, 2H, 2N-H); MS (ES) Found [Calcd.]: m/z 222.31 [222.28] ($MH^+$).

*5-(4-Methoxyphenyl)dipyrromethane* **(4)**: Pale yellow powder; mp: 98–99°C, IR (KBr) 3407, 2965, 2934, 1615, 1507, 1455, 1298, 1246, 1174, 1105, 1025, 964, 837, 775, 722, 554 $cm^{-1}$; $^1H$ NMR (400 MHz, $CDCl_3$): d 3.80 (s, 3H, OCH3), 5.44 (s, 1H, mesoH), 5.90–5.93 (m, 2H, 2C3-H), 6.16 (dd, J 2.8, 5.7, 2H, 2C4-H), 6.64–6.68 (m, 2H, 2C5-H), 6.87 (d, J

8.4, 2H, Ar-H), 7.16 (d, J 8.4, 2H, Ar-H), 7.80 (br s, 2H, 2N-H); MS (ES) Found [Calcd.]: m/z 252.33 [252.31] ($MH^+$).

*5-(2,6-Dichlorophenyl)dipyrromethane* **(5):** Yellow solid. mp 102–103°C, IR (KBr) 3412, 2925, 1612, 1495, 1447, 1282, 1175, 1101, 792, 741, 529 cm_1; 1H NMR (400 MHz, $CDCl_3$): δ 6.08 (s, 1H, mesoH); 6.15 (dd, J 2.4, 5.4 Hz, 2H); 6.49 (s, 1H); 6.70–6.71 (m, 2H), 7.14 (t, J 8.0 Hz, 2H), 7.30 (d, J 8.0 Hz, 2H), 8.29 (bs, 2H, NH); MS (ES) Found [Calcd.]: m/z 291.21 [291.17] ($MH^+$) 293.15 ($MH^{+2}$)

*5-(4-Methylphenyl)dipyrromethane* ***(6):*** Pale yellow crystals; mp: 110°C, IR (KBr) 3415, 2356, 1634, 1509, 1425, 1253, 1090, 1024, 964, 908, 791, 745, 506 cm_1; 1H NMR (400 MHz, $CDCl_3$): d 2.35 (s, 3H, CH3), 5.45 (s, 1H, mesoH), 5.90 (br s, 2H, 2C3-H), 6.15 (dd, J 2.7, 5.8, 2H, 2C4-H), 6.64 (dd, J 2.5, 4.0, 2H, 2C5-H), 7.10–7.14 (m, 4H, Ar-H), 7.84 (br s, 2H, 2N-H); MS (ES) Found [Calcd.]: m/z 236.28 [236.31] ($MH^+$).

*5-(4-Chlorophenyl)dipyrromethane* **(7):** Pale yellow powder; mp: 112–114°C, IR (KBr) 3380, 2960, 2922, 2860, 1642, 1487, 1405, 1250, 1085, 1018, 765, 722, 554, 507 $cm^{-1}$; $^1H$ NMR (400 MHz, $CDCl_3$): d 5.40 (s, 1H, mesoH), 5.88 (br s, 2H, 2C3-H), 6.15 (dd, J 2.8, 5.6, 2H, 2C4-H), 6.67 (dd, J 2.7, 4.2, 2H, 2C5-H), 7.14 (d, J 8.1, 2H, Ar-H), 7.30 (d, J 8.1, 2H, Ar-H), 7.82 (br s, 2H, 2N-H); MS (ES) Found [Calcd.]: m/z 256.75 [256.73] ($MH^+$), 257.74 ($MH^{+1}$).

(KBr) 3372, 3096, 2958, 2922, 2860, 1705, 1484, 1404, 1081, 1020, 767, 721, 645, 545, 502 $cm^{-1}$; $^1H$ NMR (400 MHz, $CDCl^3$): d 5.40 (s, 1H, mesoH), 5.92 (br s, 2H, 2C3- H), 6.15 (dd, J 2.7, 5.6, 2H, 2C4-H), 6.66 (dd, J 2.5, 4.2, 2H, 2C5-H), 7.12 (d, J 8.2, 2H, Ar-H), 7.45 (d, J 8.2, 2H, Ar-H), 7.81 (br s, 2H, 2N-H); MS (ES) Found [Calcd.]: m/z 301.15 [301.18] ($MH^+$).

## RESULTS AND DISCUSSION

We began our study by grinding the mixture of 2 mmol pyrrole and 1 mmol 4- nitrobenzaldehyde (Scheme 1) under the reaction conditions described in Table 1.

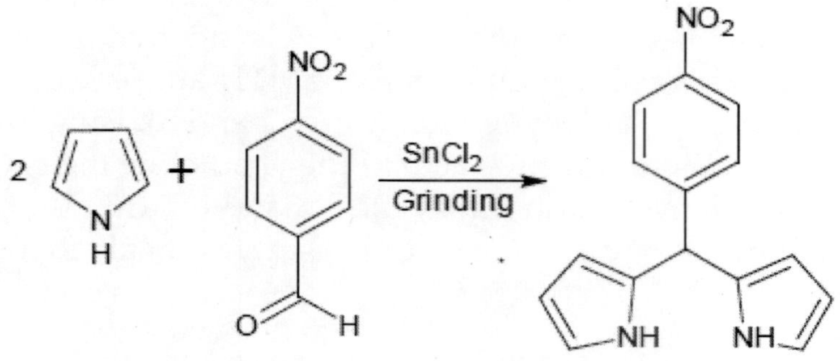

**Scheme 1**. Synthesis of 5-(4-nitrophenyl)dipyrromethane.

Initially, the mixture was ground in mortar with a pestle at room temperature under neat condition. However, result demonstrated the need of catalyst since the starting material was recovered (Entry 1). Thus, we chose 0.1 mmol $SnCl_2 \cdot 2H_2O$ as a catalyst. The result demonstrated that stoichiometric use of $SnCl_2 \cdot 2H_2O$ gives moderate yield of the product (Entry 2). The excellent yield was obtained at 0.2 mmol of $SnCl_2 \cdot 2H_2O$ (Entry 3) which is greater than that of 0.3 mmol of $SnCl_2 \cdot 2H_2O$ as a catalyst (entry 4). For evaluating the amount of catalyst, $SnCl_2 \cdot 2H_2O$ was employed in 0.4 and 0.5 mmol, however result demonstrated that moderate yield of product (Entry 5 and 6). Thus, from all above data, we confirmed that reaction gives excellent yield at 0.2 mmol of $SnCl_2 \cdot 2H_2O$. This study also confirmed that the $SnCl_2 \cdot 2H_2O$ played a great role as a Lewis-acid catalyst because its absence did not give desired product (Entry 1).

**Table 1**. Optimization for synthesis of 5-(4- nitrophenyl)dipyrromethane

| **Entry** | **$SnCl_2$ 2O (mmol)** | **Time (min)** | **Yielda (%)** |
|---|---|---|---|
| **1** | 0.0 | 20 | 00b |
| **2** | 0.1 | 2 | 62 |
| **3** | 0.2 | < 1 | 98 |
| **4** | 0.3 | < 1 | 84 |
| **5** | 0.4 | < 1 | 78 |
| **6** | 0.5 | < 1 | 78 |

[a]Isolated yield of the products. [b]The starting material was recovered.

Another interesting result in this article was found that we got excellent yield at lowest pyrrole/aldehyde ratio i. e. 2:1. This result was compared with the literatures best pyrrole/aldehyde ratio (Table 2). In these Littler et al [7c] and Naik et al [11] afforded moderate yield at high pyrrole/aldehyde ratio (Entry 2 and 3). Faugeras et al [14] gave good yield of the product but they had done their work at high pyrrole/aldehyde ratio (Entry 4). Rohand et al13 also carried out good job but they did not try at lowest pyrrole/aldehyde ratio (Entry 1).

**Table 2**. Comparison of the pyrrole/aldehyde ratio with literature

| **Entry** | **Catalyst** | **Pyrrole/aldehyde ratio** | **Yielda (%)** |
|---|---|---|---|
| **1** | HCl | 3:1 | 97 [13] |
| **2** | TFA | 25:1 | 56 [7c] |
| **3** | Cation exchange resin (T-63) | 20:1 | 61 [11] |
| **4** | I2, CH2Cl2 | 10:1 | 84 [14] |
| **5** | SnCl2 2H2O | 2:1 | 98 |

[a]Isolated yield of the 5-(4-nitrophenyl)dipyrromethane.

Thus, from above data, it was cleared that none of the above best literatures can be compared with our best results i. e. use of low pyrrol/aldehyde ratio and excellent yield of the product.

In order to confirm these interesting results, we applied this method to the synthesis of various *meso*-substituted dipyrromethanes (Scheme 2) and obtained results were compared with literature best methods (Table 3).

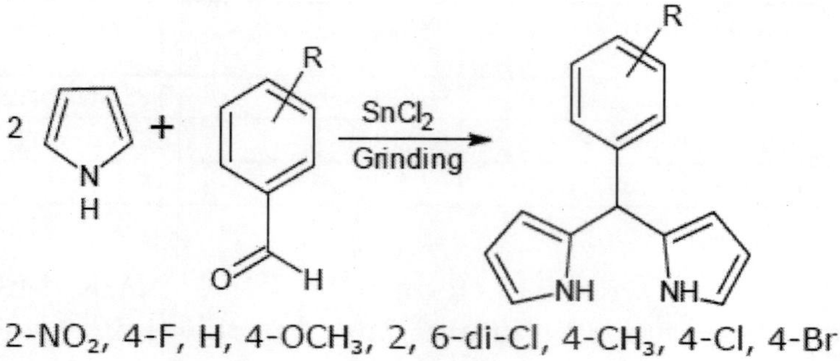

R = 2-$NO_2$, 4-F, H, 4-$OCH_3$, 2, 6-di-Cl, 4-$CH_3$, 4-Cl, 4-Br

**Scheme 2.** Synthesis of various *meso*-substituted dipyrromethanes.

Table 3 cleared that there was no influence of the electronic nature of the substituent on the reaction time or yield. Thus, it is possible to affirm that aldehydes containing electron donating or withdrawing groups reacted well in very short time (< 1 min) and gave corresponding dipyrromethanes in excellent yield (92-97%) as compared to best method found in literatures. In the case of 5-(4-metoxyphenyl)dipyrromethane, Faugeras et al. [14] gave excellent yield (90 %) in short reaction time but they performed their work by using dichloromethane as a solvent and by using microwave oven, which is always harmful to the environment and have economically higher cost.

In the another case of 5-phenyldipyrromethane and 5-(2,6-dichlorophenyl)dipyrromethane, Rohand et al. [13] gave good yields (86 and 85 %) in prolonged reaction time but they performed their work by using large quantity of strong acid which is always hazardous to the environment.

**Table 3**: Comparison of the yields with best methods found in the literatures

| **Entry** | **Aldehyde** | **Isolated Yielda (%) /Pyrrole:aldehyde ratio** | **Literature best yield (%)/Pyrrole:aldehyde ratio** |
|---|---|---|---|
| 1 | 2-nitrobenzaldehyde | 94/2:1 | 73/20:1 [11] |
| 2 | 4-fluorobenzaldehyde | 95/2:1 | 77/20:1 [11] |
| 3 | Benzaldehyde | 97/2:1 | 86/03:1 [13] |

| | | | |
|---|---|---|---|
| 4 | 4-methoxybenzaldehyde | 94/2:1 | 90/10:1 [14] |
| 5 | 2,6-Dichlorobenzaldehyde | 95/2:1 | 85/03:1 [13] |
| 6 | 4-methylbenzaldehyde | 95/2:1 | 83/40:1* [12] |
| 7 | 4-chlorobenzaldehyde | 97/2:1 | 77/20:1 [11] |
| 8 | 4-bromobenzaldehyde | 92/2:1 | 74/40:1* [12] |

[a]products were characterized by IR, $^{1}$H NMR, MS (ES) and coincided with literature data, *pyrrole/imine ratio instead ofpyrrole/aldehyde.

Thus, from all above results and discussion it was cleared that this method is superior in terms of use of inexpensive catalyst, solvent free reaction, lowest pyrrole/aldehyde ratio, very short reaction time and excellent yield of the products.

## CONCLUSION

In conclusion, we have discovered highly rapid and simple method for the quantitative synthesis of *meso*-substituted dipyrromethanes at lowest pyrrole/aldehyde ratio and by using $SnCl_2 \cdot 2H_2O$ as a catalyst. The use of this nontoxic, easily available and inexpensiveness catalyst make this protocol practical and economically attractive.

## ACKNOWLEDGMENTS

We would like to thank DST, New Delhi for financial assistance and Prof. Mohammed Tilawat Ali for providing necessary facilities for research work.

## REFERENCES AND NOTES

1. Temelli, B.; Unaleroglu, C. *Tetrahedron* **2009,** *65,* 2043. [CrossRef]
2. Gale, P. A.; Anzenbacher, P. Jr.; Sessler, J. L. *Coord. Chem. Rev.* **2001**, *222,* 57. [CrossRef]
3. Zhan, H.-Y.; Liu, H.-Y.; Chen, H.-J. *Tetrahedron Lett.* **2009**, *50,* 2196. [CrossRef]
4. Baeyer, A.; *Ber. Dtsch. Chem. Ges.* **1986**, *19,* 2184. [CrossRef]
5. Boyle, R. W.; Xie, L. Y.; Dolphin, D. *Tetrahedron Lett.* **1994,** *35,* 5377.

[CrossRef]. (b) Mizutani, T.; Ema, T.; Tomita, T. *J. Am. Chem. Soc.* **1994**, *116*, 4240. [CrossRef]

6. Setsune, J.-I.; Hashimoto, M.; Shiozawa, K. *Tetrahedron* **1998**, *54*, 1407. [CrossRef]
7. Srinivasan, A.; Sridevi, B.; Reddy, M. V. R. *Tetrahedron Lett.* **1997**, *38*, 4149. [CrossRef]. (b) Lee, C. H.; Lindsay, J. S. *Tetrahedron* **1994**, *50*, 11427. [CrossRef]. (c) Littler, B. J.; Miller, M. A.; Hung, C.-H. *J. Org. Chem.* **1999**, *64*, 1391. [CrossRef]
8. Biaggi, C.; Benaglia, M.; Raimondi, L. *Tetrahedron* **2006**, *62*, 12375. [CrossRef]
9. Gao, G. H.; Lu, L.; Gao, J. B. *Chinese Chemical Lett.* **2005**, *16*, 900.
10. Sobral, A. J. F. N.; Rebanda, N. G. C. L.; Silva, M. D. *Tetrahedron Lett.* **2003**, *44*, 3971. [CrossRef]
11. Naik, R.; Joshi, P.; Kaiwar, S. P.; *Tetrahedron* **2003**, *59*, 2207. [CrossRef]
12. Tamelli, B.; Unaleroglu, C. *Tetrhedron* **2006**, *62*, 10130. [CrossRef]
13. Rohand, T.; Dolusic, E.; Ngo, T. H. Maes, W.; Dehaen, W. *Arkivok* **2007**, *x*, 307.
14. Faugeras, P.-A.; Boëns, B.; Elchinger, P.-H. *Tetrahedron Lett.* **2010**, *51*, 4630. [CrossRef]
15. Laha, J. K.; Dhanalekshami, S.; Taniguchi, M. *Org. Process Res. Dev.* **2003**, *7*, 799. [CrossRef]
16. Rai, G.; Jeong, J. M.; Lee, Y. S. *Tetrahedron Lett.* **2005**, *46*, 3987. [CrossRef]
17. Wu, Z.; Philip, R.; Wickham, G. *Tetrahedron Lett.* **2000**, *41*, 9871. [CrossRef]
18. Shi, D. Q.; Dou, G. L.; Ni, S. N. *Heterocyclic Chem.* **2008**, *45*, 1797. [CrossRef]
19. Blass, B. E.; Srivastava, A.; Coburn, K. R. *Tetrahedron Lett.* **2003**, *44*, 3009. [CrossRef]

# Chapter 12

# TOTAL SYNTHESIS OF SIX 3, 4-UNSUBSTITUTED COUMARINS

Wenqing Gao[1], Qingyong Li [1, 2,*], Jian Chen[1], Zhichao Wang[1] and Changlong Hua[1]

[1]Key Laboratory of Forest Plant Ecology, Ministry of Education, Northeast Forestry University, Harbin 150040, China

[2]College of Pharmaceutical Science, Zhejiang University of Technology, Hangzhou 310014, China

## ABSTRACT

In this article we describe a new methodology for the total synthesis of 3,4-unsubstituted coumarins from commercially available starting materials. Six examples were prepared, including five naturally occurring coumarins—7-hydroxy-6,8-dimethoxy-coumarin (isofraxidin), 7-hydroxy-6-methoxycoumarin (scopoletin), 6,7,8-trimethoxy-coumarin, 6,7-dimethoxycoumarin (scoparone), and 7,8-dihydroxycoumarin (daphnetin) and one synthetic coumarin, 7-hydroxy-6-ethoxycoumarin. Moreover, five important o-hydroxybenzaldehyde intermediates were also obtained, namely 2,4-dihydroxy-3,5-dimethoxybenzaldehyde, 2,4-dihydroxy-5-methoxybenzaldehyde, 5-ethoxy-2,4-dihydroxy-benzaldehyde, 2-hydroxy-3,4,5-trimethoxybenzaldehyde, and 2-hydroxy-4,5-

dimethoxy-benzaldehyde. The method developed herein involves just three or four steps and allows for the rapid synthesis of these important molecules in excellent yields. This is the first synthesis of 6, 7, 8-trimethoxycoumarin and 7-hydroxy-6-ethoxycoumarin.

## INTRODUCTION

Coumarins (2H-1-chromene-2-ones) are abundant in Nature and are common motifs found in drugs, dyes, spices, and agricultural chemicals. In particular, 3, 4-unsubstituted coumarins are the most common naturally occurring coumarins, which show potential antimalarial [1], antioxidant [2–6], antimicrobial [7], anti-inflammatory [8], and antitumor [9–11] activity, however, their abundance in plants is very low and the purification processes are complex.

Therefore, various methods have been developed for the total synthesis of coumarins [11–22], including the Perkin [23], Pechmann [24], and Knoevenagel reactions [25]. These reactions mostly lead to coumarins with substituents at the 3- or 4-position.

A one-pot Wittig reaction/cyclization has been adopted by de Kimpe for the synthesis of 3, 4-unsubstituted coumarins [26]. Herein, we report the highly efficient total synthesis of five naturally occurring coumarins, as well as a synthetic 3,4-unsubstituted coumarin (Figure 1). To the best of our knowledge, this is the first report on the total synthesis of coumarins 3 and 4.

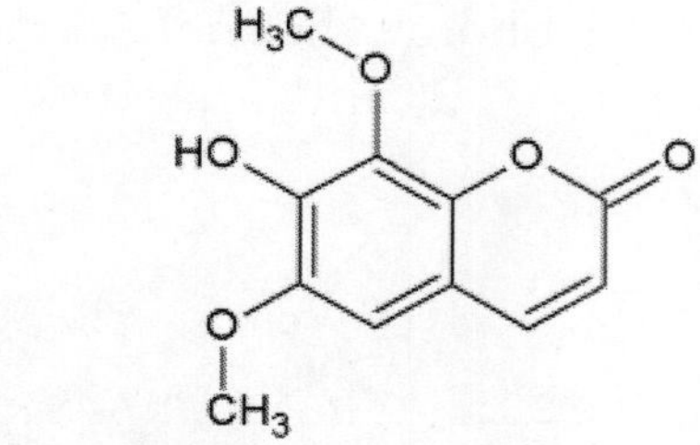

**Isofraxidin (1)**

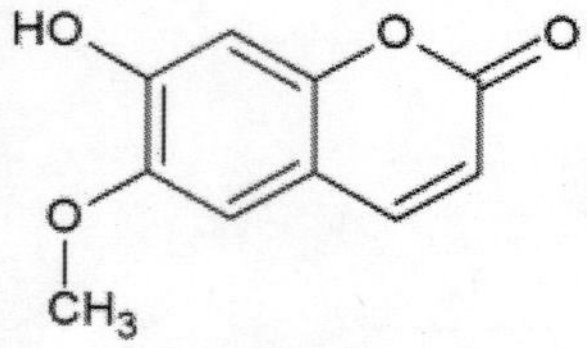

**Scopoletin (2)**

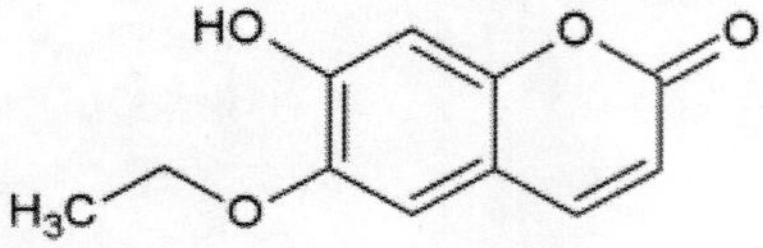

**7-Hydroxy-6-ethyloxycouamrin (3)**

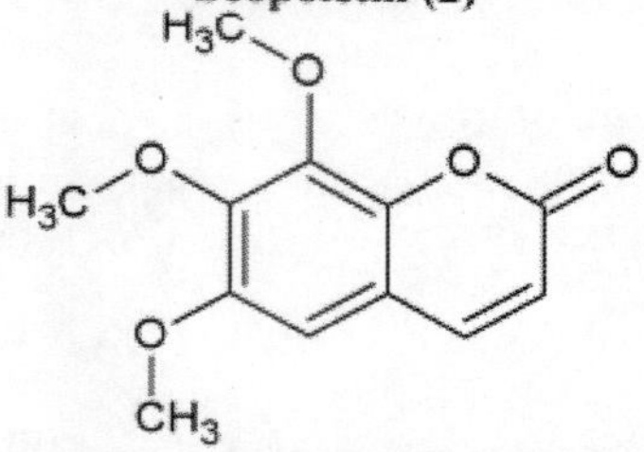

**6,7,8-Trimethoxycoumarin (4)**

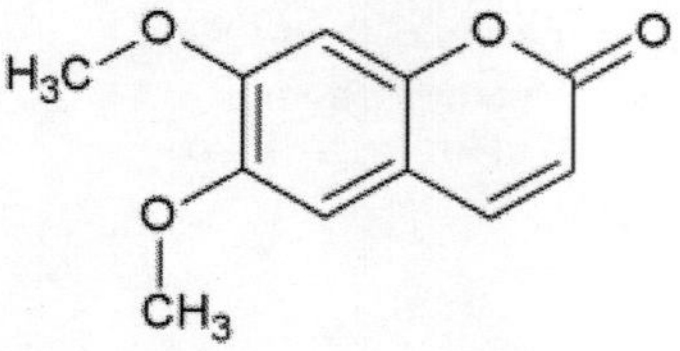

**Scoparone (5)**

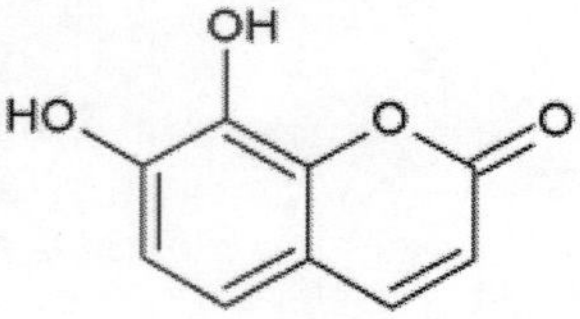

**Daphnetin (6)**

**Figure 1.** The structure of the synthesized coumarins.

## RESULTS AND DISCUSSION

Coumarins 1, 2, and 3 were synthesized from simple starting materials (Scheme 1).

The hydroxyl group was protected using pivaloyl chloride [27] to give the corresponding products in 100% yield. These compounds were then iodinated in 80%, 77%, 80% yields, respectively, using N-iodo-succinimide [28, 29], followed by hydrolysis using cuprous oxide, pyridine-2-aldoxime, tetrabutylammonium bromide, and cesium hydroxide [30]. The resulting o-hydroxybenzaldehydes were finally reacted with ethyl (triphenylphosphoranylidene) acetate in N, N-diethyl aniline, forming coumarins 1, 2, and 3 as described above [26].

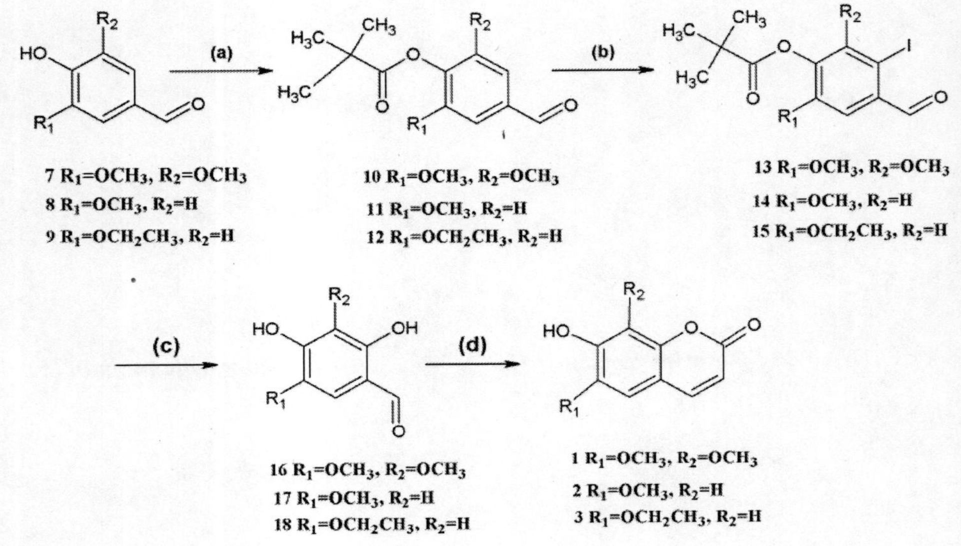

7 $R_1=OCH_3$, $R_2=OCH_3$
8 $R_1=OCH_3$, $R_2=H$
9 $R_1=OCH_2CH_3$, $R_2=H$

10 $R_1=OCH_3$, $R_2=OCH_3$
11 $R_1=OCH_3$, $R_2=H$
12 $R_1=OCH_2CH_3$, $R_2=H$

13 $R_1=OCH_3$, $R_2=OCH_3$
14 $R_1=OCH_3$, $R_2=H$
15 $R_1=OCH_2CH_3$, $R_2=H$

16 $R_1=OCH_3$, $R_2=OCH_3$
17 $R_1=OCH_3$, $R_2=H$
18 $R_1=OCH_2CH_3$, $R_2=H$

1 $R_1=OCH_3$, $R_2=OCH_3$
2 $R_1=OCH_3$, $R_2=H$
3 $R_1=OCH_2CH_3$, $R_2=H$

**Scheme 1**. Syntheses of coumarins 1, 2, and 3.

*Reagents and conditions*: (a) $(CH_3)_3COCl$, DMAP, $Et_3N$, $CH_2Cl_2$; (b) **10** NIS, $CF_3COOH$, $CH_3CN$, reflux;

**11, 12** *N*-iodosuccinimide (NIS), $CF_3SO_3H$, MeOH; (c) CsOH, *syn*-2-pyridinealdoxime, $Cu_2O$, $Bu_4NBr$, $N_2$, $H_2O$; (d) $(C_6H_5)_3P=CHCOOCH_2CH_3$, $Et_2NPh$, $N_2$, reflux.

It was envisaged that this method could be applied to the synthesis of 6, 7-dimethoxy-8- hydroxycoumarin (**34**) and 5, 6, 7, 8-tetrahydroxycoumarin (**35**) (Scheme 2).

(a) (b)

**26** $R_1$=H, $R_2$=OH
**27** $R_1$=OH, $R_2$=$OCH_3$

**28** $R_1$=H, $R_2$=$(CH_3)_3COO$
**29** $R_1$=$(CH_3)_3COO$, $R_2$=$OCH_3$

**30** $R_1$=H, $R_2$=$(CH_3)_3COO$
**31** $R_1$=$(CH_3)_3COO$, $R_2$=$OCH_3$

(c)

**32** $R_1$=H, $R_2$=OH
**33** $R_1$=OH, $R_2$=$OCH_3$

**34** $R_1$=H, $R_2$=OH
**35** $R_1$=OH, $R_2$=$OCH_3$

**Scheme 2.** Attempted syntheses of coumarins **34**, **35**.

*Reagents and conditions*: (a) **26** $(CH_3)_3COCl$, DMAP, $Et_3N$, $CH_2Cl_2$; **27** $MeSO_4$, $K_2CO_3$, $CH_3CN$; (b) **28** NIS,

$CF_3COOH$, $CH_3CN$, reflux; **29** NIS, $CF_3SO_3H$, MeOH; (c) CsOH, *syn*-2-pyridinealdoxime, $Cu_2O$, $Bu_4NBr$,

$N_2$, $H_2O$.

Protection of the hydroxyl groups using pivaloyl chloride and iodination of the resulting compounds proceeded as expected, but hydrolysis of the iodo-compounds **30** and **31** failed to afford the desired products **32** and **33**. It was presumed that both electronic effects and steric hindrance of the three methoxy groups and the hydroxyl group at the *ortho* position of **30** and **31** affected the success of these reactions. Synthesis of coumarins **4** and **5** (Scheme 3) began with the iodination of 3, 4, 5- trimethoxybenzaldehyde (**19**) and veratraldehyde (**20**), respectively, both in 99% yield. Hydrolysis of 2-iodo-3, 4, 5-trimethoxybenzaldehyde (**21**) and 2-iodo-4, 5 dimethoxybenzaldehyde (**22**) afforded 2-hydroxy-3, 4, 5-trimethoxybenzaldehyde (**23**) and 2-hydroxy-4, 5-dimethoxybenzaldehyde (**24**) in 83% and 85% yield, respectively. Finally, the *o*-hydroxybenzaldehydes **23** and **24** were converted to the corresponding coumarins **4** and **5**. The overall yield of **4** and **5** was higher compared to that of compounds **1–3**.

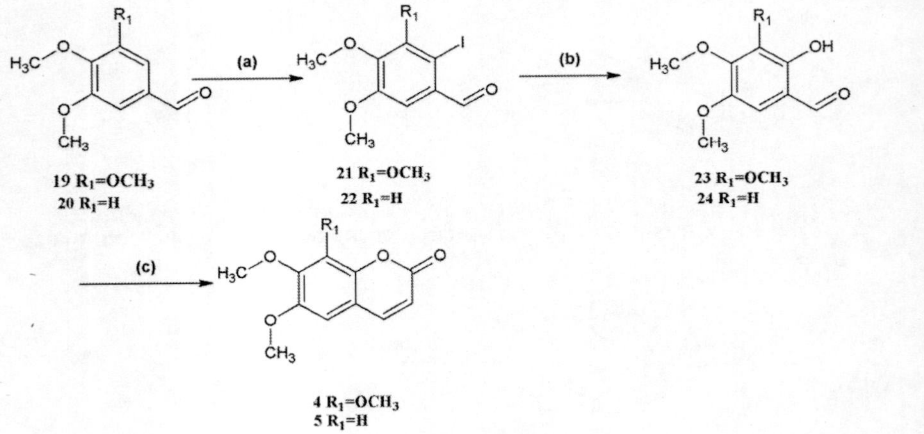

19 $R_1$=$OCH_3$
20 $R_1$=H

21 $R_1$=$OCH_3$
22 $R_1$=H

23 $R_1$=$OCH_3$
24 $R_1$=H

4 $R_1$=$OCH_3$
5 $R_1$=H

**Scheme 3.** Syntheses of coumarins 4 and 5.

Reagents and conditions: (a) 19 NIS, $CF_3COOH$, $CH_3CN$, reflux; 20 NIS, $CF_3SO_3H$, MeOH; (b) CsOH, syn-2-pyridinealdoxime, $Cu_2O$, $Bu_4NBr$, $N_2$, $H_2O$; (c) $(C_6H_5)_3P$=$CHCOOCH_2CH_3$, $Et_2NPh$, $N_2$, reflux.

Finally, daphnetin (6) was synthesized using the one-pot Wittig/cyclization reaction from commercially available 2, 3, 4-trihydroxybenzaldhyde in 65% yield (Scheme 4).

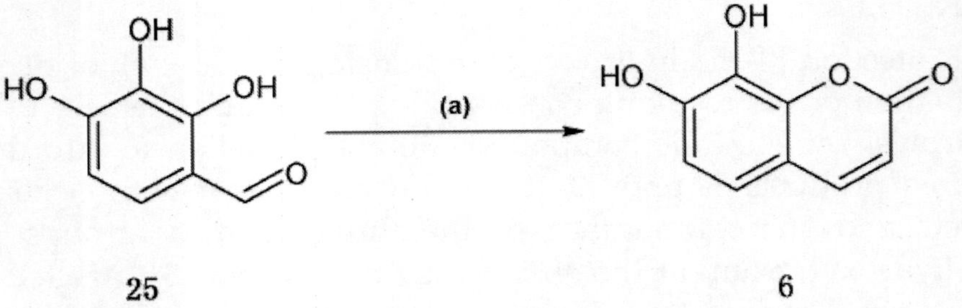

25

6

**Scheme 4.** Synthetic procedures of coumarin 6.

*Reagents and conditions*: (a) $(C_6H_5)_3P$=$CHCOOCH_2CH_3$, $Et_2NPh$, $N_2$, reflux.

This result for daphnetin (**6**) was identical compared to the previously reported synthesis [26]. Similarly for scopoletin (**2**), the overall yield of 50.4% (for four steps) is comparable to that of the previously reported synthesis (52.8%) [24]. However, for isofraxidin

(**1**) and scoparone (**5**), the yields of 57.6% and 77.4% are significantly higher compared to those obtained previously [23, 25].

The key step in this novel method is the hydrolysis of the iodinated compound, as substituents on the phenyl group have a great influence on this reaction. Activating groups can reduce the probability of hydrolysis and steric hindrance at the *ortho* position also influences the reaction.

## EXPERIMENTAL

All solvents and commercially available reagents were purchased from the suppliers and used without further purification. 1H-NMR and 13C-NMR spectra were recorded using Bruker DPX 500 (Bruker, Billerica, MA, USA) and 300 spectrometers, respectively. Spectra were recorded in $CDCl_3$ and DMSO solutions and chemical shifts are reported in parts per million (ppm) relative to tetramethylsilane (TMS) as the standard. IR spectra were recorded on an Infinity Spectrum One spectrophotometer (Shimadzu, Kyoto, Japan). Mass spectra (Applied Biosystems, Toronto, Canada) were recorded on an Agilent 1100 Series VS (ES, 4000 V) mass spectrometer. Melting points were measured using a Büchi B-450 apparatus (Shanghaishenguang, Shanghai, China). Flash chromatography was performed with ACROS silica gel (particle size 0.030–0.040 mm, pore diameter ca. 6 nm) using a glass column.

### *General Procedure for the Synthesis of Coumarins* 1–6.

The appropriate *o*-hydroxybenzaldehyde (1 mmol) and ethyl (triphenylphosphoranylidene) acetate (1.2 mmol) were dissolved in *N, N*-diethyl aniline (1.5 mL) and the resulting mixture was stirred under a $N_2$ atmosphere and reflux for 15 min. The solvent was removed under reduced pressure (1 mmHg, 52 °C) and the resulting brown oil was purified by column chromatography (petroleum ether-ethyl acetate, 3:1).

#### *7-Hydroxy-6, 8-dimethoxy-2H-chromen-2-one (Isofraxidin,* **1***).*

Yellow solid. Yield: 80%. mp (°C): 147–148.5 (lit. 148–149) [13].

$^1$H-NMR (500 MHz, $CDCl_3$): 3.95 (s, 3H, $OCH_3$), 4.10 (s, 3H, $OCH_3$), 6.16 (s, 1H, Ar), 6.29 (d, $J$ = 9.5 Hz, 1H, 3-CH), 6.66 (s, 1H, OH), 7.60 (d, $J$ = 9.5 Hz, 1H, 4-CH); $^{13}$C-NMR (75 MHz, $CDCl_3$) 56.6 ($OCH_3$), 61.2 (OCH3), 105.0 (C-5), 110.7 (C-4a), 112.5 (C 3), 135.2 (C-8), 143.5 (C-7), 144.5 (C-8a), 145.3 (C-4), 146.1 (C-6), 160.6 (C-2); IR (KBr, $cm^{-1}$) 1699 (C=O), 1570 (C=C), 3327 (br, OH); HRMS (EI): *m/z* 223 [$M+H^+$].

*7-Hydroxy-6-methoxy-2H-chromen-2-one* **(***Scopoletin*, **2)**. ***Yellow solid. Yield: 85%. mp*** *(°C):*

202.7–204.4 (lit. 204) [15]. $^1$H-NMR (500 MHz, $CDCl_3$): 3.82 (s, 3H, $OCH_3$), 6.23 (d, $J$ = 9.5 Hz, 1H, 3-CH), 6.79 (s, 1H, Ar), 7.23 (s, 1H, Ar), 7.92 (d, $J$ = 9.5 Hz, 1H, 4-CH), 10.33 (s, 1H, OH); 13C-NMR (75 MHz, $CDCl_3$) 56.4 ($OCH_3$), 103.2 (C-8), 110.0 (C-5), 111.0 (C-3), 112.1 (C-4a), 144.9 (C-4), 145.7 (C-6), 150.0 (C-7), 151.6 (C-8a), 161.1 (C-2); IR (KBr, $cm^{-1}$) 1558 (C=C), 1683 (C=O), 3334 (br, OH). HRMS (EI): *m/z* 193 [$M+H^+$].

*7-Hydroxy-6-ethoxy-2H-chromen-2-one* **(3)**.

Yellow solid. Yield: 80%. mp (°C): 138.5–142.0. $^1$H-NMR (500 MHz, $CDCl_3$): 1.37 (t, $J$ = 7.0 Hz, 3H, $CH_3$), 4.06 (q, $J$ = 7.0 Hz, 2H, $OCH_2$-), 6.23 (d, $J$ 9.5 Hz, 1H, 3-CH), 6.81 (s, 1H, Ar), 7.90 (d, $J$ = 9.5 Hz, 1H, 4-CH), 10.27 (s, 1H, CHO);$^{13}$C-NMR (75 MHz, $CDCl_3$) 15.1 ($CH_3$), 64.8 (-$CH_2$O-), 103.2 (C-8), 111.0 (C-5), 111.2 (C-3), 112.1 (C-4a), 144.8 (C-4), 144.9 (C-6), 149.9 (C-7), 151.9 (C-8a), 161.1 (C-2); IR (KBr, $cm_{-1}$) 1558 (C=C), 1699 (C=O), 1716, 3383 (br, OH);.HRMS (EI): *m/z* 207 [$M+H^+$].

*6,7,8-Trimethoxy-2H-chromen-2-one* **(***Dimethylfraxetin*, **4)**.

Yellow solid. Yield: 81%. mp (°C): 103.8–104.5 (lit. 104–105) [31]. $_1$H-NMR (500 MHz, $CDCl_3$): 3.91 (s, 3H, $OCH_3$), 4.00 (s, 3H, $OCH_3$), 4.04 (s, 3H, $OCH_3$), 6.35 (d, $J$ = 9.5 Hz,1H, 3-CH), 6.69 (s, 1H, Ar), 7.63 (d, $J$ = 9.5 Hz, 1H, 4-CH); $^{13}$C-NMR (75 MHz, $CDCl_3$) 56.3 ($OCH_3$), 61.5 ($OCH_3$), 61.8 ($OCH_3$), 103.7 (C-5), 114.3 (C-4a), 115.2 (C-8), 141.2 (C-6), 143.1 (C-7), 143.5 (C-4), 145.9 (C-8), 150.1 (C-8a), 160.5 (C-2); IR (KBr, $cm_{-1}$) 1566 (C=C), 1716 (C=O). HRMS (EI): *m/z* 237 [$M+H^+$].

*6, 7-Dimethoxy-2H-chromen-2-one* **(***Scoparone,* **5)**.

Yellow solid. Yield: 91%. mp (°C): 142.6–148.4 (lit. 145–146) [18]. $_1$H-NMR (500 MHz, $CDCl_3$): 3.81 (s, 3H, $OCH_3$), 3.87 (s, 3H, $OCH_3$), 6.30 (d, *J* = 9.5 Hz, 1H, 3-CH), 7.07 (s, 1H, Ar), 7.25 (s, 1H, Ar), 7.955 (d, *J* = 9.5 Hz, 1H, 4-CH); $_{13}$C-NMR (75 MHz, $CDCl_3$) 56.3 ($OCH_3$), 56.6 ($OCH_3$), 100.5 (C-8) 109.3 (C-5), 111.6 (C-4a), 113.1 (C-3), 144.8 (C-6), 146.3 (C-4), 149.9 (C-8a), 153.0 (C-7), 161.0 (C-2); IR (KBr, cm$_{-1}$) 1516 (C=C), 1558, 1708 (C=O), 1616, 3566 (br, OH). HRMS (EI): *m/z* 207 [M+ $H^+$].

*7, 8-Dihydroxy-2H-chromen-2-one* **(***Daphnetin,* **6)**.

Yellow solid. Yield: 65%. mp (°C): 265.7–267.2 (lit. 265–268) [21].$^1$H-NMR (500 MHz, $CDCl_3$): 6.20 (d, *J* = 6.5 Hz, 1H, 3-CH), 6.81 (s, 1H, Ar), 7.03 (s, 1H, Ar), 7.91 (d, *J* = 7.0 Hz, 1H, 4-CH), 9.79 (d, *J* = 7.0 Hz, 1H, OH), 9.80 (d, *J* =7.0Hz, 1H, OH); $_{13}$C-NMR (75 MHz, $CDCl_3$) 111.7 (C-3), 112.5 (C-4a), 112.9 (C-6), 119.3 (C-5), 132.6 (C-8), 144.2 (C-8a), 145.6 (C-4), 150.1 (C-7), 160.9 (C-2); IR (KBr, cm$^{-1}$) 1674 (C=O), 1508 (C=C), 1589. HRMS (EI): *m/z* 179 [M+$H^+$].

### General Procedure for the Synthesis of Compounds 10–12

Thecorrespondingphenol(3mmol)wasdissolvedindichloromethane (5 mL) and 4-dimethylamiopyridine (0.1 mmol) was added. The reaction was stirred for 0.5 h before addition of pivaloyl chloride (6 mmol) followed by dropwise addition of triethylamine (6 mmol). The reaction mixture was stirred at room temperature for 2 h. The reaction solution was poured into dichloromethane (100 mL) and washed with saturated sodium chloride solution (2 × 100 mL) and saturated sodium bicarbonate solution (2 × 100 mL). The organic phase was collected, dried over anhydrous magnesium sulfate, and filtered, and then the solvent was removed *in vacuo* to obtain compounds **10–12**.

*4-Formyl-2, 6-dimethoxyphenyl-2,2-dimethylpropanoate* **(10, $C_{14}H_{18}O_5$)**.

Yellow solid. Yield 100%. mp (°C): 108–110.9. $_1$H-NMR ($CDCl_3$, 500 MHz): 1.39 (s, 9H, C $(CH_3)_3$), 3.88 (s, 6H, $OCH_3$), 7.14 (s, 2H, Ar), 9.91 (s, 1H, CHO); $_{13}$C-NMR (75 MHz, $CDCl_3$) 27.2 ($CH_3$), 29.7 ($CH_3$

*2,4-Dihydroxy-5-methoxybenzaldehyde* **(17, $C_8H_8O_4$)**.

Yellow solid. Yield: 77%. mp (°C): 152.1−153.2. $^1$H-NMR (500 MHz, $CDCl_3$): 3.92 (s, 3H, $OCH_3$), 6.41 (s, 1H, Ar), 6.53 (s, 1H, Ar), 6.89 (s, 1H, OH), 9.68 (s, 1H, OH), 11.34 (s, 1H, CHO); $_{13}$C-NMR (75 MHz, $CDCl_3$) 56.5 ($OCH_3$), 103.2 (C-3), 112.9 (C-1), 113.3 (C-6), 140.7 (C-5), 154.4 (C-4), 159.6 (C-2), 193.8 (CHO); IR (KBr, $cm^{-1}$) 3292 (br, OH), 1635 (C=O); MS (EI): *m/z* 169 [M+H$^+$].

*5-Ethoxy-2, 4-dihydroxybenzaldehyde* **(18, $C_9H_{10}O_4$)**.

Yellow solid. Yield: 77%. mp (°C): 131.8−134.4. $^1$H-NMR (500 MHz, $CDCl_3$): 1.47 (t, $J$ = 7.0 Hz, 3H, $CH_3$), 4.12 (q, $J$ = 7.0 Hz, 2H, $OCH_2$-), 6.49 (s, 1H, Ar), 6.53 (s, 1H, Ar), 6.88 (s, 1H, OH), 9.65 (s, 1H, OH), 11.33 (s, 1H, CHO); $^{13}$C-NMR (75 MHz, $CDCl_3$) 14.8 ($CH_3$), 65.3 (-$CH_2O$), 103.1 (C-3), 113.3 (C-1), 113.9 (C-6), 139.9 (C-5), 154.6 (C-4), 159.5 (C-2), 193.8 (CHO); IR (KBr, $cm_{-1}$) 1627 (C=O), 1699, 3367 (br, OH); MS (EI): *m/z* 183 [M+H$^+$].

*2-Hydroxy-3, 4, 5-trimethoxybenzaldehyde* **(23, $C_{10}H_{12}O_4$)**.

Yellow solid. Yield: 83%. mp (°C): 46.0–46.6. $^1$H-NMR (500 MHz, $CDCl_3$): 3.87 (s, 3H, $OCH_3$), 3.94 (s, 3H, $OCH_3$), 4.05 (s, 3H, $OCH_3$), 6.78 (s, 1H), 9.78 (s, 1H, OH), 11.01 (s, 1H, CHO); $_{13}$C-NMR (75 MHz, $CDCl_3$) 61.4 ($OCH_3$), 61.1 ($OCH_3$), 56.5 ($OCH_3$), 109.2 (C-6), 115.2 (C-1), 141.0 (C-3), 146.3 (C-4), 150.3 (C-2), 151.8 (C-5), 194.9 (CHO); IR (KBr, $cm_{-1}$) 1641 (C=O), 3240 (br, OH); MS (EI): *m/z* 213 [M+H$^+$].

*2-Hydroxy-4,5-dimethoxybenzaldehyde* **(24, $C_9H_{10}O_3$)**.

Yellow solid. Yield: 85%. mp (°C): 105.5–112.8. $^1$H-NMR (500 MHz,

$CDCl_3$): 3.88 (s, 3H, $OCH_3$), 3.94 (s, 3H, $OCH_3$), 6.48 (s, 1H, Ar), 6.91 (s, 1H, Ar), 9.71 (s, 1H, OH), 11.40 (s, 1H, CHO);$^{13}$C-NMR (75 MHz, $CDCl_3$) 56.4 ($OCH_3$), 56.4 ($OCH_3$), 100.1 (C-3), 112.9 (C-1), 113.2 (C-6), 143.0 (C-5), 157.2 (C-2), 159.4 (C-4), 194.0 (CHO); IR (KBr, $cm^{-1}$) 1624 (C=O), 3446 (br, OH); MS (EI): *m/z* 183 [$M+H^+$].

## CONCLUSIONS

A novel method for the total synthesis of six coumarins and five important intermediates has been developed. In most cases, the overall yields of these products are higher than those previously reported. Furthermore, this is the first reported total synthesis of 7-hydroxy-6-ethoxycoumarin (**3**) and 6, 7, 8-trimethoxycoumarin (**4**). All of these coumarins and advanced intermediates will be evaluated for their activity, and the results will be reported in due course.

## ACKNOWLEDGMENTS

This work was financially supported by Fundamental Research Funds for the Central Universities (DL12EA01) and the Heilongjiang Postdoctoral Scientific Research Fund (LBH-Q11182).

## REFERENCES

1. Cubukcu, B.; Bray, D.H.; Warhurst, D.C.; Mericli, A.H.; Ozhatay, N.; Sariyar, G. *In vitro* antimarlarial activity of crude extracts and compounds from *Artemisia abrotanum* L.*Phytother. Res.* **1990**, *4*, 203–204.
2. Huang, H.-Y.; Ko, H.-H.; Jin, Y.-J.; Yang, S.-Z.; Shih, Y.-A.; Chen, I.-S. Dihydrochalcone glucosides and antioxidant activity from the roots of *Anneslea fragrans* var. *lanceolata*. *Phytochemistry* **2012**, *78*, 120–125.
3. Shafiullah, K.; Naheed, R.; Nighat, A.; Abdul, M.; Aziz-ur-Rehmana; Lubna, I.; Mehreen, L. Antixoidant constituent from Cotoneaster racemiflora. *J. Asian Nat. Prod. Res.* **2009**, *11*, 44–48.
4. Shaw, C.-Y.; Chen, C.-H.; Hsu, C.-C.; Chen, C.-C.; Tsai, Y.-C. Antioxidant properties of scopoletin isolated from *Sinomonium acutum*. *Phytother. Res.* **2003**, *17*, 823–825.
5. Thuong, P.T.; Hung, T.M.; Ngoc, T.M.; Ha, D.T.; Min, B.S.; Kwack, S.J.; Kang, T.S.; Choi, J.S.; Bae, K. Antioxidant activity of coumarins from Korean medicinal plants and their structure-activity relationships. *Phytother. Res.* **2010**, *24*, 101–106.

6. Deianaa, M.; Rosaa, A.; Casua, V.; Cottigliab, F.; Bonsignoreb, L.; Dessìa, M.A. Chemical composition and antioxidant activity of extracts from *Daphne gnidium* L. *J. Am. Oil Chem. Soc.* **2003**, *80*, 65–70.

7. Smyth, T.; Ramachandran, V.N.; Smyth, W.F. A study of the antimicrobial activity of selected naturally occurring and synthetic coumarins. *Int. J. Antimicrob. Ag.* **2009**, *33*, 421–426.

8. Niu, X.; Xing, W.; Li, W.; Fan, T.; Hu, H.; Li, Y. Isofraxidin exhibited anti-inflammatory effects *in vivo* and inhibited TNF-α production in LPS-induced mouse peritoneal macrophages *in vitro* via the MAPK pathway. *Int. Immunopharmacol.* **2012**, *14*, 164–171.

9. Jiménez-Orozco, F.A.; Rosales, A.A.R.; Vega-López, A.; Domínguez-López, M.L.; García-Mondragón, M.J.; Maldonado-Espinoza, A.; Lemini, C.; Mendoza-Patiño, N.; Mandoki, J.J. Differential effects of esculetin and daphnetin on *in vitro* cell proliferation and *in vivo* estrogenicity. *Eur. J. Pharmacol.* **2011**, *668*, 35–41.

10. Liu, H.; Li, Z.; Yu, L.; Zhang, Y. Antitumor activity and mechanisms of scoparone. *J. Zhong Guo Yao Li Tong Xun* **2005**, *22*, 40–41.

11. Liu, W.; Hua, J.; Zhou, J.; Zhang, H.; Zhu, H.; Cheng, Y.; Gust, R. Synthesis and *in vitro* antitumor activity of novel scopoletin derivatives. *Bioorg. Med. Chem. Lett.* **2012**, *22*, 5008–5012.

12. Rouesac, F.; Leclerc, A. An efficient synthesis of isofraxidin. *Synth. Commun.* **1993**, *23*, 1147–1153.

13. Chen, W. Total synthesis of isofraxidin. *Chin. J. Med. Chem.* **1996**, *6*, 269–271.

14. Chen, W.A. Convenient total synthesis of isofraxidin, a natural coumarin. *Indian J. Chem. Sect. B Org. Chem. Incl. Med. Chem.* **1996**, *35*, 1085–1087.

15. Demyttenaere, J.; Syngel, K.V.; Markusse, A.P.; Vervisch, S.; Debenedetti, S.; de Kimpe, N. Synthesis of 6-methoxy-4*H*-1-benzopyran-7-ol, a character donating component of the fragrance of *Wisteria sinensis*. *Tetrahedron* **2002**, *58*, 2163–2166.

16. Crosby, D.G. Improved synthesis of scopoletin. *J. Org. Chem.* **1961**, *26*, 1215–1217.

17. Hauer, H.; Ritter, T.; Grotemeier, G. An improved and large scale synthesis of the natural coumarin scopoletin. *Arch. Pharm.* **1995**, *328*, 737–738.

18. Hao, S.; Zong, G.; Wei, Y.; Meng, Z. Synthesis and pesticidal activity of 6,7-dimethoxy coumarin. *J. Qingdao Agric. Univ.* **2007**, *24*, 241–244.

19. Zhang, S.; Ma, J.; Chen, S.; Li, H.; Xin, J. Improved synthesis techniques of 6,7-dimethoxy coumarin. *J. Hebei Univ. Sci. Technol.* **2007**, *28*, 24–28.

20. Tanaka, H.; Kato, I.; Ito, K. Total synthesis of daphnetin, a coumarin lignold. *Chem. Pharm. Bull.* **1986**, *34*, 628–632.

21. Ahmed, B.; Khan, S.A.; Alam, T. Synthesis and antihepatotoxic activity of some heterocyclic compounds containing the 1,4-dioxane ring system. *Pharmazie* **2003**, *58*, 173–176.

22. Curini, M.; Epifano, F.; Maltese, F.; Marcotullio, M.C.; Gonzales, S.P.;

Rodriguez, C. Synthesis of collinin an antiviral coumarin. *Aust. J. Chem.* **2003**, *56*, 59–60.

23. Aslam, K.; Khosa, M.K.; Jahan, N.; Nosheen, S. Synthesis and application of coumarin. *Pak. J.Pharm.* **2010**, *23*, 449–454.

24. Mahesh, K.P.; Swapnil, S.M.; Manikrao, M.S. Coumarin synthesises via Pechmann condensation in Lewis acidic chloroaluminate ionic liquid. *Tetrahedron Lett.* **2001**, *42*, 9285–9287.

25. Shaabani, A.; Ghadari, R.; Rahmati, A.; Rezayan, A.H. Coumarin synthesis via Knoevenagel Condensation Reation in 1,1,3,3-*N,N,N'N'*-tetramethylguanidinium trifluoroacetate ionic liquid. *J. Iran Chem. Soc.* **2009**, *6*, 710–714.

26. Harayama, T. Synthtic studies on aromatic heterocyclic compounds. *Pharm. Soc. Japan* **2006**, *126*, 543–564.

27. Kraus, G.A.; Cui, W.; Seo, Y.H. The reaction of Aryl triflates and Aryl pivalates with electrophiles. The triflate as a meta-directing group. *Tetrahedron Lett.* **2002**, *43*, 7077–7078.

28. Castanet, A.-S.; Colobert, F.; Broutin, P.-E. Mild and regioselective iodination of electron-rich aromatics with *N*-iodosuccinimide and catalytic trifluoroacetic acid. *Tetrahedron Lett.* **2002**, *43*, 5047–5048.

29. Olah, G.A.; Wang, Q.; Sandford, G.; Surya, P.G.K. Iodination of deactivated aromatics with *N*-iodosuccinimide in trifluoromethanesulfonic acid (NIS-CF3SO3H) via *in situ* generated superelectrophilic iodine (I) trifluoromethanesulfonate. *J. Org. Chem.* **1993**, *58*, 3194–3195.

30. Yang, D.; Fu, H. A simple and practical copper-catalyzed approach to substituted phenols from Aryl halides by using water as the solvent. *Chem. Eur. J.* **2010**, *16*, 2366–2370.

31. Qi, S.; Wu, D.; Luo, X. Coumarins and ellagic acids from sapium chihsinianum S. Lee. *Nat. Prod. Res. Dev.* **2004**, *16*, 297–299.

# Chapter 13

# LABORATORY EVALUATION OF ACOUSTIC BACKSCATTER AND LISST METHODS FOR MEASUREMENTS OF SUSPENDED SEDIMENTS

Ramazan Meral

Kahramanmaras Sutcu Imam University, Faculty of Agriculture, Department of Agricultural Structures and Irrigation. 46060, Kahramanmaras, Turkey

## ABSTRACT

The limitation of traditional sampling method to provide detailed spatial andtemporal profiles of suspended sediment concentration has led to an interest in alternativedevices and methods based on scattering of underwater sound and light . In the presentwork, acoustic backscatter and LISST (the Laser In Situ Scattering Transmissometry) devices, and methodologies were given. Besides a laboratory study was conducted tocompare pumping methods for different sediment radiuses at the same concentration. Theglass spheres (ballotini) of three different radiuses of 115, 137 and 163 μm were used toobtain suspension in the sediment tower at laboratory. A quite good agreement wasobtained between these methods and pumping results with the range at 60.6-94.2% forsediment concentration and 91.3-100% for radius measurements. These results and theother studies show that these methods have potential for research tools for sedimentstudies. In addition further studies are needed to determine the ability of these methods forsediment measurement under different water and sediment material conditions.

## INTRODUCTION

The transport of suspended sediments has a growing interest because of environmental pollution, siltation problem in water storage structures and channels, and the removing of fine soil from agricultural lands. Measuring suspended sediment is difficult as it changes temporally and spatially with the effects of flow and turbulence.

The traditional sediment measurement method is to take periodic water samples for water analyses. Application of this method is simple, but labor intensive to collect and process. Also, this method is restrictive in its capability to provide detailed spatial and temporal profiles of suspended sediment concentration. Over the past decade, technological advances in electronics and computer sciences have led to new devices that produce accurate measurements of suspended sediment concentration in water. In particular, instruments based on the scattering of underwater sound [1-4], and light [5-7], have been investigated in some detail for suspended sediment measurements.

The acoustic backscatter system (ABS) involves the propagation of sound around a megahertz through the water column. Short bursts of high frequency sound, 0.5 MHz -5.0 MHz, those transmitted from a transducer are directed towards the measurement volume. Sediment in suspension directs a portion of this sound back to the transducer. The strength of the backscattered signal allows the calculation of sediment concentration. The backscatter amplitude depends on the concentration, particle size, and acoustic frequency. This can be exploited by using multiple frequencies to determine both particle size and concentration. The strong signal from the bed can be used to measure the bed forms. Acoustic devices measure the concentration in a range-gated vertical profile of 1-2 m in depth [8-10].

This method has been demonstrated and utilized successfully under laboratory and field conditions by several investigators. Results showed that sediment concentration and particle size can be measured relatively non-intrusive, with high spatial and temporal resolution with acoustic backscatter systems. But acoustic measurement has some limitations; the signal is susceptible to absorption and scattering by bubbles, from the biological materials and flow depth limitation for shallow rivers [11].

In recent years, the application of laser scattering techniques have opened new possibilities for the measurement of sediment concentration and size distribution. The LISST-100 (the Laser In Situ Scattering Transmissometry) which is developed by Sequoia Scientific Inc. for laser scattering technique, directs a laser beam through a sample of water where particles in suspension will scatter, absorb and reflect the beam. The diffracted laser beam is received by ring detectors that allow measurement of the scattering angle of the beam. Particle size and volumetric concentration can be calculated from knowledge of this angle [12, 13]. Gartner et al., (2001) have reported that laboratory and field measurements indicate the potential of the LISST as a powerful research tool for sediment measurement studies and the instrument is capable of determining size distribution and volume concentration within acceptable limits [7]. Traykovski et al. (1999) performed several experiments with natural sediments in the laboratory to compare LISST results to traditional sieving, filtering and weighing techniques [14]. They found the LISST was able to adequately determine the particle volumetric size distribution of two different natural sediments. Van Wijgaarden and Roberti (2002) have used LISST for the measurement of the particle size and settling velocity of suspended sediment in calm fresh water [15] and Thonon et al. (2005) have used LISST on river floodplains [16]. Both of these authors report a greatly expanded capacity for the collection of the spatial and temporal variability of suspended sediment in these conditions.

A disadvantage of the LISST instrument is its large size, which causes a significant flow obstruction. Another disadvantage is the design of the LISST laser mount, which is sensitive to impacts that can easily throw it out of alignment. The high energy regime of the surf zone could easily provide impacts of such magnitude [11].

These methods are still in an ongoing developmental phase and the studies are needed to introduce and explore of methodology details and to compare with each other. The objectives of this paper are to describe of ABS and LISST devices, to give an explanation of the basic physics which is used to interpret the backscatter signal methodology, and to determine methods of mean mass concentration and mean particle size using LISST results of volumetric particle size distribution. Finally a laboratory study was conducted to compare ABS and LISST methods with pumping method for different sediment radius.

## Acoustic methodology; backscattering

The concentration of suspended sediments in water can be determined from backscattered signal from suspensions using following approach [17, 18, 10, 9, 2, 19, 1]. The suspended sediment concentration *M*, can be expressed as:

$$M = \left\{ \frac{V_{rms} \psi \, r}{k_s k_t} \right\}^2 e^{4r(\alpha_w + \alpha_s)} \qquad (1)$$

*M* is the suspended sediment concentration, $V_{rms}$ is recorded voltage from the transducer, ψ, accounts for the departure from spherical spreading within the transducer near field.

$$\psi = \frac{1 + 1.35z + (2.5z)^{3.2}}{1.35z + (2.5z)^{3.2}} \qquad (2)$$

where;

$$z = r/rn, \; r_n = \pi \, a_t^{\,2} / \lambda$$

$a_t$ is the effective radiating radius of the transducer, λ is the sound wavelength and *r* is range from transducer. $k_t$ is a constant and found from calibration, $a_w$ is the attenuation coefficient due to water absorption; its dependence upon water temperature, depth and salinity can be obtained from tables or by formulae. The absorption of sound for freshwater at atmospheric pressure can be written as [20]:

$$\alpha_w = (55.9 - 2.37T + 4.77*10^{-2}T^2 - 3.84*10^{-4}T^3)10^{-3} f_r^2 \quad (3)$$

$T$ is degrees Celsius, $f_r$ frequency in MHz. αs is the attenuation due to scattering in suspension and can be described by

$$\alpha_s = \frac{1}{r}\xi(r)M(r) \quad (4)$$

$$\xi = \frac{3}{4a_s\rho}\chi \quad (5)$$

χ, is the normalized total scattering cross-section. $a_s$ is the radius of the particles in suspension. is the density of the particles in suspension (~2650 kg.m-3 for sandy sediments).

$$k_s = \frac{f}{\sqrt{a_s\rho_s}} \quad (6)$$

*f* is form function and describes scattering properties of the particle. The normalized total scattering cross-section and form function values are related with particle radius and wave number of the sound (*k*) in water. This relation was given by Thorne and Meral (2008) as [21];

$$\chi = \frac{0.29(ka_s)^4}{0.95 + 1.28(ka_s)^2 + 0.25(ka_s)^4} \quad (7)$$

$$f = \frac{(ka_s)^2(1-0.35e^{-((ka_s-1.5)/0.7)^2})(1+0.5e^{-((ka_s-1.8)/2.2)^2})}{1+0.9(ka_s)^2} \quad (8)$$

Calibration constant, $k_t$, *is* determined from the acoustic backscatter strength of homogenous suspended sediment of known particle size and concentration level.

Suspended sediment equation (Eq. (9)) contains two *M* values. If the all other equation parameters are known the following procedures are used for obtaining suspended sediment concentration and particle size.

$$M = \left\{ \frac{V_{rms} \psi \, r}{k_s k_t} \right\}^2 e^{4r\left[\frac{1}{r}\int \xi(r) M(r)\right]} \tag{9}$$

The second $M$ (on right of equation) is accepted as zero and the first $M$ is computed which is called $M_0$. After that the final $M$ value is computed using $M_0$ with Eq.(9) .This procedure is applied for different particle radiuses and each range($r$) values, and β coefficients are computed for selected particle radius.

$$\beta = \frac{M_{std}}{M_{mean}} \tag{10}$$

where; $M_{std}$ is standard deviation and $M_{mean}$ is mean of computed $M$ values. The particle radius is chosen as real one which gives a minimum β value [1].

## The LISST-100 devices and principle of laser scattering; diffraction

The LISST-100, with diameter of 13 cm and length of 81 cm, is a laser particle size analyzer and consists of a diode laser operating at 670 µm which is collimated to form a parallel beam of light (figure 1). The beam passes through a 5 cm sample cell of water where particles produce scattering. The forward diffracted beam is collected by a set of 32 ring detectors that are logarithmically spaced and located in the focal plane of the laser receiving lens. Each ring detector measures scattering over a sub-range of angles. The ring radius and focal plane length of the receiving lens determine the range of the angles. A photo-diode is placed behind the ring detector which detects the main laser beam. This supplies the optical transmissometer function [5].

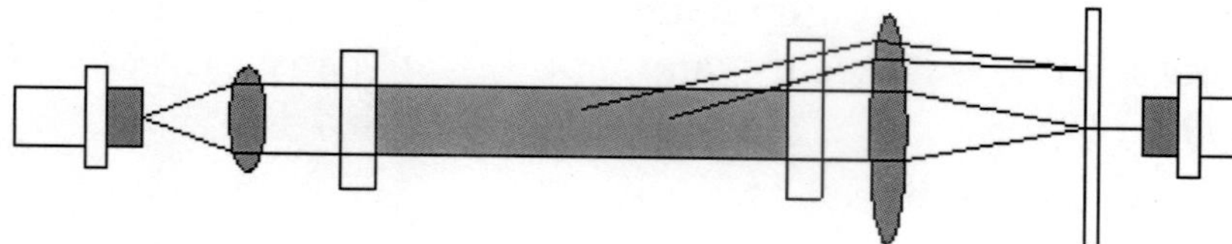

**Figure 1.** The LISST devices elements: (a) laser diode source, (b) companion focusing optics, (c) and (e) two pressure windows, (d) sample volume, (f) receiving lens, (g) concentric photodetector rings, (h) transmission detector

The light scattering measured by the LISST represents light scattering within a thin radial annulus in the focal plane of radius $r$ and with $dr$. This annulus represents an area, $dA=2\pi rdr$, and the light gathered by this annulus is all the light that is forward scattered into the solid range for distance $f$ is, $\theta+d\theta = atan(r+dr/f)$ . The scattering per unit solid angle into angle θ, from a single particle radius is defined as light power (energy) .The light power can be written in the matrix form ($K_{i,j}$) for each detector ($i= 1, 2, 3......N$) and each size class ($j =1, 2, 3......N$), and its relation between the volume distribution ($Vj$) can be expressed as;

$$E_i = K_{i,j}V_j \tag{11}$$

$$V_j = \int_{a_{j,\min}}^{a_{j,\max}} a^3 n(a)da \tag{12}$$

$a_{j,min}$ and $a_{j,max}$ are minimum and maximum particle diameter in the $j^{th}$ class and $n(a)$ is the number of particles of size [13].

The scattered power must be corrected with background scattering, $z$, the optical transmission, τ, which is recorded by optical transmissometer, and detector correction factor ($D_c$) which is provided by the instrument manufacturer.

$$E = D_c\left[\left(E_i / \tau\right) - z\right] \tag{13}$$

The background scattering is recorded to account for the optical effect of the flow-through chamber. To obtain $z$, the flow-through chamber is filled with clear water, and LISST scans are recorded and averaged.

Finally, the volume concentration;

$$V = \frac{D_c[(E_i / \tau) - z]}{Cv} \tag{14}$$

*Cv*, volume conversion constant, is determined from LISST measurement of known volume concentration particles ($V_0$). The background scattering distribution from de-ionized water, and the scattering distribution from the suspension are recorded, and $E$ is computed. Finally volume conversion constant,

$$C_v = \Sigma E / V_0 \tag{15}$$

LISST measurements were logged using standard LISST-TOP software provided by Sequoia Scientific and further processed in the MATLAB.

## MATERIALS AND METHODS

This experiment was carried out under laboratory conditions and all measurements were conducted in the vertical sediment tower shown in figure 2. The tower was built to obtain a homogeneous suspension of sediments within the central region of the tower, using the re-circulation bilge pumps and a mixing unit composed of a turbulence grid impellor and propeller driven by a 12 V dc motor. The ABS transducers are located around the central shaft of the grid shaft [22].

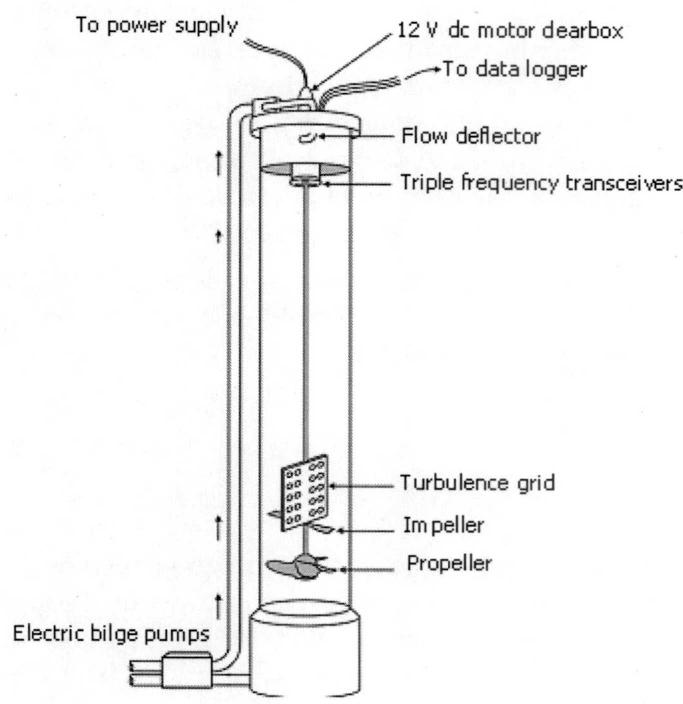


**Figure 2.** The sediment tower used for ABS measurement (Adopted from Thorne and Hanes, 2002 [1]).

Initially the tower was filled with water and the system was run to remove bubbles in the water for fifteen minutes every fifteen minutes during 3 days which is known as degassing. The ABS background measurements were taken during one hour using 3 different transducers operating at 1, 2 and 4 MHz frequencies.

Before starting of LISST measurement, the background scattering distribution was measured using clear water to take into scattering and attenuation due to water itself.

The glass spheres (ballotini) of three different radiuses of 115, 137 and 163 μm with a 10% variation in the mean size, were used to prepare of suspended sediment because all the acoustic parameters were known and they provided an unambiguous condition of the prediction of sediment concentration. 120 gr wetted and degassed ballotini was added to the tower for each radius application and homogeneous suspension was obtained with the circulation of water with pump and mixer unit which consist of turbulence grid, impeller and propeller. ABS and LISST measurements were taken during 1 hour for each radius, and pumping samples were taken three times during this period. The average suspended sediment concentration was determined with gravimetric methods from these samples. The sediment was removed from the tower using a conical mesh, of aperture size 20 µm, placed on the top of the tower, for another experiment.

The voltage values are calculated for background ($V_b$) and suspended measurement ($V_s$) and finally backscatter voltages are obtained using the final expression;

$$V = \sqrt{V_s^2 - V_b^2} \tag{16}$$

LISST-100 Particle Size Analyzer program gives results as particle size distribution and cumulative volumetric concentration for each radius. In addition a script was written to obtain mass sediment concentration, and mean particle size as follows. The LISST takes measurement for each second and each radius classes. The individual mass concentrations for each radius,,( kg.m$^{-3}$);

$$M_{c_{ind}} = \rho V_c \tag{17}$$

: particle density (2480 kg.m$^{-3}$), $V_c$: volumetric concentration

The total mass concentration;

$$M_{c_{total}} = \rho \sum_{1}^{n} V_c \tag{18}$$

$n$; radius range number (1-32)

The mean individual mass concentrations ($M_{c\ mean\_ind}$) for each radius class were obtained with considering all measurements;

$$M_{c_{mean\_ind}} = \sum_{1}^{m} M_{c_{ind}} / m \tag{19}$$

*m*; record range number

The mean sediment radius (d50) value is equal to radius versus %50 cumulative mass value

$$\%cumulative\_mass = \frac{\sum_{1}^{n} M_{c_{mean\_ind}}}{M_{c_{total}}} \quad (20)$$

# RESULTS AND DISCUSSION

The backscatter voltage values, sediment concentration and radius results for each measurement of ABS are shown figures 3-5. The backscatter voltages drop with increasing range from transducer. Similar results are given by Thorne at all (1991) [23] and are explained with water and particle attenuation and spreading of the beam. The nearfield region of the transducer was seen in the first 0.2 m range, and the reflections from the paddle were seen at about 0.9 m and 1.25 m. Therefore, sediment concentrations and radius were determined for 0.2-0.8 m range.

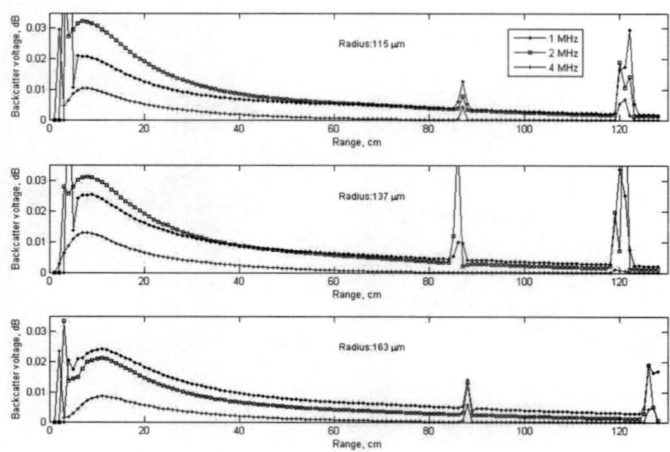


**Figure 3.** The backscatter voltages from suspension for each frequency and radius.

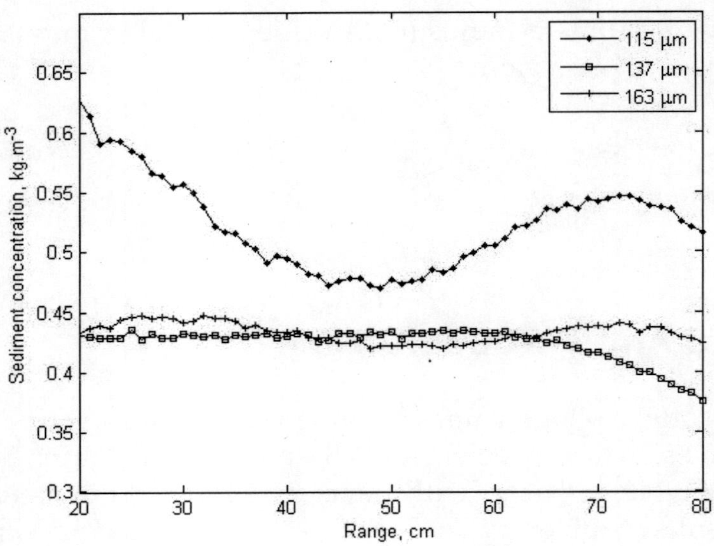


**Figure 4.** The mean sediment concentration results of ABS measurements for 3 different frequency.

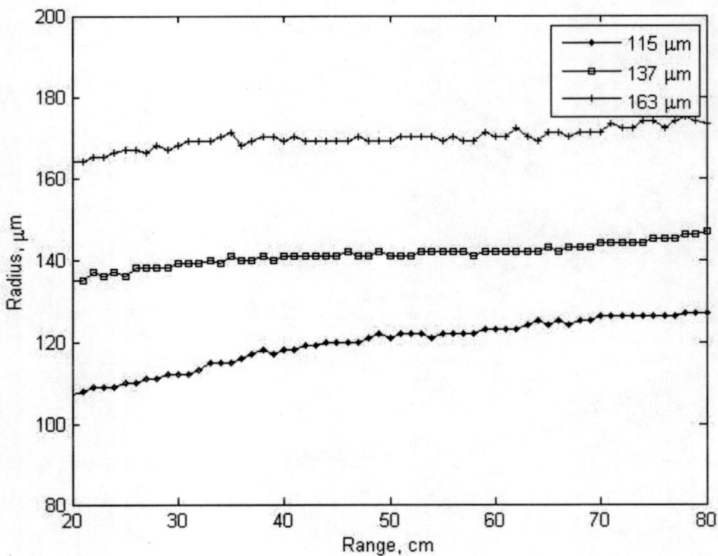


**Figure 5.** Sediment radius results of ABS measurements.

The first 100 seconds were considered because sediment concentration decreased in the course of time due to settling of ballotini. The homogenised suspension condition was obtained exactly during this time. Because when the LISST devices was put in the tower the mixing unit was stopped due to physical limitation and re-circulation was obtained just by pumps. The laser transmission, scattering distribution, sediment concentration and radius results are given each measurement of LISST in figures 6-9. Figure 6 and 8 show that the high laser transmission represents low sediment concentration, and the inverse relation is seen between radius and scattering (Figure 6 and 9).

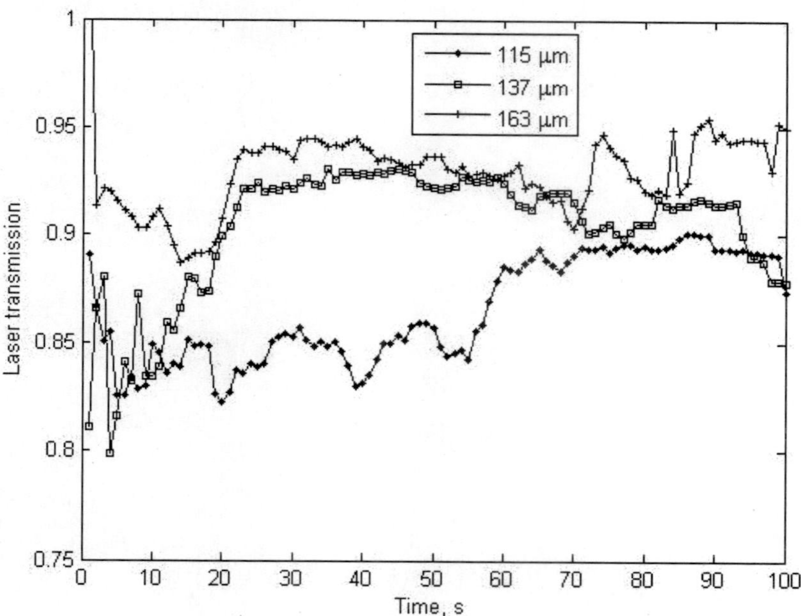


**Figure 6.** The laser transmission results of LISST measurements.

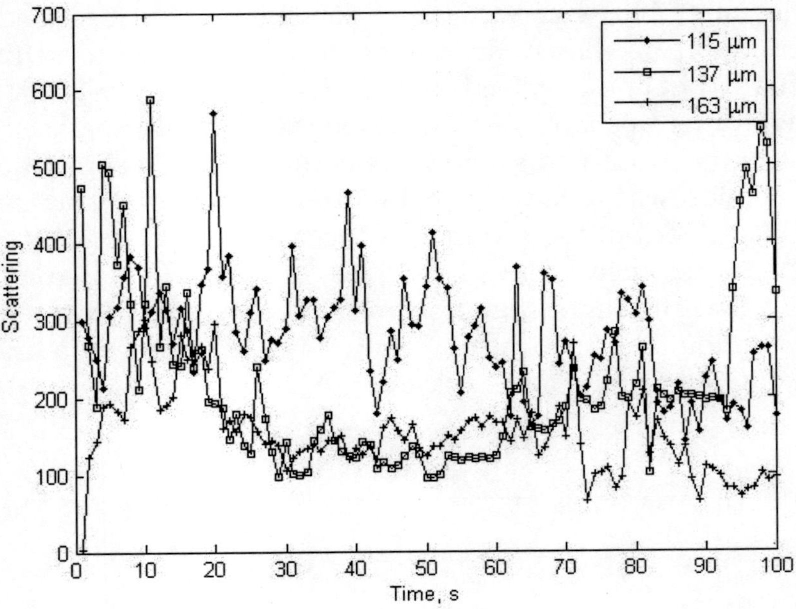


**Figure 7.** The scattering distribution of LISST measurements.

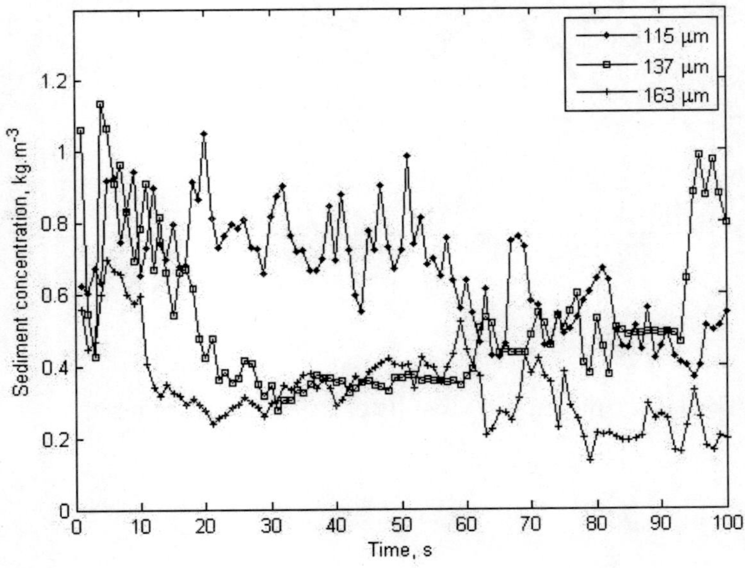


**Figure 8.** Sediment concentration results of LISST measurements.

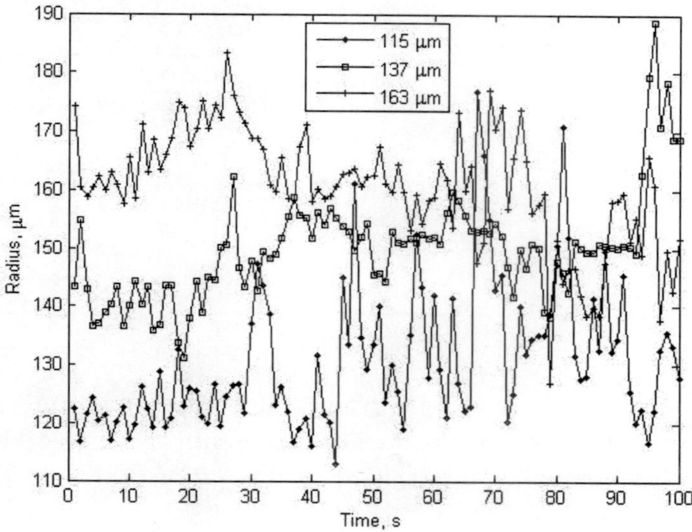


**Figure 9.** Sediment radius results of LISST measurements

The statistical analyses of results are tabulated in Tables 1 and 2. ABS measurements were considered for 45 cm range and 100 s duration to compare LISST measurements under similar conditions.

**Table 1.** Sediment concentration results and statistical evaluation.

| | | | mean | standard error | agreement with pumping,% |
|---|---|---|---|---|---|
| First measurement | | 1 MHz | 0.497 | 0.017 | 90.2 |
| (115 µm ) | ABS | 2 Mhz | 0.481 | 0.015 | 87.3 |
| | | 4 MHz | 0.385 | 0.021 | 69.9 |
| | LISST | | 0.660 | 0.016 | 80.2 |
| Second measurement | | 1 Mhz | 0.508 | 0.011 | 92.2 |
| (137 µm) | ABS | 2 Mhz | 0.494 | 0.011 | 89.7 |
| | | 4 MHz | 0.275 | 0.011 | 49.9 |
| | LISST | | 0.510 | 0.020 | 92.6 |
| Third measurement | | 1 Mhz | 0.519 | 0.012 | 94.2 |
| (163 µm) | ABS | 2 Mhz | 0.486 | 0.009 | 88.2 |
| | | 4 MHz | 0.291 | 0.008 | 52.8 |
| | LISST | | 0.334 | 0.012 | 60.6 |
| Pumping | | | 0.551 | 0.014 | |

**Table 2.** Sediment radius results and statistical evaluation.

| | | mean | standard error | agreement with pumping, % |
|---|---|---|---|---|
| First measurement | ABS | 110 | 1.314 | 95.6 |
| (115 μm) | LISST | 125 | 0.808 | 91.3 |
| Second measurement | ABS | 132 | 1.176 | 96.4 |
| (137 μm) | LISST | 131 | 1.237 | 95.6 |
| Third measurement | ABS | 163 | 1.296 | 100.0 |
| (163 μm) | LISST | 161 | 0.993 | 98.8 |

Both methods have good agreement with pumping results in determining of radius. Similarly, Thosteson and Hanes (1998) found less than 20% error in laboratory test in determining sediment concentration and size [24].

In concentration measurements, the best agreement was obtained in ABS measurements for 1 and 2 MHZ frequency. Generally, 4 MHZ frequency measurements estimated lower values than the observed values and it had low agreement with pumping. This may be due to the length of transducer connection cable. It is known that signal attenuation occurs in high frequency depending upon cable length [25].

The lowest agreement (60.6%) was obtained in LISST concentration measurements for 163 μm radius. It was observed that LISST measurements were more sensitive to sediment radius. It can be said that accuracy decreases as sediment radius increases. Because LISST makes estimation based on sediment radius and radius-based scattering angle.

## CONCLUSION

In both methods (ABS and LISST), quite good results were obtained for measurement of sediment concentration and radius in laboratory conditions. These results and other former studies show that ABS and LISST methods have potential for research tools for sediment studies. It is an advantage that LISST method can make measurement without calibration and can determine particle size distribution. This method can be preferred to optical backscattering sensors or single frequency acoustic sensors which suffer from calibration difficulties but has difficulties in representing a profile. ABS method determines mean sediment radius, but its important advantage is making measurement along a profile. Acoustic measurement is still in an ongoing developmental phase and there are limitations and shortcomings; signal highly susceptible to absorption and scattering by air bubbles, evaluation difficulties of backscatter signal from the biological materials. However the continuous and non-intrusive measurement with the centimetre interval along a profile using backscattering data, is a useful addition to the measurement of suspended sediment.

## ACKNOWLEDGEMENTS

The work described here was carried out during a visit to the Proudman Oceanographic Laboratory (POL), Liverpool, UK in 2006. I would like to thank Professor Peter D. Thorne for his interest, time and encouragement during this visit. Also thanks to Dr. Alejandro J. Souza for LISST measurements opportunities and, to Dr. Richard D. Cooke and Dr. Kyle F.E. Betteridge for their helps on laboratory studies. Financial support was provided by The Scientific & Technological Research Council of Turkey and POL for this study.

## REFERENCES

1. Thorne, P.D; Hanes, D.H. A review of acoustic measurement of small-scale sediment processes. Continental Shelf Research **2002**, *22*, 603-632.
2. Thorne, P.D.; Hardcastle, P.J. Acoustic measurements of suspended sediments in turbulent currents and comparison with in-situ samples. Journal of the Acoustical Society of America **1997**, *101* (5)(Pt. 1), 2603–2614.
3. Zedel, L.; Hay, A.E.; Cabrera, R.; Lohrmann, A. Performance of a single-beam pulse-to-pulse coherent Doppler profiler. EEE Journal of Oceanic Engineering **1996**, *21*, 290–297.
4. Hay, A.E., Wilson, D. Rotary sidescan images of nearshore bedform evolution during a storm. Marine Geology **1994**, *119*, 57–65.
5. Agrawal, Y.C.; Pottsmith, H.C. Instruments for particle size and settling velocity observations in sediment transport. Marine Geology **2000**, *168*, 89–114.
6. Styles, R. Laboratory evaluation of the LISST in a stratified fluid. Marine Geology **2006**, *227*, 151– 162

7. Gartner, J.W.; Cheng, R.T.; Wang, P.; Richter, K., 2001. Laboratory and field evaluations for the LISST-100 instrument for suspended particle size determinations. Marine Geology 175, 199–219.

8. Wren, D.G.; Roger A.; Kuhnle R.A. Surrogate techniques for suspended-sediment measurement, Turbidity and Other Sediment Surrogates Workshop **2002,** April 30 – May 2, Reno, NV

9. Thorne, P.D.; Hardcastle, P.J.; Soulsby, R.L. Analysis of acoustic measurements of suspended sediments. Journal of Geophysical Research **1993**, *98* (1), 899–910.

10. Hay, A.E.; Sheng, J. Vertical profiles of suspended sand concentration and size from multifrequency acoustic backscatter. Journal of Geophysical Research **1992**, *97*(10), 15661-15677.

11. Battisto, G.M. Field measurement of mixed grain size suspension in the nearshore under waves. **2000**, A Thesis presented to The Faculty of the School of Marine Science, The College of William and Mary in Virginia.

12. Agrawal, Y.C.; Pottsmith, H.C. Laser diffraction particle sizing in STRESS. Continental Shelf Research **1994**, *14* (10/11), 1101–1121.

13. Pedocchi, F.; Garcia, M.H. Evaluation of the LISST-ST instrument for suspended particle size distribution and settling velocity measurements. Continental Shelf Research **2006**, *26*, 943–958

14. Traykovski, P.; Latter, R.J.; Irish, J.D. A laboratory evaluation of the laser in situ scattering and transmissometery instrument using natural sediments. Marine Geology **1999**, *159*, 355–367.

15. Van Wijngaarden, M.; Roberti, J.R. In situ measurements of settling velocity and particle distribution with the LISSTST. In: Winterwerp, J.C., Kranenburg, C. (Eds.), Fine Sediment Dynamics in the Marine Environment. Elseiver Science B.V., Amsterdam, **2002**, pp. 295–311.

16. Thonon, I.; Roberti, H.; Middelkoop, H.; van der Perk, M.; Burrough, P. In situ measurements of sediment settling characteristics in floodplains using a LISST-ST. Earth Surf. Processes **2005**, *30* (10), 1327–1343.

17. Sheng, J.; Hay, A.E. An examination of the spherical scatterer approximation in aqueous suspensions of sand. Journal of the Acoustical Society of America **1988**, *83*, 598–610.

18. Thorne, P.D.; Campbell, S.C. Backscattering by a suspension of spheres. Journal of the Acoustical Society of America **1992**, *92*, 978–986.

19. Smerdon, A.M. AQ59:C- ABS system user manual. Aquatec Electronics Limited, Hartley Wintney **1996**.

20. Fisher, F.H.; Simmons, V.P. Sound adsorption in sea water. J.Acoust. Soc. Am. **1977**, *62*(3), 558-564

21. Thorne, P.D.; Meral, R. Formulations for the scattering properties of suspended sandy sediments for use in the application of acoustics to sediment transport. Continental Shelf Research **2008**, *28* (2), 309-317

22. Thorne, P.D.; Buckingham, M.J. Measurements of scattering by suspensions

of irregularly shaped sand particles and comparison with a single parameter modified sphere model. Journal of the Acoustical Society of America **2004**, *116* (5), 2876–2889.

23. Thorne, P.D.; Vincent, C. E.; Hardcastle, P. J.; Rehman, S.; Pearson, N. Measuring Suspended Sediment Concentrations Using Acoustic Backscatter Devices. Marine Geology **1991**, *98*, 7-16.

24. Thosteson, E.D.; Hanes, D.M. A Simplified method for determining sediment size and concentration from multiple frequency acoustic backscatter measurements. Journal of the Acoustical Society of America **1988**, *104* 2(1), 820–830.

25. Smerdon, A. A personal communication from the Aquatec group. See http:// www.aquatecgroup.com/ download/ datasheet/aquascat/ AQUAscat1000. pdf **2007**.

# Chapter 14

# IMPROVED CALIBRATION FUNCTIONS OF THREE CAPACITANCE PROBES FOR THE MEASUREMENT OF SOIL MOISTURE IN TROPICAL SOILS

Ali Fares [1,*], Farhat Abbas [1], Domingos Maria [2] and Alan Mair [3]

[1]Natural Resources and Environmental Management Department, University of Hawaii-Manoa, Honolulu, HI 96822, USA; E-Mail: farhat@hawaii.edu

[2]School of Engineering and Computer Sciences, New York Institute of Technology, New York, NY 10023, USA; E-Mail: dmaria@nyit.edu

[3]Geology and Geophysics Department, University of Hawaii-Manoa, Honolulu, HI 96822, USA;

*Author to whom correspondence should be addressed; E-Mail: afares@hawaii.edu.

## ABSTRACT

Single capacitance sensors are sensitive to soil property variability. The objectives of this study were to: (i) establish site-specific laboratory calibration equations of three single capacitance sensors (EC-20, EC-10, and ML2x) for tropical soils, and (ii) evaluate the accuracy and precision of these sensors. Intact soil cores and bulk samples, collected from the top 20 and 80 cm soil depths at five locations across the Upper Mākaha Valley watershed, were analyzed to determine their soil bulk density (ρb), total porosity (θt), particle size distribution, and electrical conductivity (EC). Laboratory calibration equations were established using soil packed columns at six water content levels (0–0.5 $cm^3$ $cm^{-3}$).

Soil bulk density and θt significantly varied with sampling depths; whereas, soil clay content (CC) and EC varied with sampling locations. Variations of ρb and θt at the two depths significantly affected the EC-20 and ML2x laboratory calibration functions; however, there was no effect of these properties on calibration equation functions of EC-10.

There was no significant effect of sampling locations on the laboratory calibration functions suggesting watershed-specific equations for EC-20 and ML2x for the two depths; a single watershed-specific equation was needed for EC-10 for both sampling depths. The laboratory calibration equations for all sensors were more accurate than the corresponding default equations. ML2x exhibited better precision than EC-10, followed by EC-20. We conclude that the laboratory calibration equations can mitigate the effects of varying soil properties and improve the sensors' accuracy for water content measurements.

## INTRODUCTION

The water content of surface soils is more dynamic than that of deeper soil layers because of the continuous water loss due to evapotranspiration and the periodical water inputs from rainfall and irrigation events. Variations in water content within the vadose zone are also due to variations in soil texture, ρb, θt, CC, and EC [1-4].

Soil water content is directly measured with the thermo-gravimetric method and/or indirectly with commercially available soil water content monitoring sensors, i.e., capacitance, TDR and neutron scattering. The thermo-gravimetric method is labor intensive, time consuming, destructive, and discrete for repetitive measurements. Conversely, most of the indirect measurement techniques are logged manually, or in real-time on site with data loggers or remotely via cellular and satellite phones.

Details on design, operations, and application of different water content monitoring sensors can be found in Fares and Polyakov [5] and Robinson et al. [6]. Site-specific calibration of these sensors is recommended for accurate monitoring of soil water content. ECH2O [7] sensors including EC-5, EC-10 and EC-20, and ThetaProbe® [8]

sensors including ML2x are single capacitance water content devices, which have been calibrated for different soils in the laboratory [9-11] and in the field [12]. Czarnomski et al. [13] tested the EC-20 and found that the default calibration equation under-estimated the actual water content by up to 0.12 $cm^3 cm^{-3}$, and measurements weren't sensitive to $\rho_o$. Overduin et al. [14] tested seven different water content sensors, including the ECH2O and ML2x ones, for monitoring of the water content of a feather moss stored in different layers. They concluded that the readings of most of the sensors were affected by the spatial variability of the moss bulk density. Logsdon and Hornbuckle [15] compared the performance of ML2x, the updated CS616, and the Stevens Hydra probe. They reported that the larger measurement volume of the CS616 resulted in less spatial variability of its measured water content than that of the ML2x, which has a relatively smaller measurement volume. Foley and Harris [12] assessed the performance of the EC-20 and ML2x in a Black Vertosol from southeast Queensland (Australia) and found considerable over- and under-estimations of water content when using the default calibration equations of these sensors. They also reported a significant impact of $\rho_o$ on the sensors' performance and concluded that the site-specific laboratory calibrations can significantly improve the accuracy of both sensors. Bogena et al. [16] evaluated the EC-20 and EC-5 (the latest in the ECH2O series) in the laboratory and field. They concluded that the sensors' performance was affected by variability in various soil properties. Mendes et al. [17] tested the performance of EC-5 sensors in a pile of poultry manures compacted at five densities (0.32, 0.35, 0.38, 0.42, and 0.47 g $cm^{-3}$). They reported a significant effect of the manure bulk density, temperature, and salinity on sensor performance. Fares et al. [11] evaluated the effect of media temperature and salinity on the apparent water content measured with the EC-20. They concluded that ignoring the media temperature and salinity might cause significant errors of up to 0.23 $cm^3 cm^{-3}$, particularly in the lower water content range.

An ideal sensor, e.g., soil water monitoring sensor, is accurate and precise. Accuracy and precision of most of the water content sensors vary, as does their calibration, with various soil properties [13, 18]. Accuracy and precision are the major quantitative assessments of sensor performance [19]; they define how well a sensor's output represents the actual water content [20]. Accuracy and precision are sometimes incorrectly thought to have the same meaning [21]. Accuracy is the degree of conformity with a standard [22] and; therefore, is the ability of a water content sensor to estimate the actual water content [13]. The root mean square error (RMSE) is a reasonable indicator of a sensor accuracy [23, 24]. A smaller RMSE indicates better accuracy.

Precision is an indication of the uniformity or reproducibility of a result and as such relates to the quality of an operation to obtain a result [25]. It is the degree of refinement in the performance of an operation, or the degree of perfection in the instruments and methods used to obtain a result [22]; thus, precision describes the repeatability of a measurement. Precision can also be a measure of the variability of an observation around a statistical true value [23].

Thus, a measure of the precision of an estimate is given by the standard deviation from a true mean [24] or by the variance in the multiple sensor readings simultaneously taken at the same water content level of a uniform medium [21]. Lesser precision is reflected by a larger variance.

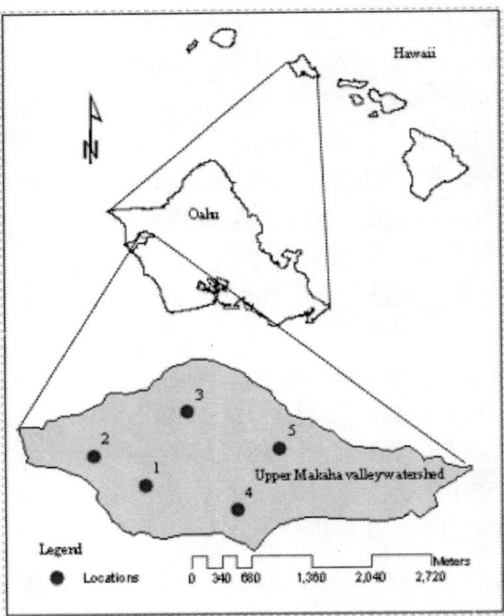


**Figure 1.** The map of the Upper Mākaha Valley sub-watershed showing the five laboratory calibration study locations where the sensors are installed and from which soil samples were collected and used for this work.

The Upper Mākaha Valley watershed is located on West O'ahu (HI, USA), which is the dry leeward side of this tropical island (Figure 1). The watershed has been home to a long-term hydrologic study aiming at determining the effects of rainfall variability, groundwater pumping, and invasive species on the hydrology of the watershed [26]. The watershed has been instrumented with EC-10, EC-20, and ML2x sensors, and other equipment's for real-time monitoring of water budget components including recharge below the root zone, changes in soil water storage within the root zone, and actual evapotranspiration. Our hypothesis was that the varying rb, θt, CC, and EC of the watershed soils will affect the performance of these sensors. Therefore, the objectives of this study were to:

- Establish site-specific laboratory calibration equations of ec-10, ec-20, and ml2x for tropical soils and
- Evaluate the accuracy and precision of these sensors.

## MATERIALS AND METHODS

### The Study Site

The soil samples were taken from five long-term monitoring locations across the Upper Mākaha Valley watershed (Figure 1); we refer to them as locations 1 through 5 from here onward. These locations were chosen to represent the spatial variation of elevation, land cover, soil type and slope across the study area. Two of these locations have weather stations; however, all of them are instrumented with soil water content sensors. Soil water content is also monitored at 20 and 80 cm depths. These locations exhibit spatial variations in topography, soil series, and vegetation cover (Table 1). The soils of the lower valley are less permeable than those along the valley ridges; whereas, those of the upper valley are clay loam, silty loam, and silty clay [27].

**Table 1.** The elevation, NRCS soil series, and the vegetation of the five monitoring locations

| Location | Elevation, m | NRCS soil series | Vegetation |
|---|---|---|---|
| 1 | 343 | Mollisol | Christmas berry (*Lycium carolinianum*) |
| 2 | 477 | Inceptisol | Strawberry guava (*Psidium cattleianum*) |
| 3 | 538 | Inceptisol | Ohia (*Metrosideros polymorpha*), Strawberry guava |
| 4 | 601 | Oxisol/Ultisol | Coffee (*Coffea arabica*), Llama (*Artiodactyla camelidae*), Strawberry guava |
| 5 | 609 | Oxisol/Ultisol | Ohia, Strawberry guava, Uluhe (*Dicranopteris linearis*) |

## Soil Water Content Monitoring Sensors

ECH2O (EC-10 and EC-20) sensors operate at 5 MHz frequency [7]; whereas Theta Probe® (the ML2x sensor) operates at 100 MHz frequency [8]. The ECH2O sensors measure dielectric constant (ε) of the surrounding soil media and convert it to a single voltage that ranges between 250 and 1,000 mV, which is related to water content through a linear calibration equation [7]. ML2x generates an electromagnetic signal (at 100 MHz) that extends into the soil by an array of four rods, the impedance of which varies with that of the soil. Soil impedance has two major components: the apparent ε and the ionic (electrical) conductivity; the operating 100 MHz frequency minimizes the latter; hence, changes in impedance are mainly due to soil's apparent ε [8]. There is a linear correlation between water content and the square root of the dielectric constant ($\sqrt{\varepsilon}$) as determined by the ML2x [28,29]. The configuration of the ML2x makes it less sensitive to minor air gaps and soil variations [15].

## Soil Sampling and Analyses

Three replicates of undisturbed soil core samples (radius = 2.5 cm; height = 7.5 cm) were collected with a sludge hammer soil sampler (Soilmoisture Equipment Crop. Santa Barbara, CA, USA) at 20 and 80 cm depths from the five locations (Figure 1). The soil cores were carefully trimmed, sealed with caps, placed in labeled Ziplock plastic bags, and transported in a cooler to the laboratory where the caps from the bottom of the cores were replaced with fine nylon mesh to secure the soil inside the cores. The caps from the top of the cores were removed and the cores were then placed vertically in a tray filled with water for 24 h, letting them slowly saturating from their bottom. The saturated samples were weighed and then oven dried at 105°C for 48 h and weighed again. The values of $\rho_b$ and $\theta_t$ were calculated following the procedures described by Grossman and Reinsch [30] and Flint and Flint [31], respectively.

Three replicates of bulk soil samples were also collected from 20 and 80 cm depths at each location. The samples were thoroughly mixed to produce a representative sample for each depth at every location. These samples were air dried and sieved (<2 mm); a sub-sample was used to determine their particle size distribution using the hydrometer method [32]. The textural triangle of the United States

Department of Agriculture (USDA) classification scheme was used to determine the soil textural class. These samples were also used to prepare 1:2 soil: water solutions for measurements of EC with the corresponding electrodes connected to a multi-functional sympHony® meter (Model SB90M5; Batavia, IL, USA).

## Column Preparation for Laboratory Calibrations

Polyvinyl chloride (PVC) cylindrical columns (internal radius = 5 cm; height = 40 cm) were used for the laboratory calibration of the selected sensors. Sieved (<2 mm) and oven-dried (105°C; 48 h) representative soil bulk samples from the two depths of the five locations were separately packed in these columns. For each location and depth, starting with oven-dried soil, an incremental amount of deionized water was added to and thoroughly mixed with the dry soil to produce soil media of six water content levels (i.e., 0, 0.1, 0.2, 0.3, 0.4, and 0.5 $cm^3\ cm^{-3}$). As incremental amounts of soil were poured in columns during packing, the soil columns were gently tapped from their sides and uniformly compacted from the top to attain the field bulk density of the corresponding depths and locations (Table 2). The $ECH_2O$ sensors were vertically placed in the center of the columns during packing; whereas, the ML2x sensors were smoothly inserted in the packed columns, which were covered with a tight-fit Styrofoam lid to prevent water evaporation.

**Table 2.** The USDA soil classification, particle size distribution, bulk density ($\rho_o$), total porosity ($\theta_t$), and the values of electrical conductivity (EC) of the soil samples collected from 20 and 80 cm depths at the five monitoring locations in the Upper Mākaha Valley watershed

| Location | Depth cm | $\rho_b$ g cm$^{-3}$ | $\theta_t$ cm$^3$ cm$^{-3}$ | Clay g kg$^{-1}$ | Sand | USDA Soil texture | EC[#] µS cm$^{-1}$ |
|---|---|---|---|---|---|---|---|
| 1 | 20 | 0.93 | 0.69 | 313 | 269 | Clay loam | 2,016 |
| | 80 | 1.27 | 0.56 | 313 | 269 | Clay loam | 840 |
| 2 | 20 | 0.86 | 0.67 | 610 | 308 | Clay | 802 |
| | 80 | 1.06 | 0.60 | 690 | 250 | Clay | 480 |
| 3 | 20 | 0.83 | 0.67 | 521 | 167 | Clay | 426 |
| | 80 | 0.97 | 0.58 | 640 | 250 | Clay | 222 |
| 4 | 20 | 0.95 | 0.66 | 313 | 218 | Clay loam | 1,888 |
| | 80 | 0.93 | 0.67 | 288 | 320 | Clay loam | 1,270 |
| 5 | 20 | 0.69 | 0.72 | 263 | 421 | Loam | 1,220 |
| | 80 | 0.93 | 0.61 | 288 | 661 | Sandy loam | 1,024 |

[#] Electrical conductivity measurements were made on 1:2 soil: water solutions.

The columns filled with soils at the desired water content level with the sensors inserted in them were left for 2 h to attain equilibrium. Ten consecutive readings at 1-minute intervals were logged with data loggers and later used to calculate an average sensor reading for each sensor and for the particular water content. At the end of each calibration experiment, actual water content was determined from soil samples collected near sensor positions in the columns following the thermo-gravimetric method. These laboratory experiments were conducted at a constant room temperature of 22 ± 2°C.

## Data Analyses

Values of actual water content were plotted versus the respective readings of the ECH$_2$O (mV) and ML2x ($\sqrt{\varepsilon}$) sensors and linear calibration equations were established separately for the two sampling depths (20 and 80 cm) at every location. A factorial analysis of variance (ANOVA) was conducted to evaluate the effect of sampling depth and location on (i) $\rho_b$, $\theta_t$, CC, and EC and (ii) the slope and y-intercept functions (a and b) of the laboratory calibration equations using the Statistix software package [33]. The performance of the laboratory calibration equations were evaluated based on P and r values obtained from the regression between the calculated and the actual water contents.

Mean bias error (MBE) was used to determine under- and/or over-estimation of water content by the laboratory and default calibration equations. Positive values of MBE indicate over-estimation, whereas negative values indicate under-estimation of water content from their actual values. RMSE was used as an indicator of sensor's accuracy. Sensor accuracy was assumed very poor, poor, fair, and good for RMSE ≥ 0.1, 0.1 > RMSE ≥ 0.05, 0.05 > RMSE ≥ 0.01, and RMSE < 0.01 $cm^3\ cm^{-3}$, respectively. MBE ($cm^3\ cm^{-3}$) and RMSE ($cm^3\ cm^{-3}$) were calculated as follows:

$$\text{MBE} = \sum_{i=1}^{n} (\theta_{ci} - \theta_{ai}) / n \qquad (1)$$

$$\text{RMSE} = \sqrt{\sum_{i=1}^{n} (\theta_{ci} - \theta_{ai})^2 / n} \qquad (2)$$

where $\theta_{ci}$ and $\theta_{ai}$ are the individual values of calculated and their corresponding actual water contents in $cm^3\ cm^{-3}$, respectively, and n is the number of observations. Improvement in the sensor accuracy with the use of laboratory calibration equations over the corresponding default equations was gauged with the percent reduction in RMSE calculated as:

$$\text{Percent reduction in RMSE} = \left( \frac{\text{RMSE}_{\text{Def}} - \text{RMSE}_{\text{Lab}}}{\text{RMSE}_{\text{Def}}} \right) \times 100 \qquad (3)$$

where $\text{RMSE}_{\text{Def}}$ and $\text{RMSE}_{\text{Lab}}$ are the RMSE of default and laboratory calibration equations, respectively. Sensor precision was gauged by the variance in the multiple sensor readings that were simultaneously taken from a uniform medium at the same water content level. Larger variance indicates poorer precision.

## RESULTS AND DISCUSSION

### Soil Properties at Sampling Depths and Locations

There was a significant increase in $\rho_b$ ($P < 0.05$) and a consequent significant decrease in $\theta_t$ ($P < 0.05$) with increase in soil depth (Table 3). Larger ρb and smaller θt at 80 cm soil depth may be due to compaction from the overburden of the top soil layer. There were statistically significant larger $\rho_b$ and smaller $\theta_t$ at 80 cm than at 20 cm depth, respectively (Figure 2). At 80 cm depths, $\rho_b$ ranged between 0.93 and 1.27 g $cm^{-3}$; however, at 20 cm depths, it ranges between 0.69 and 0.95 g $cm^{-3}$ (Table 2). The smaller values of $\rho_b$ resulted in larger values of $\theta_t$ given their inverse relationship $\theta_t = 1 - \rho b/\rho s$, where $\rho_s$ is the soil particle density. At 20 cm depth, $\theta_t$ ranged between 0.66 and 0.72 $cm^3$ $cm^{-3}$; whereas, at 80 cm depth, it ranged between 0.56 and 0.67 $cm^3$ $cm^{-3}$. Sampling location had a highly significant ($P < 0.01$) effect on CC and a significant ($P < 0.05$) effect on EC values (Table 3). At locations 2 and 3, the values of CC were significantly larger and those of EC were smaller than those at other locations, respectively (Figure 3). Based on the USDA soil classification method, the soil type at locations 1 and 4 is clay loam (Table 2). Locations 2 and 3 have a clay soil; whereas, location 5 has a loam-sandy loam duplex at 20 and 80 cm depths, respectively.

**Table 3.** Values of the probability (P) and significance levels obtained from the factorial general analysis of variance for bulk density ($\rho_b$), porosity ($\theta_t$), clay content (CC), and electrical conductivity (EC) as a function of soil depths and sampling locations.

| Factors | $\rho_b$ | $\theta_t$ | CC | EC |
|---|---|---|---|---|
| Depth | 0.0393* | 0.0320* | NS | NS |
| Location | NS | NS | 0.0021** | 0.0480* |
| Interaction | NS | NS | NS | NS |

*: significant; **: highly significant; NS: not significant.

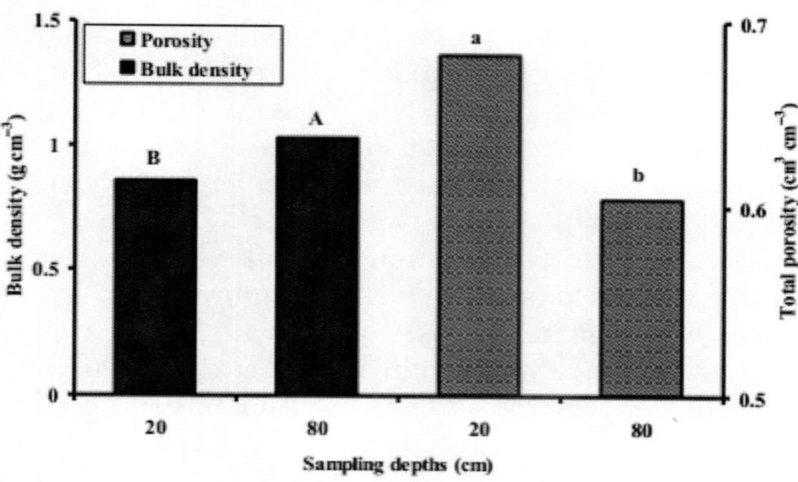


Figure 2. Effect of sampling depths on bulk density and total porosity of soil samples collected from 20 and 80 cm soil depths. Tukey's mean separation results are shown by the different letters, i.e., the two groups with two different letters were statistically different.

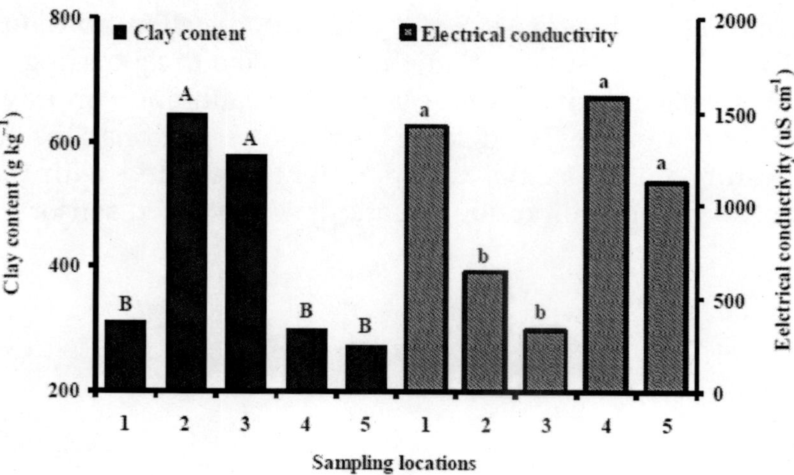


**Figure 3.** Effect of sampling locations on clay content and electrical conductivity of soil samples collected from locations 1 through 5 across the watershed. Tukey's mean separation results are shown by the different letters, i.e., the two groups with two different letters were statistically different.

The values of EC at 20 cm soil depth were almost double of those at 80 cm depth at locations 1 through 3; whereas, at locations 4 and 5, the EC values at 20 cm depth were 1.5 and 1.2 times those at 80 cm depths, respectively (Table 2). Release of nutrients due to decomposition of organic matter from tree litter in these forested watershed soils could be a reason for these larger EC values of the surface (20 cm depth) soil samples. Larger EC values at locations 4 and 5 might be due to the mineral composition of these oxisol and ulitsol that include iron and aluminum oxides, hydroxides, quartz, kaolin, clay minerals, and organic matter. Most tropical soils, including these in the study site, are acidic due to high leaching under warm temperature and intense rainfall conditions [34,35]. There was no significant effect of sampling depth on CC, and EC, and of sampling location on $\rho_b$ and $\theta_t$ (Table 3).

## Laboratory Calibration Equations

The laboratory calibration equations of EC-10, EC-20, and ML2x (Table 4) accurately ($P < 0.001$; $r > 0.95$) predicted the actual water content (RMSE $0.93 \times 10^{-2}$ to $5.59 \times 10^{-2}$ $cm^{-3}$ $cm^{-3}$) compared with their respective default equations (RMSE $3.69 \times 10^{-2}$ to $10.7 \times 10^{-2}$ $cm^{-3}$ $cm^{-3}$). The values of MBE of the laboratory calibration equations were 10 to 100 times smaller than those of their corresponding default equations. This indicates that the laboratory calibration improved the sensors' performance. The default calibration equations substantially under-estimated the actual water content compared with the site-specific laboratory calibration equations of the tested sensors.

**Table 4.** Calibration functions of EC-10, EC-20, and ML2x laboratory calibration equations and statistical indicators for their accuracy in estimating actual water content. The values of RMSE and MBE in parentheses are from the comparison of actual water content with that calculated with the manufacturer calibration equations.

| Station | Depth | Sensor | *a* | *b* | P | *r* | RMSE | MBE |
|---|---|---|---|---|---|---|---|---|
| | | | $cm^{-3}\ cm^{-3} \times 10^{-2}$ | | | | $cm^{-3}\ cm^{-3} \times 10^{-2}$ | |
| 1 | 20 | EC-10 | 0.045 | −26.020 | 1.6E-04 | 0.99 | 2.18 (4.70) | −0.010 (1.390) |
| | | EC-20 | 0.040 | −23.487 | 7.0E-04 | 0.98 | 3.13 (3.95) | 0.033 (−2.167) |
| | | ML2x | 10.83 | −14.550 | 2.4E-03 | 0.96 | 4.25 (4.55) | −0.030 (−0.787) |
| | 80 | EC-10 | 0.062 | −23.090 | 1.4E-03 | 0.97 | 4.42 (4.64) | 0.016 (0.008) |
| | | EC-20 | 0.051 | −39.035 | 3.6E-04 | 0.95 | 5.59 (6.27) | −0.041 (−5.08) |
| | | ML2x | 14.63 | −28.512 | 3.5E-05 | 0.99 | 1.77 (3.81) | 0.000 (−5.44) |
| 2 | 20 | EC-10 | 0.056 | −27.669 | 1.5E-03 | 0.99 | 2.60 (9.67) | −0.026 (−9.312) |
| | | EC-20 | 0.041 | −29.963 | 6.7E-05 | 0.99 | 0.93 (8.49) | 0.041 (−7.960) |
| | | ML2x | 12.96 | −14.989 | 7.2E-03 | 0.97 | 4.37 (8.58) | 0.001 (−7.260) |
| | 80 | EC-10 | 0.061 | −32.060 | 3.6E-03 | 0.95 | 5.25 (10.7) | 0.043 (−9.312) |
| | | EC-20 | 0.055 | −34.963 | 1.7E-05 | 0.99 | 1.40 (8.90) | 0.031 (−7.877) |
| | | ML2x | 13.44 | −16.906 | 4.0E-03 | 0.95 | 5.43 (9.08) | −0.002 (−7.037) |
| 3 | 20 | EC-10 | 0.051 | −24.630 | 2.6E-03 | 0.96 | 4.34 (8.60) | −0.012 (−7.227) |
| | | EC-20 | 0.043 | −21.547 | 2.6E-03 | 0.96 | 4.33 (9.55) | −0.019 (−8.510) |
| | | ML2x | 10.39 | −11.833 | 4.2E-03 | 0.95 | 4.93 (5.73) | 0.000 (−2.067) |
| | 80 | EC-10 | 0.060 | −31.592 | 5.8E-04 | 0.98 | 3.43 (9.61) | 0.012 (−8.893) |
| | | EC-20 | 0.053 | −31.423 | 1.3E-04 | 0.99 | 2.35 (10.1) | −0.026 (−9.153) |
| | | ML2x | 12.43 | −16.718 | 1.2E-03 | 0.97 | 4.07 (5.83) | 0.001 (−4.113) |
| 4 | 20 | EC-10 | 0.042 | −22.701 | 2.3E-04 | 0.99 | 2.38 (5.82) | 0.042 (1.002) |
| | | EC-20 | 0.037 | −21.342 | 7.9E-04 | 0.98 | 3.25 (4.18) | −0.011 (−1.740) |
| | | ML2x | 9.072 | −13.183 | 6.7E-03 | 0.97 | 3.35 (6.14) | 0.000 (4.186) |
| | 80 | EC-10 | 0.047 | −24.557 | 5.4E-04 | 0.98 | 2.97 (5.16) | 0.018 (2.800) |
| | | EC-20 | 0.040 | −21.146 | 1.5E-03 | 0.97 | 3.83 (6.29) | −0.114 (−4.895) |
| | | ML2x | 10.55 | −13.378 | 2.5E-03 | 0.96 | 4.36 (4.82) | 0.000 (0.860) |
| 5 | 20 | EC-10 | 0.049 | −26.874 | 3.6E-04 | 0.98 | 2.45 (4.44) | −0.018 (−3.012) |
| | | EC-20 | 0.043 | −25.819 | 3.4E-04 | 0.99 | 2.40 (4.76) | −0.011 (−4.090) |
| | | ML2x | 10.70 | −17.344 | 1.8E-03 | 0.99 | 1.94 (3.69) | −0.348 (2.347) |
| | 80 | EC-10 | 0.057 | −31.216 | 7.6E-04 | 0.98 | 3.52 (7.48) | −0.025 (−6.603) |
| | | EC-20 | 0.048 | −28.525 | 1.3E-03 | 0.97 | 4.05 (8.38) | 0.024 (−7.048) |
| | | ML2x | 12.18 | −17.743 | 7.9E-04 | 0.98 | 3.56 (4.23) | 0.001 (−2.248) |

a = slope of calibration equation; b = y-intercept of calibration equation; P and r: probability and coefficient of correlation values obtained from the regression between the calculated and the actual water contents.

Under-estimation of the actual water content, either by laboratory or by default calibration equations for ML2x, may be attributed to the high CC [37]. Soils that have high CC contain larger amounts of bound water due to the large surface areas of clay particles compared to silt and sand particles. Bound water, has lower ε than free water [38] and its proportion in a soil positively correlates with the soil surface area. With the increase of CC in a soil, the proportion of bound water to free water increases. Consequently, the smaller ε of the bound water in clay soils causes under-estimation of actual water content in high CC soils [39]. This is especially true at low water content where the ratio of bound water to free water substantially increases [18].

## Effect of Soil Depth and Location on the Calibration Equations

The slope and y-intercept functions (a and b) of the laboratory calibration equations of the EC-10, EC-20, and ML2x sensors for the two depths and five locations were evaluated for the effect of soil sampling depth and location. There was no statistically significant effect of sampling locations on them; but there was a significant ($P < 0.05$) effect of soil depths on a and b of the EC-20 and ML2x equations. The effect of sampling depth on the calibration equations of the EC-20 and ML2x was maybe due to the significant differences ($P < 0.05$) in the values of $\rho_b$, $\theta_t$, and CC at the two depths (Tables 2, 3). Huang et al. [36] and Foley and Harris [12] also reported that the EC-20 and ML2x were sensitive to varying $\rho_b$. For the EC-20 and ML2x, there was no significant effect of sampling locations on slope and y-intercept functions of the laboratory calibration equations. Therefore, one calibration equation per depth can be used for the entire watershed for each sensor. The sampling depths or the locations did not affect the laboratory calibration equation functions of the EC-10 suggesting that it needs one calibration equation for the entire watershed, irrespective of depth.

## Watershed-Specific Calibration Equations

One watershed-specific calibration equation for the EC-10 and two for the EC-20 and ML2x (one for each depth) were established (Table 5). Calculation of the water content using these watershed-specific calibration equations resulted in smaller RMSE and MBE than with their corresponding default equations. The watershed-specific laboratory calibration equation of the EC-10 was more accurate than its corresponding default equation as its MBE is five times smaller than that of the default equation. Similarly, there was an improvement in the accuracy of the EC-20 with the watershed-specific laboratory calibration equations for the two depths. The ML2x watershed-specific laboratory calibration equations for the two depths were more accurate (MBE $0.001 \times 10^{-2}$ and $-1 \times 10^{-6}$ $cm^{-3}$ $cm^{-3}$ for 20 and 80 cm, respectively) than the ML2x default equation (MBE $-0.565 \times 10^{-2}$ and $-2.96 \times 10^{-2}$ $cm^{-3}$ $cm^{-3}$ for 20 and 80 cm, respectively).

The accuracy of the laboratory calibration equations for ECH2O sensors was the highest for the water content between 0.2 and 0.5 $cm^3$ $cm^{-3}$ [Figure 4(A,B)]. There was slight decrease in the accuracy of the watershed-specific laboratory calibration equations of the EC-20 at 20 and 80 cm depths for the water content $\leq 0.2$ $cm^3$ $cm^{-3}$ [Figure 4(B)]. For the ML2x, the laboratory calibration equations performed better at water content $\geq 0.25$ $cm^3$ $cm^{-3}$ [Figure 4(C)]. However, there was no difference between the two calibration equations of ML2x for water content $\leq 0.15$ $cm^3$ $cm^{-3}$ [Figure 4(C)]. Overall, the default calibration equations of the ML2x performed better than those of the $ECH_2O$ sensors across the tested range of water content. The EC-10 had the highest variation in its readings among the three sensors; it had an absolute error as high as 20% as shown in Figure 4(A), 830 mV corresponds to 0.10 and 0.30 $cm^3$ $cm^{-3}$ actual water contents. The EC-20 and ML2x showed absolute errors of up to 10 to 12% and 10 to 15%, respectively, especially at medium and high soil water content ranges [Figure 4(B,C)].

**Table 5.** Calibration functions of the EC-10, EC-20, and ML2x site-specific laboratory calibration and default equations and statistical indicators for their accuracy in estimating actual water content.

| Sensor / Calibration | Depth, cm | *n* | *a* | *b* | RMSE | MBE |
|---|---|---|---|---|---|---|
| | | | $cm^{-3}$ | $cm^{-3} \times 10^{-2}$ | $cm^{-3}$ | $cm^{-3} \times 10^{-2}$ |
| EC-10 | 20–80 | 59 | 0.0496 | −25.654 | 5.54 | −0.643 |
| Default | 20–80 | 59 | 0.0571 | −37.597 | 7.40 | −4.390 |
| EC-20 | 20 | 29 | 0.0408 | −22.418 | 4.43 | 0.009 |
| | 80 | 30 | 0.0467 | −27.942 | 5.39 | 0.026 |
| Default | 20 | 29 | 0.0424 | −28.997 | 6.55 | −4.79 |
| | 80 | 30 | 0.0424 | −28.997 | 8.12 | −5.90 |
| ML2x | 20 | 28 | 10.302 | −12.961 | 5.54 | −0.001 |
| | 80 | 30 | 12.272 | −17.357 | 5.04 | −1E−04 |
| Default | 20 | 28 | 11.900 | −19.050 | 5.99 | −0.565 |
| | 80 | 30 | 11.900 | −19.050 | 5.87 | −2.96 |

a = slope of calibration equation; b = y-intercept of calibration equation.

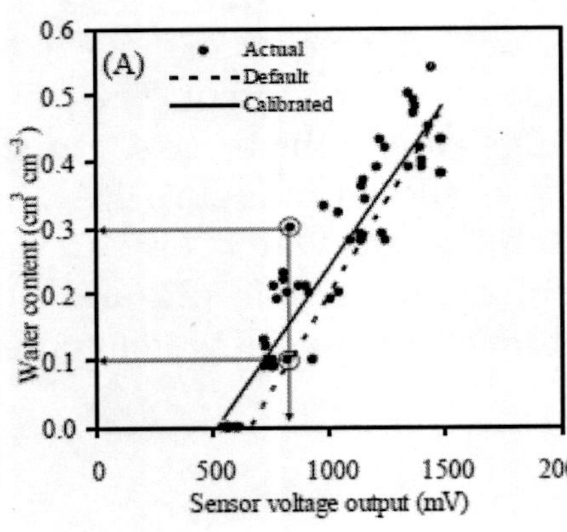


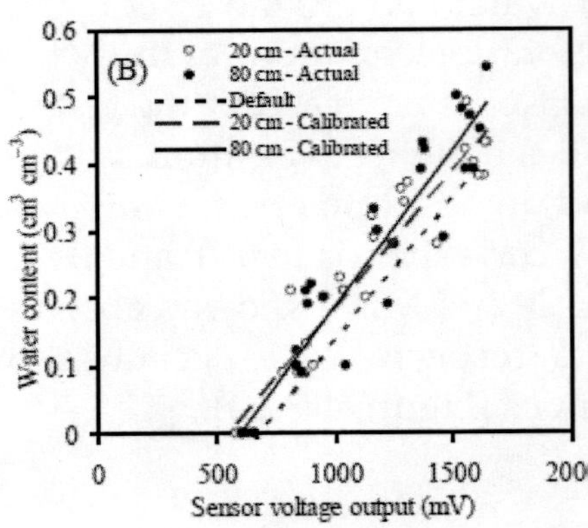

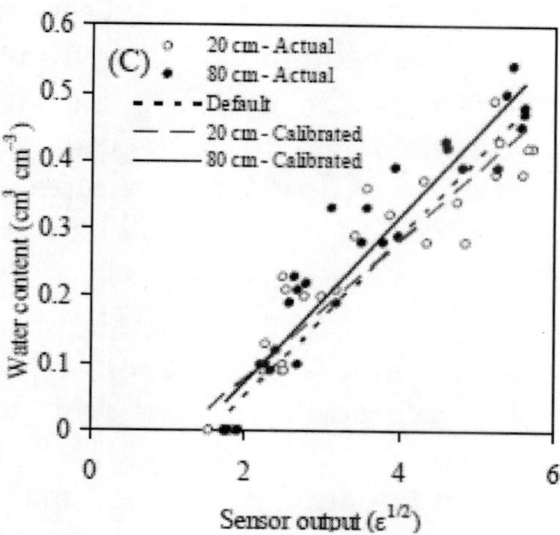


**Figure 4.** Actual soil water content as a function of the readings of the EC10 (A), EC-20 (B), and ML2x (C) using soil samples from 20 and 80 cm depths across the watershed. Arrows connecting actual water content data points (circled blue) with x and y axes in 4A show an example of absolute error from EC-10 as it gave same reading (ca. 830 mV) for 0.1 and 0.3 cm3 cm−3 water contents resulting in 20% error.

## Sensor Accuracy and Precision

Significant variations of $\rho_b$ and $\theta_t$ across the sampling depths and of CC and EC across the sampling locations had a significant effect on sensors' accuracy (large RMSE values in Tables 4 and 5).

The EC-20 exhibited poor accuracy with laboratory (RMSE $5.59 \times 10^{-2}$ cm$^3$ cm$^{-3}$) and default (RMSE $6.27 \times 10^{-2}$ cm$^3$ cm$^{-3)}$ calibration equations at 80 cm depth of location 1 which may be due to the large value of $\rho_b$, i.e., 1.27 g cm$^{-3}$ (Table 3). Likewise, a small value of $\rho_b$, i.e., 0.69 g cm$^{-3}$ at 20 cm depth of location 5 resulted in a fair accuracy of the EC-20 with the laboratory (RMSE $2.45 \times 10^{-2}$ cm$^3$ cm$^{-3}$) and default (RMSE $4.76 \times 10^{-2}$ cm$^3$ cm$^{-3}$) calibration equations.

The accuracy of the EC-10 sensor was fair with the use of laboratory calibration equation and poor with the use of default equation, except at the two depths of location 1 and at the 20 cm depth of location 5 where the default equation had similar accuracy to that of laboratory equations. Overall, the laboratory calibration equations improved the accuracy of the tested sensors as compared to their corresponding default equations except under large value of $\rho_b$ at 80 cm of locations 1 (EC-20; $\rho_b$ 1.27 g cm$^{-3}$) and 2 (EC-10; $\rho_b$ 1.06 g cm$^{-3}$). The use of default calibration equations of all sensors resulted in poor to very poor sensor accuracy especially at locations 2 and 3 (Table 4); the sensors' accuracy varied between fair and poor at the remaining locations.

The percent reduction in RMSE, calculated from Equation 3, reflected that the laboratory calibration equations of each sensor improved their accuracy compared with the use of their corresponding default equations (Table 6). The percent reduction in RMSE, calculated from Equation 3, reflected that the accuracy of each sensor improved with the use of their laboratory calibration equations compared with the corresponding default equations (Table 6). Percent reduction in RMSE ranged from 11 to 89% for EC-20, 4.7 to 64% for the EC-10, and 7 to 47% for the ML2x.

These results reflect improvement in the sensors' accuracy with the use of laboratory calibration equations over their corresponding default equations in the ascending order EC-20 > EC-10 > ML2x. The watershed-specific calibration equations also improved the accuracy of the EC-10 by 25%; whereas, the accuracy of the EC-20 and ML2x was improved by 32 and 44% for 20 cm depth and by 7.5 and 14% for 80 cm soil depth. Jones et al. [40] attributed the poor performance of $ECH_2O$ sensors to EC effects. Such effects dominate the output of sensors that operate at frequencies < 100 MHz [41], i.e., $ECH_2O$ sensors, which operate at 5 MHz. Poor performance of the EC-20 can also be attributed to the sensor's functionality of averaging its readings over its plane of interface with the soil (i.e., 20 cm) than the EC-10 (i.e., 10 cm).

**Table 6.** Percent improvement in the accuracy of the EC-10, EC-20, and ML2x sensors based on the root mean square error (RMSE) values of the laboratory and default calibration equations to predict the actual soil water content

| Location | Depth, cm | EC-10 | EC-20 | ML2x |
|---|---|---|---|---|
| 1 | 20 | 54 | 21 | 7 |
| | 80 | 4.7 | 11 | 54 |
| 2 | 20 | 73 | 89 | 49 |
| | 80 | 51 | 84 | 40 |
| 3 | 20 | 50 | 55 | 14 |
| | 80 | 64 | 77 | 30 |
| 4 | 20 | 59 | 22 | 45 |
| | 80 | 42 | 39 | 10 |
| 5 | 20 | 45 | 50 | 47 |
| | 80 | 53 | 52 | 16 |
| Watershed scale | 20 | 25 | 32 | 7.5 |
| 1–5 | 80 | - | 34 | 14 |

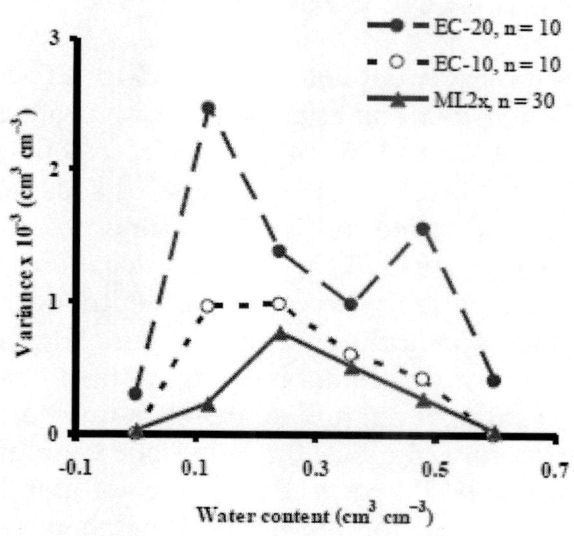


**Figure 5.** Precision of the EC-10, EC-20, and ML2x sensors judged from the variance of sensor's readings repeatedly (n = 10 for the $ECH_2O$ sensors; and n = 30 for the ML2x, where n is number of repeated measurements) taken at the same water content level.

The ML2x exhibited better precision (mean variance; MV 0.31 × 10−3 cm3 cm−3) than the EC-10 (MV 0.49 × $10^{-3}$ $cm^3$ $cm^{-3}$) and EC-20 (MV 1.18 × $10^{-3}$ $cm^3$ $cm^{-3}$) (Figure 5). The variance from the true mean of the actual water content ranged from 0.03 × 10−3 to 0.77 × $10^{-3}$, 0 to 0.98 × $10^{-3}$, and 0.29 × $10^{-3}$ to 2.47 × $10^{-3}$ $cm^3$ $cm^{-3}$ for the ML2x, EC-10, and EC-20, respectively. The high precision of the ML2x was may be due to: (i) its rods' perfect contact with the surrounding media and (ii) its impedance of 100 MHz sinusoidal signal, which is supposed to minimize the effect of EC on sensor's readings [8,41]. Overall, the sensors showed better precision at low (<0.1 $cm^3$ $cm^{-3}$) and high water content (>0.5 $cm^3$ $cm^{-3}$) levels than at medium water content (0.15–0.45 $cm^3$ $cm^{-3}$) where all the sensors' exhibited poorer precision. Rosenbaum et al. [42] reported higher variances (low precision) than our results from the repeatability experiments of the $ECH_2O$ and other sensors. Extensively repeated measurements might result in more realistic variances and better representation of sensors' precision.

## SUMMARY AND CONCLUSIONS

Laboratory calibration equations of The EC-10, EC-20, and ML2x sensors were established and evaluated for accurate measurement of water content at 20 and 80 cm depths across five locations of the forested Upper Mākaha Valley watershed soils of varying $\rho_b$, $\theta_t$, CC, and EC. None of the studied soil properties, except CC and EC, was significantly affected by sampling locations. Bulk density and $\theta_t$ significantly varied with sampling depth and consequently affect the laboratory calibration equation functions of the EC-20 and ML2x; however, the calibration equation functions of the EC-10 had no effect of spatial variability neither as a function of depth nor location. Consequently, the EC-10 needed one calibration equation for the entire watershed, irrespective of the soil depths. However, one calibration equation for the entire watershed per depth was needed for EC-20 and ML2x to capture the spatial variations encountered on this tropical watershed. The laboratory calibration equations improved the sensors' measurement ability as compared to that with their corresponding default equations. The maximum improvement was for the EC-20, followed by the EC-10 and ML2x. Moreover, the ML2x exhibited the highest precision, most probably due to its higher operating frequency, followed by the EC-10 and EC-20. These results reinforce the need for site-specific calibration equations specifically for fields with large spatial variability.

## ACKNOWLEDGEMENTS

This project was supported by two grants from the US Department of Agriculture: (1) Cooperative State Research, Education and Extension Service grant number 2004-34135-15058, (2) McIntire-Stennis formula grant number 2006-34135-17690. The authors wish to thank the Honolulu Board of Water Supply for its cooperation, and Amjad Ahmad and Nghia Dai Tran for their help in soil sample collection.

## REFERENCES

1. Henninger, D.L.; Peterson, G.W.; Engman, E.T. Surface soil moisture within a watershed: variation, factors influencing, and relationships to surface runoff. Soil Sci. Soc. Am. J. 1976, 40, 773-776.
2. Gómez-Plaza, A.; Martínez-Mena, M.; Albaladejo, J.; Castillo, V.M. Factors regulating spatial distribution of soil water content in small semiarid catchments. J. Hydrol. 2001, 253, 211-226.
3. Hu, W.; Shao, M.A.; Wang, Q.J.; Reichardt, K. Soil water content temporal-spatial variability of the surface layer of a loess plateau hillside in china. Sci. Agric. 2008, 65, 277-289.
4. Pan, Y.-X.; Wang, X.-P.; Jia, R.-L.; Chen, Y.-W.; He, M.-Z. Spatial variability of surface soil moisture content in a re-vegetated desert area in Shapotou, Northern China. J. Arid Environ. 2008, 72, 1675-1683.
5. Fares, A.; Polyakov, V. Advances in crop water management using capacitive water sensors. Adv. Agron. 2006, 90, 43-77.
6. Robinson, D.A.; Campbell, C.S.; Hopmans, J.W.; Hornbuckle, B.K.; Jones, S.B.; Knight, R.; Ogden, F.; Selker, J.; Wendroth, O. Soil moisture measurement for ecological and hydrological watershed-scale observatories: A review. Vadose Zone J. 2008, 7, 358-389.
7. Decagon Devices, Inc. ECH2O Utility Mobile For Windows Handheld PCs Operator's Manual; 2002. Available online: http://www.decagon.com/literature/manuals/Ech2oUtilityMobileManual.pdf (accessed on 12 March 2009).
8. Dynamax Inc. ThetaProbe Type ML2x; 2009. Available online: ftp://ftp.dynamax.com/ DynamaxPDF/irrigation-controls/ML2x.pdf (accessed on 29 April 2011).

9. McMichael, B.; Lascano, R.J. Laboratory evaluation of a commercial dielectric soil water sensor. Vadose Zone J. 2003, 2, 650-654.

10. Fares, A.; Hamdhani, H.; Jenkins, D.M. Temperature-dependent scaled frequency: Improved accuracy of multisensor capacitance probes. Soil Sci. Soc. Am. J. 2007, 71, 894-900.

11. Fares, A.; Safeeq, M.; Jenkins, D.M. Adjusting temperature and salinity effects on single capacitance sensors. Pedosphere 2009, 19, 588-596.

12. Foley, J.L.; Harris, E. Field calibration of ThetaProbe (ML2x) and ECHO probe (EC-20) soil water sensors in a Black Vertosol. Aus. J. Soil Res. 2007, 45, 233-236.

13. Czarnomski, N.M.; Moore, G.W.; Pypker, T.G.; Licata, J.; Bond, B.J. Precision and accuracy of three alternative instruments for measuring soil water content in two forest soils of the Pacific Northwest. Can. J. For. Res. 2005, 35, 1867-1876.

14. Overduin, P.P.; Yoshikawa, K.; Kane, D.L.; Harden, J.W. Comparing electronic probes for volumetric water content of low-density feathermoss. Sen. Rev. 2005, 25, 215-221.

15. Logsdon, S.D.; Hornbuckle, B. Soil moisture probes for dispersive soils. In Proceedings of TDR 2006: Third International Symposium and Workshop on Time Domain Reflectometry for Innovative Soils Applications, West Lafayette, IN, USA, September 2006; Available online: https://engineering.purdue.edu/TDR/Papers/13_Paper.pdf (accessed on 29 April 2011).

16. Bogena, H.R.; Huisman, J.A.; Oberdorster, C.; Vereecken, H. Evaluation of a low-cost soil water content sensor for wireless network applications. J. Hydrol. 2007, 344, 32-42.

17. Mendes, L.B.; Li, H.; Xin, H. Evaluation and Calibration of a Soil Moisture Sensor for Measuring Poultry Manure or Litter Moisture Content. In Proceedings of 2008 ASABE Annual Meeting, Providence, RI, USA, 29 June–2 July 2008; Paper Number: 084438.

18. Wraith, J.M.; Or, D. Soil water characteristic determination from concurrent water content measurements in reference porous media. Soil Sci. Soc. Am. J. 2001, 65, 1659-1666.

19. Vaughan, B. How to determine an accurate soil testing laboratory. In Site Specific Management Guidelines SSMG 4; Potash and Phosphate Institute: Norcross, GA, USA, 1999; p. 2.

20. Adamchuk, V.I.; Hummel, J.W.; Morgan, M.T.; Upadhyaya, S.K. On-the-go soil sensors for precision agriculture. Comp. Elect. Agri. 2004, 44, 71-91.

21. Williams, J.; Sinclair, D.F. Accuracy, bias and precision. In Soil Water Assessment by the Neutron Method; Greacen, E.L., Ed.; CSIRO: Melbourne, VIC, Australia, 1981; pp. 35-49.

22. Fares, A.; Buss, P.; Dalton, M.; El-Kadi, A.I.; Parsons, L.R. Dual field calibration of capacitance and neutron sensors in a shrinking-swelling clay soil. Vadose Zone J. 2004, 3, 1390-1399.

23. Allmaras, R.R.; Kempthorne, O. Errors, variability and precision. In Methods

of Soil Analysis, Part 4; Dane, J.H., Topp, G.C., Eds.; SSSA Book Ser. 5; SSSA: Madison, WI, USA, 2002; pp. 15-44.

24. Evett, SR.; Tolk, J.A.; Howell, T.A. Soil profile water content determination: sensor accuracy, axial response, calibration, temperature dependence, and precision. Vadose Zone J. 2006, 5, 894-907.

25. Kendall, M.G.; Buckland, W.R. Dictionary of statistical terms. Oliver and Boyd: London, UK, 1957.

26. Fares, A. Determining the Impacts of Water Pumping and Alien Species Invasion on Stream Flow for a Sustainable Water Resource Management in Makaha Valley, Hawaii; Project Report; Submitted to the Board of water supply, Honolulu, HI, USA, 2007; p. 16.

27. Foote, D.E.; Hill, E.L.; Nakamura, S.; Stephens, F. Soil Survey of the Islands of Kauai, Oahu, Maui, Molokai, and Lanai, State of Hawaii; United States Department of Agriculture, Washington D.C. United States Printing Office: Washington, DC, USA, 1972.

28. Whalley, W.R. Considerations on the use of time-domain reflectometry (TDR) for measuring soil moisture content. J. Soil Sci. 1993, 44, 1-9.

29. 29. White, I.K.; Zegelin, J.H.; Topp, G.C. Comments on 'Considerations on the use of time-domain reflectometry (TDR) for measuring soil water content' by W R Whalley. Eur. J Soil Sci. 1994, 45, 503-508.

30. Grossman, R.B.; Reinsch, T.G. Bulk density and linear extensibility. In Methods of Soil Analysis, Part 4; Dane, J.H., Topp, G.C., Eds.; SSSA Book Ser. 5; SSSA: Madison, WI, USA, 2002; pp. 201-254.

31. Flint, L.E.; Flint, A.L. Porosity. In Methods of Soil Analysis, Part 4; Dane, J.H., Topp, G.C., Eds.; SSSA Book Ser. 5; SSSA: Madison, WI, USA, 2002; pp. 241-254.

32. Gee, G.W.; Or, D. Particle-size analysis. In Methods of Soil Analysis, Part 4; Dane, J.H., Topp, G.C., Eds.; SSSA Book Ser. 5; SSSA: Madison, WI, USA, 2002; pp. 255-293.

33. 33. Analytical Software. User's manual: STATISTIX8; Analytical Software: Tallahassee, FL, USA, 2003.

34. Fox, R.L.; Hue, N.V.; Jones, R.C.; Yost, R.S. Plant-soil interactions associated with acid, weathered soils. Plant Soil 1991, 134, 65-72.

35. Hunter, D.J.; Yapa, G.G.; Hue, N.V. Effects of green manure and coral lime on corn growth and chemical properties of an acid Oxisol in Western Samoa. Biol. Fertil. Soils 1997, 24, 266-273.

36. Huang, Q.; Akinremi, O.; Rajan, R.S.; Bullock, P. Laboratory and field evaluation of five soil water sensors. Can. J. Soil Sci. 2004, 84, 431-438.

37. Little, K.M.; Metelerkamp, B.; Smith, C.W. A comparison of three methods of soil-water content determination. S. Afr. J. Plant Soil. 1998, 15, 80-89.

38. Dobson, M.C.; Ulaby, F.T.; Hallikainen, M.T.; Elrayes, M.A. Microwave dielectric behavior of we soil 2. Dielectric mixing models. IEEE T. Geosci. Rem. Sen. 1985, 23, 35-46.

39. Seyfried, M.S.; Murdock, M.E. Response of a new soil water sensor to variable soil, water content, and temperature. Soil Sci. Soc. Am. J. 2001, 65, 28-34.

40. Jones, S.B.; Blonquist, J.M., Jr.; Robinson, D.A.; Rasmussen, V.P.; Or, D. Standardizing characterization of electromagnetic water content sensors. Part 1. Methodology. Vadose Zone J. 2005, 4, 1048-1058.

41. Hilhorst, M.A. Dielectric Characterisation of Soil. Ph.D. Thesis, Wageningen Agricultural University, Wageningen, The Netherlands, 1998; p. 141.

42. Rosenbaum, U.; Huisman, J.A.; Weuthen, A.; Vereecken, H.; Bogena, H.R. Quantification of sensor-to-sensor variability of the ECH2O EC-5, TE and 5TE sensors in dielectric liquids. Vadose Zone J. 2010, 9, 181-186.

# Chapter 15

# FROM LABORATORY STUDIES TO THE FIELD APPLICATIONS OF ADVANCED OXIDATION PROCESSES: A CASE STUDY OF TECHNOLOGY TRANSFER FROM SWITZERLAND TO BURKINA FASO ON THE FIELD OF PHOTOCHEMICAL DETOXIFICATION OF BIORECALCITRANT CHEMICAL POLLUTANTS IN WATER

S. Kenfack,[1] V. Sarria,[2] J. Wéthé,[3] G. Cissé,[4] A. H. Maïga,[3] A. Klutse,[1] and C. Pulgarin[5]

[1]Département de Recherche et Projets de Démonstration (REPRODEM), Centre Régional pour l'Eau potable et l'Assainissement (CREPA), 03BP 7112 Ouagadougou 03, Burkina Faso

[2]Departamento de Química, Universidad de los Andes, Cra 1 No 18A-10, Bogota, Colombia

[3]Unité Thématique d'Enseignement et de Recherche en Gestion et Valorisation de l'Eau et de l'Assainissement (UTER-GVEA), Institut International d'Ingénierie de l'Eau et de l'Environnement, 01 BP 594 Ouagadougou 01, Burkina Faso

[4]Centre Suisse de Recherche Scientifique de Côte d'Ivoire (CSRS), 01 BP 1301 Abidjan 01, Cote d'Ivoire

[5]Ecole Polytechnique Fédérale de Lausanne, Institute of Chemical Sciences and Engineering, GGEC, Station 6, CH, 1015 Lausanne, Switzerland

## ABSTRACT

The Fenton and photo-Fenton detoxification of non-biodegradable chemical pollution in water was investigated under simulated UV light in the laboratory and under direct sunlight in Ouagadougou, Burkina Faso. The laboratory experiments enable one to make a systematic diagnosis among three types of wastewaters, identifying a biorecalcitrant wastewater containing the Chloro-hydroxy-Pryridine (CHYPR). The application of the photo-Fenton process on effluent containing the CHYPR showed not to stimulate the generation of biodegradable by-products. Optimal conditions for detoxification of effluent containing the CHYPR were found at , mM, initial mM, for an effluent concentrated at 2.2 mM of CHYPR. The application of the photochemical process on a field pilot solar photoreactor for the detoxification of water polluted with a pesticide made with Endosulfan showed very promising results, with potential biodegradable effluents obtained at the end of the photochemical treatment. Optimal conditions of the applied study were found at . mM and mM for an initial concentration of 0.36 mM of Endosulfan.

## INTRODUCTION

Anestimation of the global worldwater pollution reveals that 60%–70% of the total pollution is due to the agricultural activities, 25% to 30% by industries, and the remaining 5% to 10% due to the domestic uses [1]. The report showed that, in the most optimistic cases, the water pollution from agricultural origin is mostly consisted of the residues of pesticide and contributes to at least 50% on the deterioration of the quality of natural waters (surface and subsoil waters). This is closely followed by the pollution of chemical industrial origin. In fact, more than 1000 new substances are marketed each year in the world whereas up to now the toxicological information on only 1000 to 2000 of these products is accessible [2].

Most of these substances are xenobiotic (c.a. foreign to life), and very often they are synthetic products. Indeed, xenobiotic substances are in general not easily biodegradable, and they cross the biological water treatment systems without being completely degraded or even sometimes not degraded at all.

Some of these substances such as the pesticides are recognized as cause of carcinogenic, mutagen effects, or of hormonal disruptions to the fauna, wild life, and human beings.

To face these problems, waste treatment techniques such as incineration, wet oxidation, and activated carbon adsorption are often used. Unfortunately, these techniques are very expensive, and the environmental virtues of some of them are discussed.

However, biological processes remain the most economical and environmentally compatible alternative for wastewater treatment. Thus, it would be important to confirm the biorecalcitrancy of a wastewater before the application of a photocatalytic treatment. Moreover, biorecalcitrant wastewater could only be photocatalytically treated up to the point where its biodegradability is enough to let the phototreated water follows a biological treatment. Two case studies are reported in this paper to illustrate the strategy of enhancing the biodegradability of biorecalcitrant wastewaters throughout a photocatalytic process. These studies are representative of two problem solving approaches: one in an industrial developed country (Switzerland) and the other in a tropical developing country (Burkina Faso).

## The Laboratory Solving Problem Approach in Switzerland versus the Field Applied Approach in Burkina Faso

### *Case Study 1: The Chemical Pollution in Wastewater in a Swiss Manufacture*

In June 2004, a Swiss chemical company started the manufacture of a new significant chemical, following a three-step process. Three types of effluent of a specific composition were generated at each step, and all the three effluents were mixed to other effluents of the manufacture and treated in an activated sludge wastewater treatment plant. Before the new production, the yield of the wastewater treatment plant was efficient, and it could respect the legal rejection norms. But since the production of the new chemical has started, the yield has fallen down and the plant could not any more respect the standards, even when the hydraulic capacity overshooting of the treatment plant was still very low. The laboratory study aimed at making a diagnosis identifying the biorecalcitrant effluent and studied its degradation using the photoFenton treatment process.

## *Case Study 2 The Problem of Persistent Organic Pollutants and Obsolete Pesticides in Burkina Faso*

In order to implement the national plan of the Stockholm's Convention on the persistent organic pollutants (POPs), the government of Burkina Faso carried out in 2001 and 2004 two inventories of stocks of POPs pesticides available on the extent territory of Burkina Faso. Table 1 reports the results of these inventories.

Table 1: Results of the surveys of POPs and obsolete pesticides in Burkina Faso (2001 and 2004) [3].

| Form | Quantity | | | |
|---|---|---|---|---|
| | 2001 | 2004 | Nature of the pesticide in 2004 | |
| | | | Cypermethrin | Endosulfan |
| Contaminated empty containers (—) | 1450000 | 120'000 | | |
| Powder (kg) | 26000 | 3000 | | |
| Liquid (Liters) | 250000 | 130000 | 85700 | 4000 |

It was noticed that 90% of these pesticides were formulated with Cypermethrin (67%) and Endosulfan (3%) which are classified in the categories II and III (from the toxic to very toxic) by the WHO. The same report revealed that 13 contaminated sites of major importance are available in Burkina Faso. Since there is neither incinerator in Burkina Faso nor a specialized hazardous waste treatment plant, the experimental approach developed at the EPFL, Switzerland, was tested for the degradation of a pesticide containing Endosulfan, one of the chemicals which is known to generate toxic metabolites after a first step biological degradation [4].

# EXPERIMENTAL

## Materials and Apparatus

At the LBE-EPFL, the photocatalytic studies were performed using a 50 mL Pyrex flask with a cut-off at nm placed into a Hanau Suntest (Figure 1).

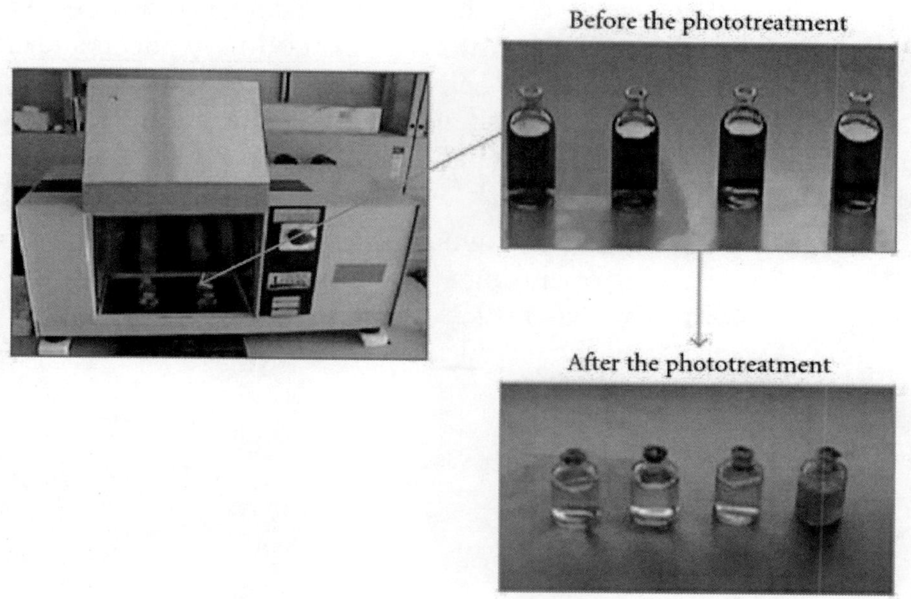


Figure 1: A Hanau Suntest Simulator containing 04 Pyrex flask samples during a laboratory processing at the BE-EPFL, Lausanne, Switzerland.

The radiation source was a Xenon lamp where the radiant flux (80 ) was measured with a power meter of YSI Corporation. The lamp had a regular distribution of wavelengths with about 0.5% of the emitted photons at wavelengths shorter than 300 nm (UV-C range) and about 7% between 300 and 400 nm (UV-B range). The profile of the photons emitted between 400 and 800 nm (UV-A range) followed the solar spectrum.

Experiments were done at the light intensity of 560, which corresponds to a relative UVA (300–400 nm) intensity of . All the chemicals ($FeSO_4$, $7H_2O$), HCl and $H_2O_2$ (30% w/w) were bought from Fluka (Switzerland), and all were analytical grade (p.a.).

Three types of wastewaters (see Table 1) containing pyridine compounds were received from DLK Technology S.A, a Swiss company specialized in the developing specific technical plants for water and wastewater treatment.

The pilot plant, a SOLARDETOX ACADUS-2003 device model delivered by Ecosystem SA, Barcelona, Spain, is a one plate Compound Parabolic Concentrator (CPC) module (collector useful surface: 2.12 $m^2$, photoreactor active volume 15.1 L within a total volume of 16.07 L) made of sixteen borosilicate cylindrical glass tubes. Each tube lies on a CPC aluminum mirror with one sun of concentration.

The reactor is mounted on a two-position fixed platform inclinable at 10° and 35° allowing operating at the approximate local latitude of Ouagadougou-Burkina Faso (12.2° N), at 10° angle position most of the time. A picture and the technical design of the CPC photoreactor used are shown in Figure 2.

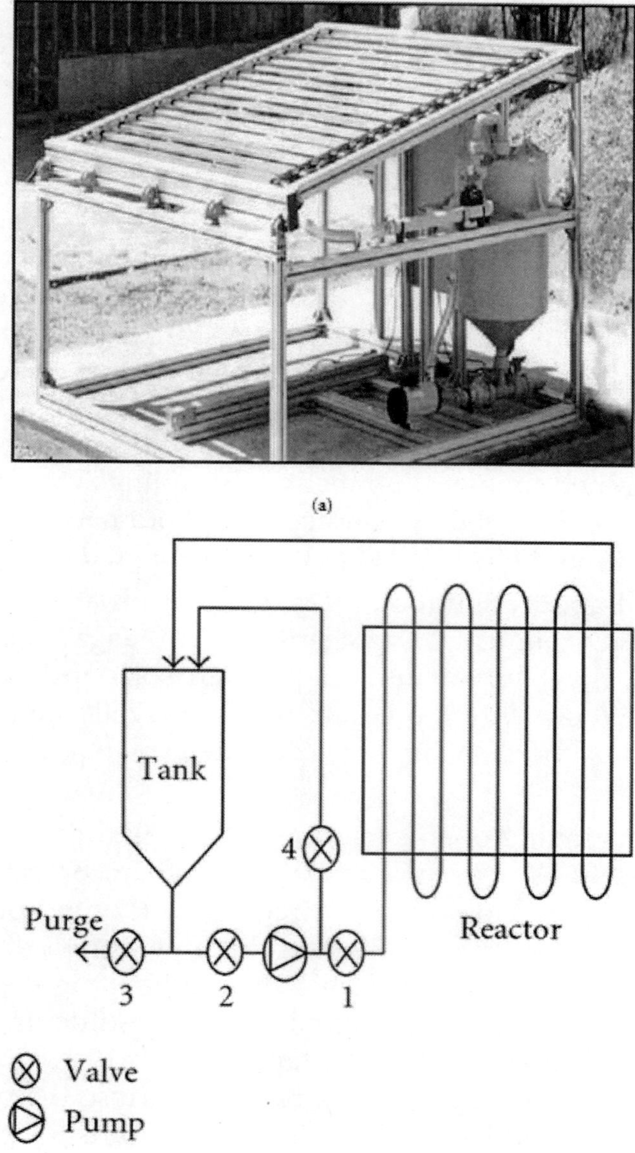


Figure 2: A Picture and the technical design of the CPC photoreactor.

From the operational point of view, the plant behaves a plug-flow reactor in the collector zone, connected to a 50 L polypropylene stirred tank (recirculating tank or buffer tank) for wastewater preparation: pH adjustments, catalyst, and oxidant feeding. The ensemble constituted by a plug-flow reactor in the collector zone plus a tank and a recirculation system is equivalent to a batch photoreactor.

## Procedures

In the laboratory, a systematic diagnosis including the characterization of each of the three effluents received was first made, followed by the photoassisted studies on the effluent containing chloro-2-pyridin. 2 liters of the raw effluent were first acidified at pH = 3. For the photodegradation processing, 50 mL of this preparation was put into a 60 mL of a Pyrex flask glass, and reactants (iron and peroxide) were added. The flask glass was placed into the Suntest photoreactor for a determined time of illumination. Samplings were made every 10 minutes for TOC, HPLC, and spectrophotometric analyses, and during the Zahn Wellens tests sampling were made daily.

The field application took place at the International Institute of Water and Environmental Engineering (2iE), Ouagadougou, Burkina Faso (12° N, 1° W). It is an ideal place where solar applied researches can be carried out with more than 300 days—2500 sunny hours per year (e.g., in Alméria, Spain, it is estimated up to 3000 sunny hours per year) [3].

At the beginning of an experiment, the mixture is homogenized to achieve a fixed concentration in the system, by circulating the preparation in a closed loop circuit (Tank-pump-photoreactor-tank). A desired stock of pesticide solution was first dissolved in 2 liters of tap water and stirred for at least 2 hours in the laboratory. Afterward, the solution was poured into the conditioning tank of the photoreactor in which 20 liters of tap water were previously added. Three more liters of tap water were used to rinse the predissolved Pyrex flask and added to the conditioning tank. The pH of the solution was adjusted to 3 with HCl, and the catalyst ($FeCl_3$) was added. The entire system was homogenized by closing the valves (1) and (3) and circulating the polluted water in a closed loop circuit for 15 minutes (see Figure 2).

Once the solution is ready, the required amount of hydrogen peroxide was added, and the feeding valve (1) was opened according to the water circulation speed needed for the operation. The water flow was varied by regulating the valve (1) to: 0.3, 0.4, and 0.5 .

The solar ultraviolet radiation was determined during the experiments through a UVA radiometer ACADUS 85 UV mounted on the reactor, at the same angle (10°) as its plate part. It provides data in terms of incident UVA radiation per area unit () and the accumulated energy  on the 2.12 $m^2$ surface of the photoreactor (in Wh).  is linked to the instantaneous irradiance flux (), the reactor total active area (A), the water volume (V), and the experiment duration (t) by the following relation:

$$E_{(t)} = E_{(t_0)} + \Delta t \times \overline{UV \times \frac{A}{V}}. \tag{1}$$

## Analytical Methods

At the LBE, HPLC analyses were carried out in a Varian 9065 unit provided with a Varian 9012 solvent delivery system, an automatic injector 9100 and a Varian Pro Star Variable (200–400 nm) diode array detector: 9065 Polychromic. All modules were piloted with a computer on which the Varian Star 5.3 software is installed for liquid chromatography data delivery. A reverse phase spherisorb silica column ODS-2 and a (70/30) (v/v) mixture of acetonitrile/water mobile phase were used to run the chromatography in isocratic mode at a flow rate of 1 . This technique allows the measuring of the main pollutant concentration in the solution and to follow the overall formation and decay of the aromatic and aliphatic by-products during the process.

AHitachiUV-visU-3010andaBiomate3modelspectrophotometers were used, respectively, at the LBE and at the 2iE, with a 1.0 cm quartz cell. The Hitachi UV-vis U-3010 spectrophotometer allowed generating the characteristic spectra of phototreated samples in the wavelength range of 200–600 nm, so that all the chemicals and/or complexes in the solutions can be signalled at their maximum absorption wavelength. Once we identified the specific absorption wavelengths through the spectra, we chose these particular wavelengths for the analyses on the multiwavelengths Biomate 3 spectrophotometer during the field's application at the 2iE at Ouagadougou, Burkina Faso.

The biological Zahn Wellens tests were carried out on the phototreated samples after 60% degradation of the TOC, following the OCDE standard method [5].

The Chemical Oxygen Demand (COD) analyses were carried out via a Hach-2000 spectrophotometer using the dichromate solution as the oxidant in a strong acid medium. 2 mL of the samples were put into the low range kits (0–150 ) and digested at 150°C for two hours. Once the digested samples were cooled, their optical density was determined at = 430 nm, in comparison to a blank, prepared with 2 mL of distilled water. The Biological Oxygen Demand for five days ($BOD_5$) was measured by means of an Hg-free WTW 2000 Oxytop unit thermostated at 20°C according to the standard method.

# RESULTS

## A Systematic Study of the Photocatalytic Removal of the Chemical Pollution of the Industrial Wastewater in Switzerland

### *Diagnosis: Physicochemical and Biological Characterization of Different Types of Effluents Resulting from the Processing of LQV.*

Table 2 presents the characteristics of three effluents ($WW_i$) from the production of the LQV (its molecular formula is confidential), and Figure 3 shows the results of the Zahn wellens tests carried out on the three effluents.

Table 2: Synthesis of the characteristics of the three effluents.

| Type of effluent | Main pollutant | Molecular structure | pH | TOC $(g \cdot L^{-1})$ |
|---|---|---|---|---|
| $WW_1$ | CTFEP | 2-Chloro-3-(2,2,2-trifluorethoxy)-pyridin | 8.3 | 45 |
| $WW_2$ | LQV | Confidential | 6 | 43.5 |
| $WW_3$ | CHPYR | 2-Chloro-3-hydroxypyridin | 5.8 | 6.5 |

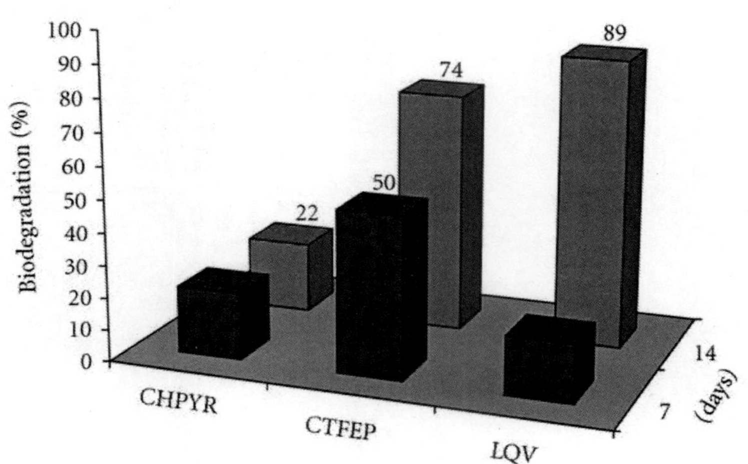


Figure 3: Percentage of biodegradation respectively 07 (■) and 14 days (■).

From these results, one can conclude the following.

(i)The wastewater mainly containing CTFEP is easily biodegradable and that with LQV is biodegradable after a certain period for bacteria adaptation.(ii)The wastewater mainly containing CHPYR is biorecalcitrant. No change in the degradation rate was observed at the , and up to the day. The Fenton and photoFenton processes were applied to $WW_3$, and the experimental conditions were optimized for the catalyst ($Fe^{2+}$) and the electron acceptor ($H_2O_2$).

## Fenton and PhotoFenton Treatment of the CHPYR Polluted Wastewater.

Figure 4 shows the chromatograms of theCHPYR effluent carried out at the beginning and after 24 hours of Fenton treatment on 50% diluted effluent on the one hand compared to the photoFenton treatment on the same effluent for 1 h30. The UV spectra of three significant peaks are observed on the raw effluent (nontreated). The two most important peaks of the chromatograms completely disappeared in case of the photoFenton treatment, and the one at 275.2 nm is 90% degraded; all the other peaks regressed relatively for 50% in case of the Fenton process after 24 hours.

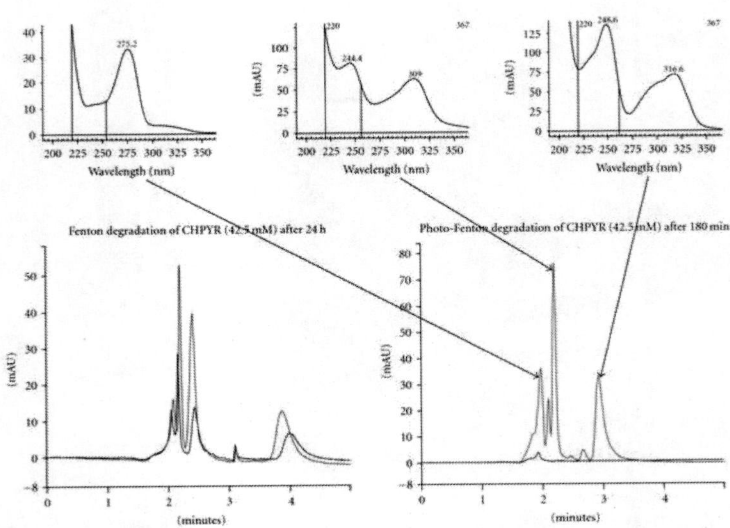


Figure 4: HPLC chromatograms of the CHPYR effluent before and after the Fenton (a) and the photoFenton (b) processes.

### *Optimization of the $H_2O_2$ and $Fe^{2+}$ Concentrations*

Figure 5 shows that the optimal conditions for the photoFenton treatment of the CHPYR effluent are obtained in 2 hours with an initial concentration of the solution of TOC = 2200 , mM, mM. Under these conditions, 60% of the organic pollutants (main pollutants and their degradation intermediates) are mineralized when the CHPYR is completely exhausted in the solution within the two hours of photoFenton treatment.

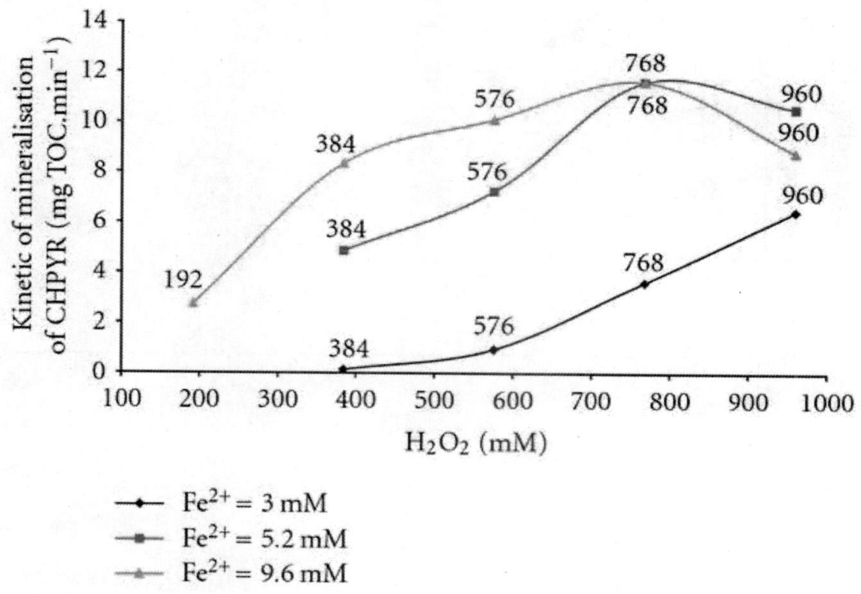


Figure 5: Kinetic of mineralisation of the CHPYR at various $H_2O_2$ concentration: (■) for 3mM; (▲) for 5.2 mM and (◆) for 9.6 mM, as a function of the $H_2O_2$ concentration during the PhotoFenton process in a 50% diluted effluent (42.5 mM).

## Biodegradability of the PhotoTreated Effluent

Figure 6 shows the results of the OECD's Zahn Wellens test carried out on the phototreated CHPYR affluent under the optimal photoFenton degradation conditions ( mM; $H_2$ mM; pH = 2.8), with a decrease of 60% of total organic carbon. This shows that, relatively less than 15% to 18% of biological degradation is reached at the end of the first weeks followed by a fold down to 12% and stagnation until the  day. It is assumed that the photogenerated intermediates of degradation of the CHPYR are also biorecalcitrant or toxic. Hence a total mineralization following the Fenton or photoFenton process is necessary.

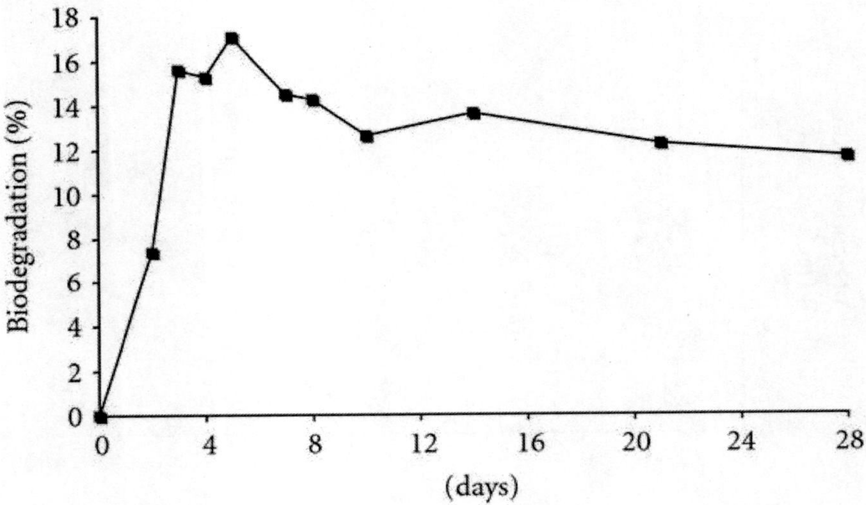


Figure 6: Evolution of the biodegradability following the Zahn Wellens process carried out on the CHPYR phototreated effluent.

## Application of the Helio-PhotoFenton Process for the Degradation Wastewater Polluted with Endosulfan

The monitoring of the biodegradability of the solar-treated effluent was conducted on 30 minutes sampled effluents. The results in Figure 7 show that the $BOD_5$ increases during the first 30 minutes of the treatment and decreases during the next hour; after that it remains constant. At the same time, the COD is continuously decreasing during one and half hour after which it remains constant. As a consequence of the evolution of these two parameters, the intrinsic biodegradability defined as the ratio

$BOD_5/COD$ is continuously increasing up to 50% where it became constant when the energetic treatment factor of the process is .

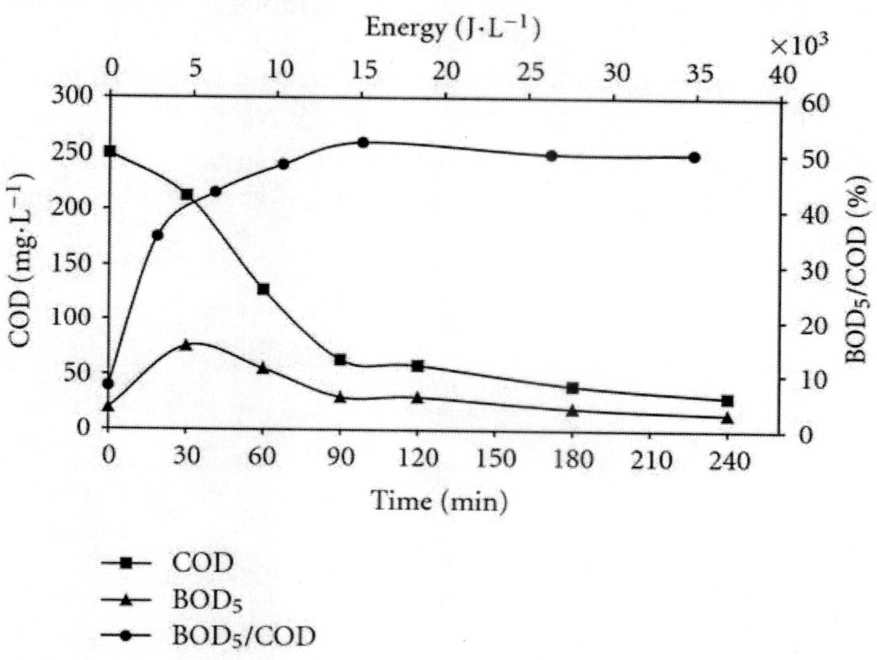


Figure 7: Evolution of biochemical parameter of the Endosulfan polluted water during the helio-photoFenton treatment, (■) for the COD, (▲) for $BOD_5$, and (•) for the $BOD_5/COD$ ratio.

## DISCUSSIONS

### Study of the Management of the Chemical Industry Effluents

Figure 4 confirmed the presence of biorecalcitrant pollutant (CHPYR) with maximum UV absorption at 309 nm. The studies carried out on the three types of effluents also showed that the effluents mainly containing LQV and of the CTFEP were biodegradable and only that of CHPYR was biorecalcitrant. However, the CHPYR effluent can be degraded through the Fenton and photoFenton processes, within the optimal conditions: [] = 5.2 mM; $[H_2O_2]$ = 768 mM for effluent concentrated at 2.2 of CHPYR. However, the Fenton process was not timely efficient.

The Zahn Wellens tests carried out on the phototreated effluent up to 60% TOC reduction did not show any significant biodegradability meaning that the remaining carbonic components in the treated effluent still were biorecalcitrant. It would be preferable to phototreat the effluent totally (i.e., up to more than 90% mineralization).

For a better management of the effluents resulting from that manufacture, the following diagram in Figure8 would be recommendable.

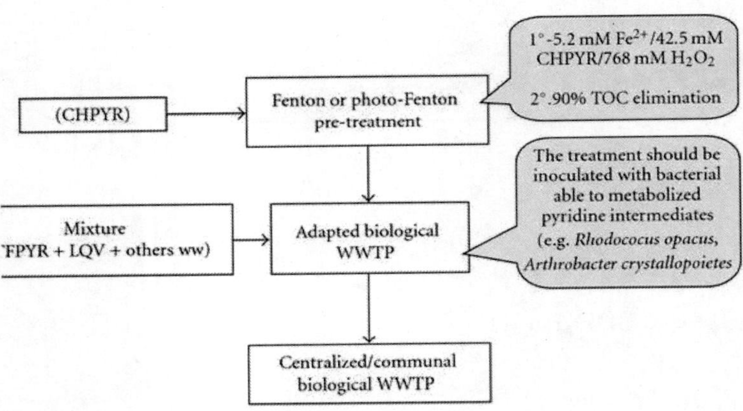


Figure 8: A proposed scheme of a diagram for the management of the studied effluents.

## The Scaling up Approach from Laboratory Studies to the Field Application

Studies in the laboratory and within very high-tech experimental device potential, made it possible to highlight the applicability of the Fenton and photoFenton processes for the detoxification of non biodegradable pollution in water. However, since the aim of this research is to undertake this process for solving real problems, its scaling up applicability (on more significant volumes) still to be evaluated, particularly when operating using the sunlight and not the lamps which of course, are cost expensive. The example of the artificial contamination of water by Endosulfan in Burkina Faso approaches more the reality since cases of contamination of water of the ground water by the pesticides had been reported in Burkina [6].

It's possible that, by submitting the commercial pesticide which composed of the active component: the endosulfan (water solubility (0.325) and other additives such as emulsifiers and adhering agents (which stabilizes the product on the plant once it is pulverized) to the Fenton and photoFenton processes, some of these additives would be destroyed by photolysis and/or oxidation with oxygen from air [7]. The decrease of 40% of the absorbance at 218 nm as observed in Figure 9, does not inevitably mean that the Endosulfan is degraded, since it was shown that the molecule of Endosulfan is stable when exposed to solar radiation and the tests of stabilization for the commercial formulation of pesticides, taking into account the sunlight exposition aspects [8]. The effect of photolysis and aeration on the Endosulfan solution could be put into account of the degradation of the additives (emulsifiers and adherents).

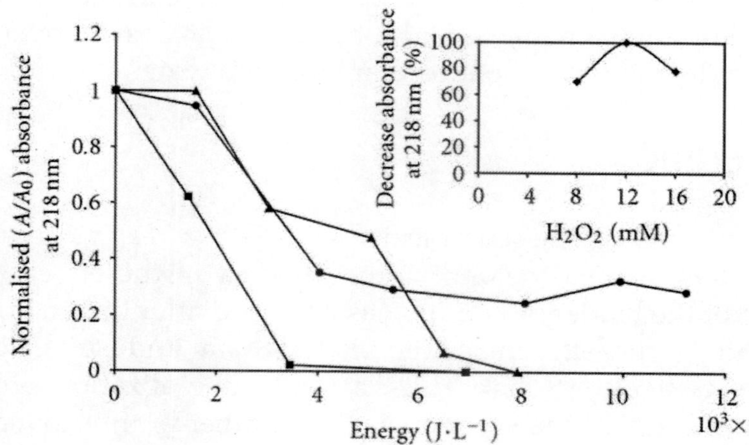


Figure 9: Relative absorbance of the phototreated Endosulfan contaminated water as a function of the specific energy. The $H_2O_2$ is varied from: (●) 8 mM, (■) 12 mM to (▲) 16 mM. The insert is percentage of relative degradation of the Endosulfan (monitored at 218 nm) when energy is accumulated in the photoreactor for the three $H_2O_2$ values tested.

Since the photocatalytic treatment is not considered as a cost effective process when compared to biological treatment [9–12], when taken into account the cost of hydrogen peroxide on the one hand, the technical choice of a treatment, total photochemical mineralization or coupling photochemical-biological processes, on the other hand, it is necessary to make a compromise over the optimal concentration of and the duration of the treatment, by considering the biodegradability of the phototreated effluent. In the case of this study, the choice was made on the lowest concentration of the peroxide (8 mM) with 0.18 $Fe^{2+}$ mM for an initial concentration of 0.36 mM of Endosulfan, conditions within which, is that 50% of the biodegradability of the phototreated effluent is reached after1 h30.

Assuming the relatively high value of the $BOD_5$/COD ratio noticed in Figure 7, one can conclude that the phototreated pesticide effluent containing Endosulfan can become biodegradable in the natural environment. However, it would be necessary to send such effluents into a biological wastewater treatment plant before they could be reused for any purpose or be rejected in the natural media (river, dam, or lake) following the scheme proposed in Figure 8. Otherwise, more studies need to be carried out, that is, by coupling photochemical and a biological process to oversee the complete degradation of endosulfan polluted water.

## CONCLUSION

The difficulty and the complexity of treating real biorecalcitrant wastewaters were observed within the two studied cases. The increase of the biodegradability was observed after the photoFenton treatment of the effluent contaminated with Endosulfan but not with the CHPYR effluent. Thus, a total mineralization process is recommendable in the second case and rather an improvement of the biodegradability of the Endosulfan effluent.

The study on the pesticide effluent enables one to notice that the photolysis could lower by 30% the COD of the effluent (Figure 9) by contrast to the absence of COD decrease for a simple aeration.

In the two case studies, the helio-photoFenton process is effective proportionally to the amount of hydrogen peroxide added; but for economic reasons, a compromise should be made between the highest kinetics of the treatment process of the Endosulfan effluent and the overall objective (or strategy) of the treatment, namely, the coupling of photochemical and biological processes which is a cost minimization option in the treatment strategies of biorecalcitrant wastewaters.

Is the high $DBO_5$/DCO ratio in the case of the photodegradation of Endosulfan effluent necessarily a good indicator to confirm the biotreatability of the phototreated effluent? Such a request could be looked out through very concise chemical analyses of the components of the effluents.

Eventually, this study shows a soft approach of transferring a laboratory high-tech context study to a field applied context within a North-South cooperation.

## REFERENCES


1. Population.Information.Program, "Solutions for a water-short world," 1–31, 1998.
2. K. Toepfer, "Statement on the occasion of the Stockholm Convention on Persistent Organic Pollutants Coming into Force," UNEP, 2004.
3. J. Blanco and S. R. Malato, "Solar detoxification," Alméria, UNESCO/PSA, 2001.
4. G. S. Kwon, H. Y. Sohn, K. S. Shin, E. Kim, and B. I. Seo, "Biodegradation of the organochlorine insecticide, endosulfan, and the toxic metabolite, endosulfan sulfate, by Klebsiella oxytoca KE-8,"Applied Microbiology and Biotechnology, vol. 67, pp. 845–850, 2005.
5. OECD, "Guidelines for testing of chemicals," test 302 B, 1996.
6. H. K. Tapsoba and Y. L. Bonzi-Coulibaly, "Production cotonnière et pollution des eaux par les pesticides au Burkina Faso," Journal de la Société Ouest-Africaine de Chimie, no. 21, pp. 87–93, 2006.
7. N. Sethunathan, M. Megharaj, Z. Chen, N. Singh, R. S. Kookana, and R. Naidu, "Persistence of endosulfan and endosulfan sulfate in soil as affected by moisture regime and organic matter addition," Bulletin of Environmental Contamination and Toxicology, vol. 68, no. 5, pp. 725–731, 2002. View at Publisher · View at Google Scholar
8. K. P. Bentson, "Fate of xenobiotics in foliar pesticide deposits; reviews of environmental contamination and toxicology," CODEN RCTOE4, vol. 114, pp. 125–161, 1990.
9. D. F. Ollis, "Process economics for water purification: a comparative assessment," in Photocatalysis and Environment, pp. 663–677, Kluwer Academic Publishers, Boston, Mass, USA, 1988.

10. C. Pulgarin, M. Invernizzi, S. Parra, V. Sarria, R. Polania, and P. Péringer, "Strategy for the coupling of photochemical and biological flow reactors useful in mineralization of biorecalcitrant industrial pollutants," Catalysis Today, vol. 54, no. 2-3, pp. 341–352, 1999.

11. J. Blanco and S. R. Malato, "Solar detoxification," Alméria, UNESCO/PSA, 2003.

12. V. Sarria, S. Kenfack, O. Guillod, and C. Pulgarin, "An innovative coupled solar-biological system at field pilot scale for the treatment of biorecalcitrant pollutants," Journal of Photochemistry and Photobiology A, vol. 159, no. 1, pp. 89–99, 2003. View at Publisher · View at Google Scholar

# Citations

## CHAPTER 1

Zhou, R.; Guo, L.; Peng, C.; He, G.; Ouyang, L.; Huang, W. Diastereoselective Three-Component Reactions of Chiral Nickel(II) Glycinate for Convenient Synthesis of Novel α-Amino-β-Substituted-γ,γ-Disubstituted Butyric Acids. Molecules 2014, 19, 826-845.

## CHAPTER 2

Synthesis, Characterization And Application Of Novel Bisazo Reactive Dyes On Various Fibers, Divyesh R. Patel, Nikul S. Patel, Hemant S. Patel and Keshav C. Patel*, Synthesis, Characterization and Application Of Novel Bisazo Reactive Dyes On Various Fibers, ISSN 1984-6428 ONLINE

## CHAPTER 3

Nadia Balucani, Elementary Reactions and Their Role in Gas-Phase Prebiotic, Chemistry, doi:10.3390/ijms10052304

## CHAPTER 4

Ajit P. Rathod, Kailas L. Wasewar, and Chang Kyoo Yoo, "Enhancement of Esterification of Propionic Acid with Isopropyl Alcohol by Pervaporation Reactor," Journal of Chemistry, vol. 2014, Article ID 539341, 4 pages, 2014. doi:10.1155/2014/539341

## CHAPTER 5

Arvind M. Patil, Sainath B. Zangade, Yeshwant B.Vibhute, and Sarala N. Kalyankar, 2-Methoxyethanol: A Remarkably Efficient And Alternative Reaction Medium For Iodination Of Reactive Aromatics Using Iodine And Iodic Acid, ISSN: 1984-6428.

## CHAPTER 6

Jean Jacques Vanden Eynde and Annie Mayence, Synthesis and Aromatization of Hantzsch 1,4-Dihydropyridines under Microwave Irradiation. An Overview, doi:10.3390/80400381.

## CHAPTER 7

Wolfgang Stockmann, Werner Engeldinger, Albert Kunst, and Margaret McGovern, "An Innovative Approach to Functionality Testing of Analysers in the Clinical Laboratory," Journal of Automated Methods and Management in Chemistry, vol. 2008, Article ID 183747, 6 pages, 2008. doi:10.1155/2008/183747

## CHAPTER 8

Hongke Wu, Jiaqi Zhong, Haimin Shen, and Hongxin Shi, "Synthesis of Branch Fluorinated Cationic Surfactant and Surface Properties," Journal of Chemistry, vol. 2014, Article ID 460356, 5 pages, 2014. doi:10.1155/2014/460356

## CHAPTER 9

Mauro F. Rebelo; José M. Monserrat; Wanderley G. Bastos, Analysis of laboratory intercomparison data. A matter of independence, doi.org/10.1590/S0100-40422003000300020

## CHAPTER 10

Figueroa-Valverde Lauro, Díaz-Cedillo Francisco, Rosas-Nexticapa Marcela, et al., "Design and Synthesis of Two Oxazine Derivatives Using Several Strategies," Journal of Chemistry, vol. 2014, Article ID 757953, 9 pages, 2014. doi:10.1155/2014/757953.

## CHAPTER 11

Kabeer A. Shaikh, Vishal A. Patila and B. P. Bandgar, An efficient solvent-free synthesis of meso-substituted dipyrromethanes using SnCl2•2H2O catalysis, ISSN: 1984-6428.

## CHAPTER 12

Wenqing Gao1, Qingyong Li 1, 2,*, Jian Chen1, Zhichao Wang1 and Changlong Hua1, Total Synthesis of Six 3, 4-Unsubstituted Coumarins, doi: 10.3390/molecules181215613.

## CHAPTER 13

Meral, Ramazan. 2008. "Laboratory Evaluation of Acoustic Backscatter and LISST Methods for Measurements of Suspended Sediments." Sensors 8, no. 2: 979-993. doi:10.3390/s8020979

## CHAPTER 14

Ali Fares, Farhat Abbas, Domingos Maria and Alan Mair, Improved Calibration Functions of Three Capacitance Probes for the Measurement of Soil Moisture in Tropical Soils, ISSN 1424-8220, doi:10.3390/s110504858.

## CHAPTER 15

S. Kenfack, V. Sarria, J. Wéthé, et al., "From Laboratory Studies to the Field Applications of Advanced Oxidation Processes: A Case Study of Technology Transfer from Switzerland to Burkina Faso on the Field of Photochemical Detoxification of Biorecalcitrant Chemical Pollutants in Water," International Journal of Photoenergy, vol. 2009, Article ID 104281, 8 pages, 2009. doi:10.1155/2009/104281

# INDEX

## T



## W